Thin Film and Flexible Thermoelectric Generators, Devices and Sensors

Sergey Skipidarov • Mikhail Nikitin
Editors

Thin Film and Flexible Thermoelectric Generators, Devices and Sensors

 Springer

Editors
Sergey Skipidarov
RusTec LLC
Moscow, Russia

Mikhail Nikitin
RusTec LLC
Moscow, Russia

ISBN 978-3-030-45864-5   ISBN 978-3-030-45862-1   (eBook)
https://doi.org/10.1007/978-3-030-45862-1

This Springer imprint is published by the registered company Springer Nature Switzerland AG
The registered company address is: Gewerbestrasse 11, 6330 Cham, Switzerland

Contents

Part II   Prospects and Application Features of Wearable TEGs

Part III   Modeling of Thermoelectric Materials Properties

Part I
Trends in Flexible and Thin Film Thermoelectric Power Generation Technologies

Achievements and Prospects of Thermoelectric and Hybrid Energy Harvesters for Wearable Electronic Applications

Mengying Xie, Chris Bowen, Tom Pickford, Chaoying Wan, Mingzhu Zhu, Shima Okada, and Sadao Kawamura

1 Introduction

Wearable electronics are of great interest and have the potential to play an important role in next-generation electronic devices, such as health-monitoring system [1]. The trend for wearable electronics is lightweight, flexible, and self-powered systems and devices to provide autonomous operation. Nowadays, wearable electronics can be miniaturized and consume ultralow power from milli- to nanowatt levels due to the rapid development in micro- and nano-electromechanical systems (MEMS and NEMS) and devices [2, 3]. Energy harvesters [4–6] as an alternative to conventional batteries have gained significant interest over recent decades since these can deliver sustainable energy to supply low-power electronic devices using ambient forms of energy, which enable electronic devices to work without external power source and eliminate the need for replacement and management of batteries.

M. Xie (✉)
Tianjin University, School of Precision Instruments and Opto-Electronics Engineering, Tianjin University, Tianjin, China
e-mail: mengying_xie@tju.edu.cn

C. Bowen (✉)
University of Bath, Department of Mechanical Engineering, Bath, UK
e-mail: c.r.bowen@bath.ac.uk

T. Pickford · C. Wan
University of Warwick, International Institute for Nanocomposites Manufacturing (IINM), Warwick, UK

M. Zhu
Zhejiang University of Technology, Information Engineering College, Hangzhou, China

S. Okada · S. Kawamura
Ritsumeikan University, Department of Robotics, Kusatsu, Japan

© Springer Nature Switzerland AG 2021
S. Skipidarov, M. Nikitin (eds.), *Thin Film and Flexible Thermoelectric Generators, Devices and Sensors*, https://doi.org/10.1007/978-3-030-45862-1_1

TEGs can convert directly low-grade waste heat into electricity via Seebeck effect [7–9] and are ideal for miniaturized, distributed self-powered autonomous system because of the advantages of silent operation, no moving parts, high reliability, and scalability [5, 10, 11]. The human body is a stable energy source of thermal radiation energy which can continuously generate heat through metabolic functions, and by exploiting this heat, TEG is a promising technology to power wearable electronics. Since the last decade, TE energy harvesters have been exploited in several commercial products such as Seiko Thermic watch and Citizen Eco-Drive Thermo by scavenging human body heat.

TEG conversion efficiency is indicated by TE figure of merit [12]:

$$ZT = \frac{S^2 \sigma T}{\kappa} = \frac{S^2 \sigma T}{\kappa_e + \kappa_L}, \tag{1}$$

where S is Seebeck coefficient, σ is electrical conductivity, T is absolute temperature in Kelvin, and κ is thermal conductivity, which is the combination of electronic thermal conductivity κ_e and lattice thermal conductivity κ_L.

The parameter $S^2\sigma$ is called the *power factor* (*PF*):

$$PF = S^2 \sigma. \tag{2}$$

It indicates that a key challenge to improve TE performance and efficiency is to develop new materials that possess high σ to minimize Joule heating but low κ to maintain a large temperature gradient between hot and cold sides of thermoelectric couple. However, in many cases, S, σ, and κ are interdependent due to the concentration of charge carriers [13].

TEG's energy conversion efficiency η_{TEG} is proportional to Carnot efficiency η_c:

$$\eta_{TEG} = \eta_C \frac{\sqrt{1+ZT}-1}{\sqrt{1+ZT}+\dfrac{T_c}{T_h}} = \frac{T_h - T_c}{T_h} \times \frac{\sqrt{1+ZT}-1}{\sqrt{1+ZT}+\dfrac{T_c}{T_h}}, \tag{3}$$

where η_c is Carnot efficiency and T_h and T_c are temperatures of hot and cold sides of TE couple, respectively. Therefore, in addition to ZT value of TE material, temperature difference across TE legs is also critical to achieve high performance of TEGs. Figure 1 shows the temperature at various locations of the human body when at different ambient temperatures, e.g., 45, 27, and 15 °C [14]. It can be seen, that heat dissipation from the human body varies depending on the body location and surrounding conditions. Compared to industrial waste heat which is higher in temperature than the environment, typically in the range of several tens to hundreds of K [15], the maximum temperature difference between human body and ambient temperature (ΔT) is approximately 15 °C when it is cold and reduces to 8 °C when it is hot and comfortable outside. The temperature across TE legs is even less than ΔT between the human body and environment as the heat dissipates not only in TE legs but also in the surrounding environment [16]. This reduced temperature differential

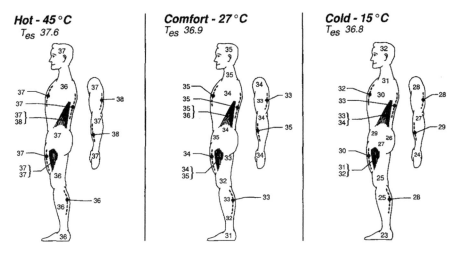

Fig. 1 Diagram of the distribution of skin temperatures at hot, comfortable, and cold ambient room temperature [14]

poses a challenge and limits severely the energy that can be scavenged from the body, resulting in power levels on the order of tens of microwatts per cm^2. Researchers attempted to improve the efficiency of TE energy-harvesting technology by optimizing TE performance of material, designing optimized device structures for improved heat transfer and even combining several energy sources, such as piezoelectric, triboelectric, electromagnetic, pyroelectric, and photovoltaic.

2 Materials

Organic and inorganic thermoelectric materials will be now described, including composite systems and use of nanostructuring.

2.1 Organic Thermoelectric Materials

For wearable applications, low weight and flexibility are important factors. Therefore, polymeric or polymer-based composites are of interest and are covered in this section. A wide variety of polymers have been investigated for a range of innovations in TE applications. In addition, the concurrent development of nanotechnology has led polymer nanocomposites being explored in depth since the turn of the century [17, 18]. The nanocomposite route enables material properties to be fine-tuned with careful control of the individual composite components and structure.

The use of polymer structures offers a range of advantages, including the ease of manufacture and abundance of the necessary materials for polymer synthesis, which are favorable for large-scale and low-cost industrial production of thermoelectrics [19]. This is also supplemented by the potential nontoxic nature of TE polymers, making installation and maintenance of devices safer. Similarly, the potential for thermal degradation of TE materials poses a lower threat in polymeric systems as they remain thermally stable at moderate temperatures. Polymer materials and nanocomposites may be constructed in a range of shapes and sizes such as thin films, foams, and nanofibers. Furthermore, due to low density and low stiffness, it is possible to design lightweight, flexible, and low-cost polymer TE devices with a geometry tailored for an application [20–28]. Polymers possessing high σ, therefore, hold a unique position as being electrically conductive but thermally insulating materials that are ideal for energy harvesting via Seebeck effect should the efficiency of TEG continue to approach efficiency of inorganic counterparts [29].

2.1.1   Thermoelectric Effects in Conjugated Polymers

The discovery and production of conjugated polymers have been a major development which has paved the way for conducting polymer composite and nanocomposite systems. Typically, structures of polymer matrices are amorphous and semicrystalline in which σ is limited by interchain hopping mechanisms. If the average potential energy barrier involved in moving between chains in the matrix is high, then mobility of charge carriers in the material is lowered. This can be circumvented in conjugated polymer matrices due to $\pi - \pi$ stacking; π orbital electron clouds form interchain π bonds, enhancing the crystallinity of the matrix. Moreover, π electron cloud enhances also chain rigidity, which assists $\pi - \pi$ stacking and further improves crystallinity.

High crystallinity in polymer matrix is known to improve mobility of charge carriers and thus σ, by making the energy barrier involved in interchain hopping lower. However, due to the need for high S in TE materials, the presence of energy barrier between polymer chains can be helpful. This is due to the *energy-filtering effect*, a phenomenon in which preferential transport of higher-energy charge carriers results in higher S. Charge carrier transport in TE materials can be thought of as a collective diffusion along a thermal gradient. Charge carriers will stochastically move through the material; "hot" carriers are with energy above average, i.e., above Fermi energy, which have a net transport to cold side of TE material, and "cold" carriers are with energy below average, i.e., below Fermi energy, and will undergo net transport to hot side. However, since "hot" charge carriers diffuse more easily due to higher energy, there will be an overall net transport of charge carriers to cold side of TE material. S can be thought of as a measure of the average entropy transported per charge carrier through this thermally driven diffusion [30]. Thus, to increase S of TE material, thermovoltage, namely, the thermal energy carried per charge carrier, should be maximized by increasing in diffusion of "hot" carriers and decreasing in diffusion of "cold" carriers. The presence of energy barriers in the matrix will not significantly impact the transport of "hot" charge carriers with energy sufficient to overcome barriers, while "cold" carriers will be impeded. Hence, high crystallinity and, therefore, fewer energy barriers in the matrix will enhance σ, but this must be supplemented with enhanced or preserved S.

It should be noted that when discussing polymeric TE materials, rather than comparing ZT values of various materials, PF is often investigated. This is because κ of organic composites remains low, of the order of ~0.1 W/(m×K), while electrical properties (characterized by S and σ) can change significantly. However, to characterize polymer or composite in full, measurements of κ should supplement PF measurements.

2.1.2 Doping Conducting Polymers

Site to site hopping along σ skeleton does not provide sufficiently high σ for TE applications, and, therefore, doping is required. Doping can alter the potential energy barriers involved with the hopping mechanism, providing increase in mobility of charge carriers in polymer structure (see Fig. 2). It can simultaneously increase the concentration of charge carriers, allowing for a higher thermoelectric current to be generated in TE material for a given temperature difference. In recent studies, the resulting conducting polymers formed after doping frequently possess σ of 10^3 S/cm and higher [31].

In many cases, conducting polymers are constructed by chemical or electrochemical doping of conjugated polymers. In such processes, conjugated polymer is exposed to gas or solution initiating a redox reaction [32]. Dopants are selected to

Fig. 2 Comparison of doped and undoped potential map for polymer chain. Charge carriers find it easier to hop between backbone sites with higher level of doping; coulomb traps begin to overlap as counterions saturate the polymer chains [29]

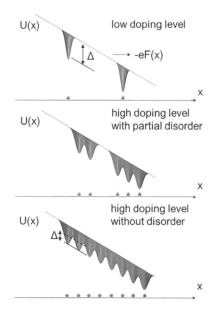

increase concentration of charge carriers by electron (or hole) donation and often simultaneously increase mobility of charge carriers [33, 34]. A counter-ion remains from the redox reaction, which settles into potential well in the matrix, such as ring structure on a polymer backbone, and retains charge neutrality of the matrix, thereby stabilizing the system. Figure 2 is a visualization of how doping affects the potential barriers observed by charge carriers moving along the backbone. Dopants are also known to offer structural modifications to the polymer matrix, notably altering how the matrix may crystallize [21, 35, 36]. This can have significant effects on PF of TE material by modifying $\pi - \pi$ stacking in the matrix and, therefore, changing the amount of interchain transport of charge carriers. In fact, dopants are often selected to increase crystallinity, enabling more metallic charge carrier transport and enhancing mobility of charge carriers significantly.

Additional concept is the use of secondary dopants. These are additives used in the process of forming conducting polymer which are not retained in the final matrix but alter the formation process to enhance TE properties [35]. Often, these dopants alter the manner in which the matrix crystallizes during formation [32]. Mengistie et al. [23] treated flexible poly(3,4-ethylenedioxythiophene):poly(styrenesulfonate) (PEDOT:PSS) hybrid polymer with several secondary dopants which included ethylene glycol (EG), polyethylene glycol, methanol, and formic acid. They found large increase in σ, while S showed little variation, which is postulated to be due to selective removal of insulating PSS from the system, thereby resulting in PEDOT chains becoming more linearly oriented. The orientation of polymer chains is a powerful tool to improve TE properties in polymer matrices as it decreases in average length of conduction pathways through TE material, increasing in σ by decreasing in number of obstacles for charge carriers to overcome. However, it should be noted that ZT will be enhanced only along the aligned axis [37].

Alternatively, post-treatments can be used to instigate changes in the matrix configuration. Chemical, thermal, and mechanical treatments of polymer materials have proven effective. Chemical treatments have similar aims to secondary doping – alter the matrix structure by chemical modification of the polymer chains [32, 38]. For example, Fan et al. [28] optimized TE properties of PEDOT:PSS films by sequential acid and base treatment, which increased electron mobility and tuned oxidation level of PEDOT, respectively. Similarly, Liu et al. [39] used co-solvent of EG and dimethyl sulfoxide (DMSO) for post-treatment of PEDOT:PSS thin films at 120 °C. Such chemical treatment has a similar effect to the study from Fan et al. [28], although the compounding thermal treatment leads to more pronounced chemical reduction of PEDOT, thereby enhancing S. Stretch alignment of polymer chains is another popular post-treatment for thin films, often resulting in similar results to alignment via chemical treatment [40]. O'Connor et al. [41] showed that poly(3-hexylthiopene) (P3HT) films have increased chain alignment upon stretching, leading to higher hole mobility and higher ZT along the direction of alignment [40, 41].

Since organic thermoelectrics are now rivaling ZT values of inorganic structures, a number of homopolymers, copolymers, and hybrid polymers are now being investigated with a breadth of methods and treatments in order to gain greater understanding of how to build efficient TE devices using organic components [18, 19,

Table 1 Notable measurements of thermoelectric properties of conducting polymers without the addition of nanoparticles

Polymer	Dopant/additive	σ, S/cm	S, µV/K	PF, µWm^{-1}K^{-2}	ZT	Reference
1	2	3	4	5	6	7
PA	I$_2$	4.42E4	15	990	–	[43]
PPy	LiClO$_4$	15	51	3.9	0.0068	[44]
PANI (p-type)	HCl	5.25	8.5	0.038	0.4E-4	[45]
PEDOT	Tos/TD AE	~80	~200	324	0.25	[46]
PEDOT:PSS	DMSO	880	73	469	0.42[a]	[47]
P3HT	TFSI$^-$	8.9E-3	5400	20	0.04[b]	[48]
FBDPPV	N – DMBI	14	−141	28	–	[49]
P(NDIOD-12)	N – DMBI	8E-3	−850	0.6	–	[50]
PDI	Imidazole	0.5	−170	1.4	–	[33]
PNDTI-BBT-DP	N – DMBI	5.0	−169	14.2	–	[51]

[a]Determined from the in-plane thermal conductivity of thin films
[b]Measured at 340 K

42]. A list of notable polymers studied is provided in Table 1, which will be discussed in the following sections.

Many conducting polymers are now known, each possessing own advantages for use in TE devices [17, 18, 31]. With the notable exception of polyacetylene (PA), conducting polymers almost always contain ring structures in the backbone which contribute electrons to π cloud, as seen in Fig. 3. The more prevalent p-type conducting polymers will be discussed first, followed by less extensive n-type materials.

2.1.3　p-Type Thermoelectric Polymers

There are many promising candidates for p-type organic TE materials. Many of these polymers have own merits and drawbacks, and many continue to be improved and innovated. For example, Kaneko et al. [43] observed large σ of 40,000 S/cm in stretched films of iodine-doped PA, and to date this is one of the highest σ observed in conducting polymer. However, S of PA is low and difficult to improve that result in PF more similar to other p-type polymers, and its poor air stability and insolubility have led to a lack of interest in attempting to increase its low S.

Polypyrrole (PPy) has also been investigated as TE polymer as its high mechanical and chemical stability and its low toxicity have led to interest in biological applications including wearable electronics [52]. Unfortunately, pure PPy films tend to have low PF values (<10 µW/(m×K^2), leading to attempts to use nanostructuring approach by fabricating nanostructures such as PPy nanowires and incorporating those into composite structures with other polymers [53–55]. Wu et al. [56] achieved PF of 0.31 µW/(m×K^2) in PPy nanotube films, which exhibited superior TE properties when aspect ratio (the ratio of length to diameter) was large. This was postulated to lead to higher degree of molecular ordering. However, Culebras et al. [44] achieved higher PF of 3.9 µW/(m×K^2) through fine control of the potential in

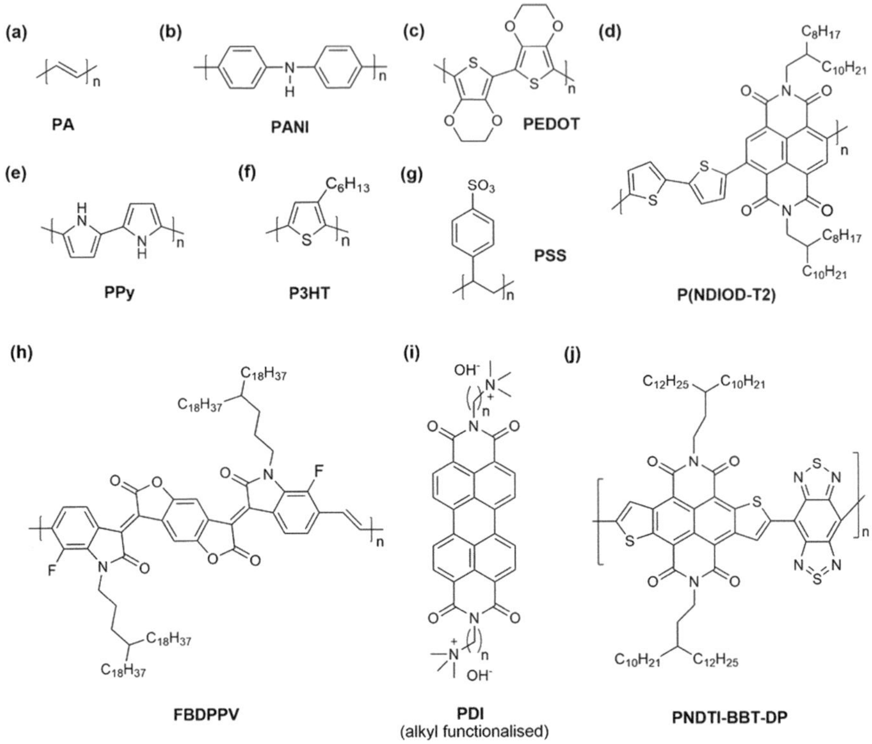

Fig. 3 Structures of polymers given in Table 1. In PEDOT:PSS hybrid systems, one of the central double bonds on PEDOT monomer donates a negative charge to SiO_3 group on PSS

$LiClO_4$ electrochemically doped PPy films without the need for any nanostructuring or incorporation of other polymers or nanofillers.

Polyaniline (PANI) remains one of the most versatile TE polymers due to its ability to act as both p-type and n-type material. Intrinsically, PANI is a weak p-type semiconductor, but this can be readily altered by doping to modify its oxidation state [57–61]. Unfortunately, PANI cannot achieve PF values of other TE polymers at room temperature (RT), with the best PF below 1 µW/(m×K^2) due to low σ [58]. The largest ZT observed was 0.4×10^{-4} in HCl-doped bulk PANI by Li et al. [45]. In contrast, PANI composites have received a large amount of attention due to versatility, making such composites ideal material for investigating different methods for optimizing TE properties [53, 62, 63]. The highest ZT for any conducting polymer was achieved at low temperature (15 K) by Nath et al. [64], who reported ultra-high S of 6×10^5 µV/K, resulting in high ZT of 2.17. This is higher than inorganic thermoelectrics where ZT is typically ~1 [19].

To date, the most successful group of conducting TE polymers is polythiophene (PTh) family. For example, P3HT has had some success as p-type TE material [31,

48, 65]. Its notable flexibility has attracted interest for use in wearable TE devices, as well as in field-effect transistors (FETs) [34, 66, 67]. P3HT exhibits a wide range of σ with doping level, varying from 10^{-10} to 10^3 S/cm, and has high S at low doping levels [67]. Zhang et al. [48] hold currently the record for highest PF achieved at 20 μW/(m×K^2) in P3HT film doped with triflimide anion (TFSI$^-$).

PEDOT has seen the most success among all conducting polythiophene derivatives [21, 68, 69]. It offers a variety of advantages over other conjugated polymers having high intrinsic σ, good processability, and stability in air. Moreover, PEDOT has seen success as the primary component of TE composites, being combined with many other polymers and nanoparticles, as well as dopants such as tosylates (e.g., iron tosylate, Fe(Tos)$_3$) and halogenated oxidants (e.g., hexafluorophosphate PF$_6^-$) [20, 28, 70, 71]. High ZT of 0.25 at room temperature was achieved by Bubnova et al. [68] in PEDOT-tosylate thin films, which remains the highest ZT observed in non-composite-based polymer material.

Especially, PEDOT:PSS has had success as TE polymer hybrid [72]. PSS is insulating polymer; it does not have π-conjugated backbone, and, therefore, it does not contribute to forming conductive pathway through the polymer matrix. Instead, it acts as dopant, providing a large increase in hole concentration, thereby enhancing σ and concurrently decreasing in S compared to PEDOT [73]. Furthermore, PEDOT:PSS forms often a core-shell structure, and, therefore, charge carriers must hop over the potential barrier posed by insulating PSS shell to move between conducting PEDOT domains. This can recover lowered S from increased concentration of charge carriers after adding PSS. PEDOT:PSS is also solution processable where regular PEDOT is not, opening up options for doping and treatments to optimize TE properties [29].

Solution processing of PEDOT:PSS provides a route for additives to be included into the polymer. For example, Kim et al. [47] provided a simple method for optimizing the doping level in PEDOT:PSS in which PEDOT:PSS thin films are de-doped with reducing agents of ethylene glycol (EG) or dimethyl sulfoxide (DMSO) by mixing those into a solution with PEDOT:PSS. DMSO-treated films display the most significant enhancement in PF, and due to low in-plane $\kappa \sim 0.33$ W/(m×K), ZT of 0.42 was observed at room temperature. This is the highest observed room temperature ZT in conducting polymer. Posttreatment methods involving hydrophilic solvents such as this have proven effective in PEDOT:PSS [38]. PEDOT becomes de-doped due to the removal of PSS from the polymer matrix, while charge carrier-filtering effects are retained as PSS is not completely removed and PEDOT:PSS boundaries must be crossed by charge carriers. This leads to increase in S after de-doping, which leads to record ZT. In addition, removal of some insulating PSS results in also increasing mobility of charge carriers which do overcome the barrier, thereby increasing σ despite the decrease in concentration of charge carriers by de-doping.

2.1.4 n-Type Thermoelectric Polymers

A major challenge for polymer thermoelectrics is the lack of promising *n*-type materials that have been discovered to date. Untreated conjugated polymers, where almost all are weak *p*-type semiconductors in undoped states, are doped by treatment with a reducing agent, which donates electrons into the lowest unoccupied molecular orbital (LUMO). Post-doping, new LUMO becomes close to the vacuum level, resulting in a conducting polymer with low work function [74]. Therefore, *n*-type conducting polymers tend to be unstable in air due to ease of oxidization in ambient atmospheric conditions, rendering those impractical for most applications. This is in contrast to *p*-type doping, which removes electrons from the highest occupied molecular orbital (HOMO), thereby decreasing in polymer work function and stabilizing it against oxidation. For the manufacture of feasible organic TEG, both *n*- and *p*-type components are required to ensure continuous current flow; using only one component leads to charge buildup on hot and cold sides of the material [50]. Therefore, optimization of *n*-type organic TE materials is necessary for the end goal of commercial organic TEG.

Consequently, *n*-type materials must be treated to circumvent this limitation by constructing features in the matrix which withdraw electrons and increase electron work function. Small *n*-type dopant molecules (4-(1,3-dimethyl-2,3-dihydro-1H-benzoimidazol-2-yl)phenyl,   N-DMBI)   and   organic   donors   (e.g., tetrakis(dimethylamino)ethylene, TDAE) are capable of retaining low LUMO in conducting polymers [33, 50]. In some cases, electron withdrawing groups are introduced as functional groups either as part of the polymer backbone or as a side group, e.g., in functionalized polyphenylene vinylene (PPV) materials [36, 49]. This increases in electron affinity of the material, making stability easier to retain upon doping.

Fluorine-functionalized benzodifurandione-based PPV (FBDPPV) has also seen success as *n*-type TE material. Shi et al. [49] achieved PF of 28 μW/(m$\times$K^2) in solution-processed FBDPPV films, in which a fluorine group was added to the backbone. They attribute high PF to the fluorine group's effect on increasing mobility and concentration of charge carriers due to modification of polymer's energy levels. Fluorine functionalization is observed to lower PPV LUMO (to 4.17 eV), increasing its affinity for H or H$^-$ by moving it closer to N-DMBI's HOMO (4.7 eV), thus increasing in effectiveness of doping. Zhao et al. [75] have managed even lower LUMO at -4.28 eV (albeit with lower PF) in similarly functionalized PPV-based polymer. A further decrease in *n*-type polymer LUMOs with similar approaches may lead to further enhanced TE properties and will likely be a focus for future work optimizing PF of organic *n*-type materials [36].

Additional complex polymers have seen success as *n*-type TE materials. Poly- or P(NDIOD-T2) [17] has received significant attention since Schlitz et al. [50] achieved PF of 0.6 μW/(m$\times$K^2) in N-DMBI-doped samples. P(NDIOD-T2) is solution processable, possesses of high electron mobility, but most notably is stable in air, making it highly attractive as *n*-type TE material. Schlitz et al. [50] took the increasingly popular approach of focusing on how fine-tuning the polymer matrix structure impacts its charge carriers' transport properties [37]. This strategy for optimizing TE properties has also been adopted to other n-type polymers and is a promising strategy for developing viable n-type counterpart to existing p-type materials [49, 51, 76].

Finally, self-doping of n-type conjugated polymers has led to large PF, most notably by Russ et al. [33] in perylene diimide (PDI) thin films. The authors made the striking observation that altering the alkyl chain length that links the conjugated backbones by reaction with imidazole had significant effects on TE properties. This form of structural modification is known to offer enhancements to σ through increasing mobility of charge carriers, as well as contributing to concentration of charge carriers of the matrix [36, 37]. At alkyl spacer length of $n_g = 6$ (n_g is number of groups), PF reached maximum of 1.4 $\mu W/(m{\times}K^2)$ due to increase in σ in two orders of magnitude over PDI at $n_g = 2$, while S saw mild enhancement with larger n_g. Through crystallographic observations, PF enhancement is attributed to structural changes in the polymer matrices, leading to morphological-induced mobility improvements.

2.2 Organic Thermoelectric Composite Materials

While electrically conducting polymers offer many advantages to inorganic counterparts, electrical properties remain typically poor in comparison to common TE ceramics [77]. This is largely due to amorphous structure preventing efficient charge carrier transport, limiting σ. Composite structures aim to alleviate this problem by enhancing electrical properties of the material without detracting from the thermal insulation of the polymer matrix. Electrically conductive nanofiller components such as inorganic nanowires, carbon nanotubes (CNTs), and graphene nanoplatelets (GNPs) have all seen success at enhancing PF of polymer matrices without significantly degrading the thermal insulation of the material [17]. As the manufacture of nanomaterials becomes industrially economical, organic TE nanocomposites have the potential to reach a point of market viability should TE materials properties exceed or match those of inorganic materials currently available.

Decoupling TE parameters has always been the main obstacle for TE material development. Nanocomposites offer a variety of targeted strategies to achieve this, as bulk materials do not have specialized structures designed to promote this behavior. In Eq. (1), it can be seen that three of TE materials properties σ, κ_e, and S are linked to concentration of charge carriers. In ideal TE material, the largest possible proportion of the thermal energy travelling across it should be transferred by charge carriers in order to maximize κ_e and minimize κ_L. In polymer nanocomposites, minimizing κ_L is not difficult; it should not increase as long as amorphous structure of the matrix is retained such that phonon scattering rates are still high in the matrix. Maximizing κ_e is more complex but can be achieved by ensuring large average energy/carrier diffusing along the thermal gradient in TE material, as well as high concentration of charge carriers. Additionally, the material electrical resistance is desired to be low, and net diffusion of charge carriers should be biased heavily toward the cold side of TE material; charge carrier transport should be impeded toward the hot side but promoted toward the cold side.

A variety of structurally tailored composite systems have managed to increase significantly mobility of charge carriers by forming highly electrically conductive pathways through the polymer matrix. Studies have also focused on enhancing the energy-filtering effect to increase PF of TE material by the use of interfacial energy barriers, the properties which may be more finely controlled in nanocomposite systems compared to the pure polymer matrix [19, 67, 78]. In particular, polymer nanocomposites involving carbon nanomaterials have achieved considerable amount of success in this respect [79]. A highly crystalline structure in the material can improve its σ by providing a low-resistance pathway for charge carriers via decreasing the energy required to move between molecules. Therefore, $\pi - \pi$ stacking, which both graphitic structures and conjugated polymers exhibit, can result in a structure with high electron mobility resulting in increase in σ of several orders of magnitude [80].

Formulating nanocomposite which takes advantage of low κ of the polymer and σ of the nanofiller depends ultimately on the material structure. This can be broken down into two categories:

(i) Nanoscale architecture and interface design
(ii) Microscale and macroscale structure

2.2.1 Nanocomposite Structures

Increasing loading fraction of the nanoparticle in the composite will alter TE properties and is the main factor investigated in studies looking into novel TE polymer nanocomposites. At loading fraction known as the *percolation threshold*, conductive pathways will form throughout the bulk material structure. At this point, σ will sharply increase as electron mobility becomes significantly higher [38, 79, 81, 82]. However, loading beyond the percolation threshold can have detrimental effects on PF due to how it affects interfacial transport of charge carriers. Interfaces in polymer nanocomposites have a greater impact on properties than in pure polymer matrices due to the contrasting electrical properties of nanofillers and polymer chains. The insulating polymer can act as a barrier between nanofillers, promoting energy-filtering effect by creating potential barrier that charge carriers must overcome to transfer between conductive nanofillers, ultimately increasing matrix PF. Composites loaded past percolation tend to form conductive pathways with a lack of these barriers as nanoparticles are more likely to be in direct contact.

For example, low-resistance conducting pathways formed by large length CNTs in polymer matrices barely impede charge carriers transport. However, energy barrier exists at junction between CNTs. The height of this barrier depends on many factors, the most important of which is the presence (or lack) of polymer between CNTs [38]. If CNTs are in direct contact, there are fewer effects contributing to the interface resistance. Conversely, the presence of polymer in between CNTs increases in barrier due to the large electrical resistance of the polymer compared to CNTs. Moreover, interfaces are a key site in polymer matrices for phonon scattering, and so interface morphology involving a greater number of barriers and scattering sites can further decrease κ of composite. Therefore, fine control of nanofiller content is critical to construct a composite with optimized figure of merit.

Interface Design: Enhancing the Power Factor

In polymer nanocomposites that contain a conductive nanofiller, the electrical resistance along the conductive pathway is minimal. A high density of interfaces helps to maximize the impact of the energy-filtering effect [83]. Hence, control of charge transport across interfaces is crucial for optimizing the matrix PF. For example, fine-tuning of energy band gap between polymer and nanofiller on each side of the interface was achieved by modification of doping level in P3HT-Bi_2Te_3 nanowire composites [67]. They reached peak PF of 13.6 $\mu W/(m \times K^2)$ (compared to 3.9 $\mu W/(m \times K^2)$ for pure P3HT samples) by fine-tuning doping level of $FeCl_3$-doped P3HT. Heavy doping helped to reduce the potential energy barrier between P3HT and Bi_2Te_3 to less than 0.10 eV, which helped to filter out low-energy charge carriers (room temperature charge carriers have energy of ~0.025 eV). This fine-tuning of interface band structure provides a direct method of controlling charge carriers transport and showcases the capability of nanocomposites as TE materials [84].

Taking more structure-focused approach, Wang et al. [62] investigated how the fabrication methodology of PANI/graphene composites affected PF of the nanocomposite. The highest PF is ~ 55 $\mu W/(m \times K^2)$ for in situ polymerized PANI/graphene films doped with camphorsulfonic acid (CSA), a common dopant for PANI. This work is prime example of the success of carbon nanomaterials in TE polymer nanocomposites, as graphene acts as a template for polymerization of aniline monomers due to strong $\pi - \pi$ interaction between PANI and graphene. This simple technique results in graphene network permeating through the nanocomposite matrix with PANI coating graphene surfaces. PANI acts as a potential barrier for charge transport, promoting energy-filtering effect. By comparison with pre-polymerized PANI/graphene films, an enhancement in S is observed without degrading σ significantly, leading to higher PF than observed in PANI/graphene composites known at the time.

Another approach to improving interfacial transport is to alter retroactively the matrix interfaces with post-processing treatments [28, 38, 85, 86]. Usually, these methods involve exposing the composite to a chemical which reacts with the polymer matrix, altering the structure of the interface and optimizing the barrier properties. For example, Hsu et al. [38] improved PEDOT:PSS-CNT nanocomposites by treating with polar solvents. Increased PF of composites was attributed to the removal of PSS from the matrix, which increased electron mobility by decreasing the resistance of CNT-PEDOT:PSS-CNT junctions on the conductive pathway. Since a portion of PSS remains in the matrix, S value of composites was retained after treatment, meaning filtering of "cold" charge carriers was still present and increase in σ was not compromised by decrease in S. These posttreatments are particularly promising since this does not complicate processing methodologies and have the potential for application on industrial scale.

Interface Design: Scattering Phonons

Thermal insulation in the polymer matrix relies on decreasing κ_L, which can be achieved by inhibiting phonon transport via providing scattering sites [78]. When phonon wave packet encounters an obstacle – such as boundary in the material structure or interface between polymer chains – part of the wave packet will be transmitted, while the rest is scattered. Ensuring thermal insulation is related to providing many sites for this situation to occur and that a large portion of the wave packet is scattered each time. There are two main mechanisms which contribute to this scattering, and targeted interface design can promote both simultaneously [87]. Since, nanofillers tend to have large σ, this approach to nanocomposite fabrication assists in retaining the thermal insulation of the polymer matrix despite the nanofillers offering a more effective pathway for phonon wave packets to travel through the material.

The first mechanism is to capitalize on acoustic mismatch between the materials [88]. Scattering is enhanced, and transmission decreased, when the phonon encounters a boundary between materials with different vibrational spectra. To increase the number of scattering sites via this effect, nanocomposite interfaces should be constructed such that phonons must travel from nanofillers to polymer and back to travel through the matrix, rather than directly between the more thermally conductive nanofillers. The second mechanism is direct scattering of a discontinuity at the boundary as the wave packet attempts to hop to the neighboring molecule. Lattice phonons are transported in a distribution of energies with corresponding wavelengths which determine the length scale of features of which it may be scattered. Short wavelength phonons will scatter at small features, such as polymer-nanofiller interfaces, whereas long wavelength phonons will scatter at large features, such as grain boundaries [83]. It is known that phonon scattering can be increased significantly if the characteristic size of matrix features is smaller than phonon mean free path, and so the inclusion of nanofillers provides much number of scattering sites for the matrix. As such, the nanostructure of interfaces can be helpful in scattering low-wavelength phonons which are thought to carry significant proportions of the thermal energy through the matrix [19, 89].

Yu et al. [90] investigated κ at interfaces in CNT-PEDOT:PSS nanocomposite thin films. Studied composites contain PEDOT:PSS as a barrier between the high aspect ratio CNTs, which form a percolated network through the composite. This helped to induce acoustic mismatch along the conductive network, impeding phonon transport while still allowing for charge carrier hopping. Ultimately, ZT of 0.12–0.24 was obtained, with the uncertainty arising due to difficulties in obtaining an accurate measurement of κ that is common problem in polymer thin films.

2.2.2 Nanocomposite Device Configurations

While the nanostructure of composites can have a large impact on the fundamental transport mechanisms in a composite, the microstructure has a large impact on the bulk thermal and electrical properties [91, 92]. For example, material with many discontinuities in the microstructure, such as abundance of grain boundaries in a crystalline structure, will both scatter large wavelength phonons and impede charge carriers transport. The presence of microscopic boundaries like these improves TE properties of the composite as long as nanofillers can bridge the gaps such that electrical resistance is not significantly decreased, while the long wavelength phonons are heavily scattered. Compounded with nanostructural scattering of short wavelength phonons, the highest-energy density portions of the phonon energy spectrum may be efficiently scattered by rational microstructure and nanostructure design [91].

Many works on conducting polymers and nanocomposites has focused on thin films. This restricts transport to two dimensions, simplifying obstruction of phonon transport. Oriented composite structures can take this further by creating quasi-1D structure in which nanofillers are aligned along the axis of the material, creating anisotropic TE properties [41, 60, 62, 78, 89, 93]. The shorter direct conductive pathways between the heat source and heat sink in aligned thin film composites can increase in σ of composite. These films may be fabricated with methods that promote nanofiller alignment (usually assisted by $\pi - \pi$ stacking) [93] or rather by posttreatments such as mechanical stretching which orient the nanofillers along the in-plane direction of the film [41].

However, thin films are not the only composite structure that can optimize transport mechanisms. Manipulation of microstructure by formation of porous materials can further enhance TE properties via decreasing κ of composite while retaining or improving the electrical properties of oriented thin film composite near percolation [94]. A porous microstructure restricts phonon transport to the material lining the edges of the pores, meaning phonons struggle to find a pathway along the thermal gradient after a scattering event [22]. Charge carriers transport can also be fine-tuned further in porous materials as conductive pathways must form in a smaller volume and often are more easily aligned by the fabrication process [60].

Fibers

Fibrous nanocomposite structures are a newly studied option for impeding phonon transport in the nanocomposite matrix [22, 60, 95]. Transport in fibrous materials is restricted to fibrous pathways, meaning control of charge and phonon transport mechanisms is restricted by the ability to move along individual fibers. Fine-tuning of nanostructure within the fibers can lead to highly specialized transport. For example, by ensuring much number of interfaces exist within the fibers, phonon transport becomes inefficient as phonons encounter an onslaught of scattering sites. Charge carriers transport may be similarly manipulated by ensuring a high degree of alignment of low-dimensionality nanofillers, e.g., carbon nanotube, since the fiber ensures a pathway of low resistance, enhancing σ.

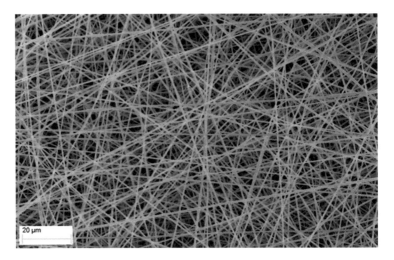

Fig. 4 Electrospun fibers of polyacrylonitrile (PAN)

Alignment of fibers is a common approach to improving PF. Wang et al. [60] produced highly aligned CNT/PANI composite fibers in which fibers displayed macroscopic alignment and CNTs exhibit alignment within fibers. They utilize the common electrospinning technique to form the nanofibers. The most successful sample was fabricated from a solution containing CNTs and aniline monomers, which underwent in situ polymerization prior to electrospinning. This method ensures nanostructure in which CNTs and PANI chains are effectively $\pi - \pi$ stacked, leading to high degree of alignment of both components of the composite within the nanofibers. They achieved maximum PF of 0.17 $\mu W/(m \times K^2)$, which is larger than any reached in pure PANI systems. Figure 4 illustrates typical structure of electrospun fibers of polyacrylonitrile (PAN)

Foams

A less commonly used approach to form TE nanocomposite microstructures is to use foams. These are highly porous materials which hold some notable advantages over other composite structures. During the foaming process, the volume of the composite increases many times, forcing the polymer chains and nanofillers into thin walls of the newly formed cells in the foam matrix. This process completely rearranges the nanostructure of the composite, and similarly to fibrous composites, this can encourage a high degree of order [26].

Foaming is essentially the forced expansion of the composite by formation of cells throughout the matrix. This may be achieved by a variety of methods; however, creation of foamed nanocomposites with conjugated polymers is not possible due to the lack of flexibility. Therefore, insulating polymers must be used, with PF of the composite needing to be supplied by the nanofiller [96, 97]. This is not necessarily a roadblock for TE polymer nanocomposite foams as κ can be lowered compared to composites with different microstructures, thereby enhancing the figure of merit without the need of exceeding PF of different composites.

The most popular methods for foaming polymer matrices involve rapid expansion of gas in the composite which, driven by strong pressure forces, escapes from the matrix while pushing it outward, forming cells [96–98]. This forces the nanofillers together in the cell walls, altering both the crystalline microstructure and the interfacial nanostructure. However, as seen in Fig. 5, the majority of nanofillers is postulated to be forced into "struts," junctions between three cells. Hence, the conductive pathway is dependent on the nanofillers forming pathways along the cell walls. The nanostructure of the cell wall conductive pathways largely determines the matrix PF as charge carriers are largely unimpeded when travelling within the struts, where direct contact between nanofillers is common.

In fact, the percolation threshold becomes the key factor in determining PF of TE foam. Due to the compression of nanofillers in the cell walls, only a small portion of those are necessary for formation of a conductive pathway. Whether more or fewer nanofillers are required to percolate the composite is unclear, and it seems to depend on microscale structural parameters such as the cell size and cell wall width. For example, Tran et al. [99] found that in poly(methyl methacrylate)/CNT foams, σ increased with the cell number density – i.e., decreased in average cell size. This was attributed to the probability of the conductive pathway breaking increasing as the cell size increased, causing the cell wall width to decrease.

Control of the percolation threshold in foam composites is important to attain the advantages of using, as super-percolated composites will have enhanced κ_L due to the abundance of thermally conductive pathways through the cell walls. Similarly, in order to maximize S in these composites, the energy-filtering effect should be induced by formation of polymer interfaces between nanofillers in cell walls, indicating the need for composites just below percolation. Aghelinejad et al. [26] found high-density polyethylene/CNT composites caused S to triple compared to thin film samples which they attribute to the aligned cell wall CNT network including a large number of junctions, although they note that the lack of understanding of this mechanism means their conclusion is fairly speculative.

2.3 Inorganic Materials

Despite the rapid development of organic and organic composite TE materials described above, inorganic TE materials continue to dominate TEG industrial sector due to excellent TE performance, especially high ZT value. The majority of inorganic TE materials works in the range of temperatures from medium to high above 500 K [100], and, to date, SnSe single crystal holds the highest ZT record of 2.6 at 923 K. This single crystal possesses ultralow $\kappa_L \sim 0.25$ W/(m×K) due to anharmonic scattering mechanism from lone s-pair electrons [101]. However, ZT of SnSe drops

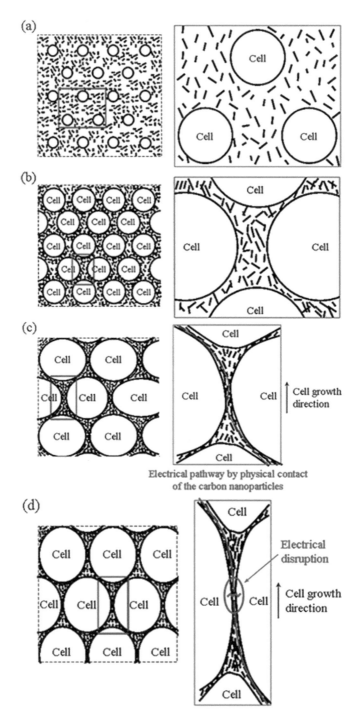

Fig. 5 Schematic of the cell formation process during foaming [97]

to 0.12 at room temperature; hence, it is not suitable for wearable applications. As this chapter has a focus on TE materials for wearable applications, here we examine TE materials that operate near room temperature. The most common room temperature inorganic TE materials are the tetradymites, alloys of general binary compounds $(Bi, Sb)_2(Te, Se)_3$ that crystallize in $R\bar{3}m$ crystal structure [100, 102, 103] due to high atomic weight. The material structure with high atomic weight can significantly reduce the speed of sound in the material and, therefore, decreases κ [13]. In addition, decrease in κ does not affect S and σ in bulk materials [104, 105].

Several strategies have been investigated to improve TE performance, including the use of nanostructured materials, electron band engineering to increase electronic transport, and phonon engineering to achieve low κ_L. When the material dimension decreases and approaches nanometer scale, it can lead to remarkable changes in electron density of states (DOS). Hicks and Dresselhaus [106] proposed that in low-dimensional TE materials, S increases with enhanced DOS. These low-dimensional TE materials include quantum dots, quantum wires, and quantum wells [107–109]. On the other hand, high-performance TE materials can be realized by developing phonon-glass electron-crystal [110] which allows sufficient charge carriers transportation in crystalline material and highly scattered phonons in amorphous materials, such as glass [105, 111]. The filled skutterudites and intermetallic clathrates exhibit such phonon-glass electron-crystal behavior and showed good TE performance. Full-spectrum phonon scattering with minimal charge carriers scattering dramatically improved ZT of p-type $Bi_{0.5}Sb_{1.5}Te_3$ alloys to 1.86 ± 0.15 at 320 K [112]. The highest $ZT \sim 2.4$ at room temperature has been achieved for p-type Bi_2Te_3/Sb_2Te_3 superlattice devices [113] by control of the transport of phonons and electrons in the superlattices. They have also achieved $ZT \sim 1.46$ for n-type $Bi_2Te_3/Bi_2Te_{2.83}Se_{0.17}$. On the other hand, $ZT \sim 1.2$ at 357 K has been realized with n-type Bi_2Te_3-based TEG, and this high TE performance was attributed to synergistic combination of reduced κ_L and high PF [114]. Some other promising TE materials that operate at room temperature have also been investigated, including MgAgSb ($MgAg_{0.965}Ni_{0.005}Sb_{0.99}$) [115–117] and transparent copper iodide (CuI) thin films [118–120]. Zhao et al. [115] doped Ni into MgAgSb to achieve high-performance $MgAg_{0.965}Ni_{0.005}Sb_{0.99}$, which has low κ of 0.5–0.6 W/(m×K) from room temperature to 570 K due to strong phonon scattering caused by the deficiencies at Ag, Sb site, and very small grain size. Furthermore, as mentioned in Sect. 2.2, organic TE composite materials and flexible inorganic/organic superlattice materials can significantly enhance PF of organic TE materials without significantly degrading thermal insulation [121].

Wan et al. [122] developed hybrid inorganic/organic superlattice $TiS_2/[(hexylammonium)_x(H_2O)_y(DMSO)_z]$ by facial electrochemical intercalation and solvent exchange methods. The organic cations in the electrolyte were intercalated into van der Waals gap in TiS_2 layers driven by coulomb force. Electrons were externally injected into inorganic layers and stabilized by organic cations, providing electrons for current and energy transport. This hybrid superlattice material possesses small κ of 0.12 W/(m×K), which is two orders of magnitude smaller than that of single layer and bulk TiS_2, with σ of 790 S/cm and ZT of 0.28 at 373 K. Du et al.

Table 2 Inorganic thermoelectric materials working at room temperature

Material	Type	ZT	PF, $\mu W/(m \times K^2)$	S, $\mu V/K$	σ, S/cm	κ, W/(m×K)	Reference
1	2	3	4	5	6	7	8
$\Upsilon - CuI$	p	0.21	3.75E2	620–890	280	0.55	[120]
$MgAg_{0.965}Ni_{0.005}Sb_{0.99}$	p	1	–	210	~380	0.7	[115]
Bi_2Te_3/Sb_2Te_3	p	2.4	–	238	–	5.8E-3	[113]
$Bi_{0.5}Sb_{1.5}Te_3$	p	1.86 ± 0.15	4E3	230	~730	0.33	[112]
$\Upsilon - Ag_2Te$	p	–	0.66	1330	3.7	0.18	[124]
WS_2	n	–	5–7	75	14	–	[125]
$NbSe_2$	p	–	26–34	14	1700	–	[125]
Yb- filled $CoSb_3$	n	0.4	4700	150	–	~3.5	[126]
$Bi_{0.5}Sb_{1.5}Te_3$	p	0.05	32.26	16	1295.21	0.2	[123]
PEDOT:PSS/TiS_2	n	0.28	450	–	790	0.12	[122]

[123] mixed different amount of $Bi_{0.5}Sb_{1.5}Te_3$ (BST) nanosheets into PEDOT:PSS and fabricated composite thin film by drop casting. They found that 4.1 wt. % BST/ PEDOT:PSS composite film has S of 16 $\mu V/K$, high σ of 1295 S/cm, and PF of 32.26 $\mu W/(m \times K^2)$. Table 2 summarizes high-performance inorganic TE materials working around room temperature, including properties, working temperature, and TEG device performance in the literature.

There are excellent articles that reviewed inorganic TE materials [100, 127], which focus on the strategies to optimize bulk TE material properties. In the following sections, we will focus on inorganic TEG with flexible substrate and self-supporting inorganic TEG. Wearable electronics require soft and flexible devices that can be attached to the human skin or clothes. Although inorganic TE materials possess high ZT values, those are restricted for the application due to non-flexible structure, non-scalable manufacturing techniques, and too expensive to mass produce. Researchers have been looking for methods to make inorganic TEG flexible without affecting significantly TE efficiency. Methods such as bonding/depositing bulk and thin film TE modules on flexible substrate have been intensively investigated. Table 3 presents performance of flexible TEGs based on inorganic TE materials.

2.3.1 Inorganic TEGs with Flexible Supporting Substrate

Commercially, inorganic TE legs can be simply physically attached to flexible substrate by thermal compression bonding with different bonding materials [128–130]. Figure 6a shows a typical TEG attached on human wrist assembled by vacuum soldering. This TEG can generate voltage about 3 mV when operator stands still, and it increases to 11.2 mV when operator is walking as the hand swing increased temperature difference during walking [128]. Researchers have also tried to manually paint thin layers of TE materials on flexible substrate [131],

Table 3 Inorganic flexible TEG devices working around room temperature

Materials	Substrate	Fabrication technique	σ, S/cm	κ, W/(m×K)	Processing temperature °C	ZT	PF, μW/(m×K²)	N of $p-n$ couples	Output voltage, V	Output power or power density	ΔT, K	Reference
1	2	3	4	5	6	7	8	9	10	11	12	13
$Bi_2Te_{2.7}Se_{0.3}$ nanoparticles	Polyimide	Ink-jet printing	6E3	–	400	–	183	–	–	139 μV K^{-1}	–	[134]
Bi_2Te_3 and Sb_2Te_3	Standard paper	PVD^a	–	2 and 2.5	RT	0.016 and 0.0056	102 and 44.6	–	–	0.5 nW	50	[133]
$\gamma - Ag_2Te$	Polyethersulfone	Spin coating	3.7E-4	0.18	RT	–	0.66	1	4.2E-3	–	2.2	[124]
$Bi_{0.5}Sb_{1.5}Te_3$ and $Bi_2Te_{2.7}Se_{0.3}$	Copper foil	Soldering	0.5E-4 and 0.7E-4	1 and 1	750 and 800	1.2 and 0.7	–	40	9E-3	80 μW	–	[146]
Bi_2Te_3 Sb_2Te_3	Silk	Manually deposit	–	–	120	–		12	10E-3	15 nW	5–35	[141]
$\gamma - CuI$	PET^b	Sputtering	280	0.55	180	0.21	3.75e2	1	–	8.2 nW	10.8	[119]
WS_2 and $NbSe_2$	PDMS^c	Contact printing	10–14, (1.5–1.7)E3	–	80	–	5–7 and 26–34	100	2.4E-3	38 nW	60	[125]
$Bi_{0.5}Sb_{1.5}Te_3$ and Bi_2Se_3	None	Cup and tube covered by fiber TE couples	769 and 1562	1.3 and 2.0	847	1.25 and 0.23	3.52E3 and 650	18 and different number	70E-3	1.46 mW cm^{-2}	60	[144]
Bi_2Te_3 Sb_2Te_3	Si	MEMS	–	1.01	–	–	–	127	6.1E-3	0.29 W m^{-2}	50	[147]
$Bi_{0.5}Sb_{1.5}Te_3$ and $Bi_2Te_{2.7}Se_{0.3}$	Flexible PCB^d	Soldering	2028 and 1651	1.73 and 1.88	–	0.68	–	200	37E-3	15 μW cm^{-2}	12	[148]

^{a}PVD physical vapor deposition
^{b}PET polyethylene terephthalate
^{c}PDMS polydimethylsiloxane
^{d}PCB printed circuit board

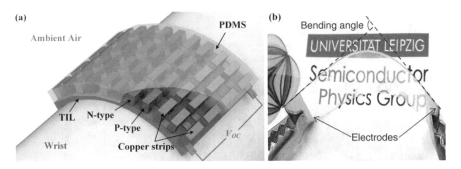

Fig. 6 (**a**) Wearable TEG on the wrist fabricated by vacuum soldering [128]; (**b**) flexible transparent Υ – CuI deposited on PET substrate [120]

and TEG with 5 $p - n$ couples achieved voltage output of 54 mV at ΔT of 50 K. In more reliable and controlled way, Fan et al. [132] deposited Al-doped ZnO (n-type) and β – Zn_4Sb_3 (p-type) on polyimide substrate by DC magnetron sputtering. The maximum output power of 10 $p - n$ couples was 246.3 μW at larger temperature difference of 180 K. Yang et al. [120] fabricated transparent and flexible Υ – CuI TEG by depositing polycrystalline CuI films on flexible polyethylene terephthalate (PET) substrates with reactive sputtering at room temperature, as shown in Fig. 6b. The resistance changed only <3% during bending of transparent thin Υ – CuI film up to 90°, and resistance change was <0.2% after repeated bending up to 400 cycles. This Υ – CuI TEG achieved ZT of 0.21 at 300 K which is the highest reported value in transparent TEG to date.

Paper has been an emerging substrate for TE devices as it is light, cheap, breathable, and foldable [132]. Affordable TEG fabricated on paper substrate by magnetron sputtering was reported by Rojas et al. [133]. This ability to be folded in diverse ways helps reduce the footprint of TE device significantly and further increase power density generation. Gao et al. [132] developed novel glass-fiber-aided cold-press method to achieve flexible n-type Ag_2Te nanowire films on copy paper substrate. It shows PF of 192.2 μW/(m×K^2) at 195 °C, and its σ has been significantly enhanced to 9.5 × 10^3 S/cm with compressive stress due to the disappearance of grain boundaries in Ag_2Te nanowires film.

For potential mass production, flexible planar thin film TEGs can be fabricated with sputtering and several printing techniques, including dispenser printing, inject printing, and screen printing. TE materials mixed with dispersing agent and conducting silver nanoparticles can be inject printed onto polyimide substrate to form thin film flexible TEG [134]. The maximum PF of printed film is ~77 μW/(m×K^2) at 75 °C, and maximum thermopower is ~139 μV/K at 50 °C, which can be controlled by adjusting the number N of $p - n$ couples. Dispenser printing technique has also been investigated to fabricate series-parallel 50 $p - n$ couples on custom-designed polyimide substrate [135]. Se-doped mechanically alloyed Bi_2Te_3 was used as n-type material, whereas Te-doped $Bi_{0.5}Sb_{1.5}Te_3$ was used as p-type material. This device achieved power output 33 μW and power density of 2.8 W/m^2.

Screen printing is a cost-effective approach [136–138] and well suited for mass production. Cao et al. [136, 137] used screen printing technique to deposit n-type

Bi_2Te_3 and p-type Sb_2Te_3 on flexible polyimide substrate and studied TE performance enhancement with different binding materials. Cold isostatic pressing process has been used to compress the printed materials to create denser film and smoother surface and further improve densification and materials properties. The single screen-printed module can generate a voltage of 6 mV and peak output power of 48 nW at temperature difference of 20 K. The maximum PF reached 2.15 μW/(m×K²) [137]. Although printable TEGs exhibit low ZT, inherent low κ makes such TEGs particularly useful for wearable applications [16]. However, it has been found that screen printing can introduce porosity to TE modules and leads to significant degradation of TE performance compared with bulk counterpart. In addition, the flexibility of polymer substrate is limited, and lowest bending radius of the film is around 30 mm, which restricts its surface conformability and application in wearable electronics [139]. Flexible polymer substrates cause thermal energy loss and tensile/compressive strain into the active region containing TE materials and electrodes during bending, resulting in performance degradation.

2.3.2 Self-Supporting Inorganic TEGs Without a Substrate

Recently, there are several studies that have investigated self-supporting TEG structure that does not require external supporting substrate. Fiber textile-based TEGs have been an emerging technique due to 3D deformation, lightweight nature, and desirability for applications in wearable electronic systems that require large mechanical deformations, high energy conversion efficiency, and electrical stability [140]. Kim et al. [52] firstly deposited n-type Bi_2Te_3 and p-type Sb_2Te_3 onto glass fabric with the aid of screen printing technique and incorporated this structure into flexible rubber sheet. Figure 7a shows Bi_2Te_3 and Sb_2Te_3 dots on glass fabric and scanning electron microscopy (SEM) images of cross section of the dots. In SEM image, glass fabric embedded in TE materials serves as a support to TE materials and thermal blocker, as well as maintains the patterns. This light and flexible TEG on glass fabric can achieve a small bending radius of as low as 20 mm and was able to generate high power output of 28 mW/g at $\Delta T = 50$ K. Similar to this work, nanostructured TE materials have been deposited onto both sides of silk fabric [141]; generated voltage and power output rose linearly with increase in ΔT and reached 10 mV and 15 nW at ΔT of 35 K. Furthermore, Siddique et al. [142] investigated manual dispenser printing technique to manufacturing TEG. The holes of the polyester fabric were manually filled with liquid TE paste with dispenser and cured in a furnace at temperature of 160 °C. TE module with 12 $p - n$ couples connected by silver conductive thread can generate voltage and power of 23.9 mV and 3.107 nW, respectively, at ΔT of 22.5 K.

Novel woven and knitted TE textiles have been fabricated by Lee et al. [143] by knitting n- and p-type TE yarns into patterns that electrically connect segments of these yarns in series and in parallel. In Fig. 7b, TE yarns were twisted; electrospun polyacrylonitrile (PAN) nanofiber cores were coated with TE semiconductor sheath, and gold-coated yarns between n- and p-type-coated segments improved electrical interconnection and improved thermal contact. TE textiles had ZT of 0.24 and 0.07

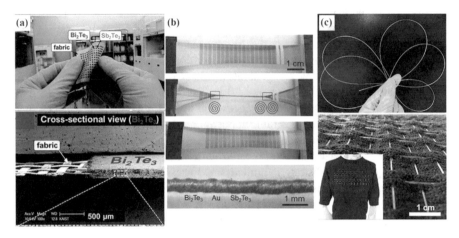

Fig. 7 (**a**) Image of 196 Bi$_2$Te$_3$ and Sb$_2$Te$_3$ dots on glass fabric of 40 mm × 40 mm and SEM images of screen-printed Bi$_2$Te$_3$ thick films on glass fabric [52]. (**b**) Image of Sb$_2$Te$_3$/PAN yarn [143]. (**c**) Single TE fibers with a length of 1 m and TE fiber woven into a large area fabric [144]

for *p*- and *n*-type TE materials, respectively, and were able to generate 0.52 Wm^{-2} at $\Delta T = 50$ K. Furthermore, another freestanding flexible TEG with no substrate has been developed by Kim et al. [145]: *p*-type Bi$_{0.3}$Sb$_{1.7}$Te$_3$ and *n*-type Bi$_2$Se$_{0.3}$Te$_{2.7}$ thick film were screen printed on SiO$_2$/a − Si/quartz substrate, and laser multi-scanning lift-off method has been used to fully separate the rigid quartz substrate from original TEG by selective reaction of XeCl excimer laser with the exfoliation a − Si layer. This freestanding TEG array with 72 modules can generate 4.78 mW/cm^2 and 20.8 mW/g at temperature difference of 25 K. Figure 7c shows flexible 1 m long crystalline TE fiber, which has TE core (diameter = 33 μm) and outer glass cladding layer (400 μm) [144]. The fiber was thermally drawn from TE material rod preform to hundreds of meters in a vertical tube furnace with the hot zone temperature of 1323 K [144]. The fiber exhibits excellent flexibility; fiber with diameter of 50 μm can reach a minimum bending radius less than 1 cm.

Table 3 shows the inorganic material based flexible TEGs working around room temperature which have been discussed in the chapter. Inorganic TE materials have advantage of high *ZT* value but require curing and annealing at high-temperature which limits applications.

3 Structural Design

Currently most wearable TEG studies have focused on the performance improvement by optimizing material engineering and single device design. In Eq. (3), TEG efficiency is proportional to the temperature difference, which is the temperature

Fig. 8 Thermal
circuit for TEG

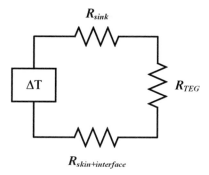

difference across TE legs. For practical wearable application using human body heat, only a small temperature gradient is applied across TEG because the surroundings have much larger thermal resistance than TEG between the human body and ambient temperature; therefore, generated power depends largely on the system design. In particular, thermal matching is important for TEGs to maximize the generated power [149]. The temperature difference across TEG is limited by the very large thermal resistance of the surroundings and the small difference between body temperature and ambient temperature. Figure 8 represents a simplified thermal circuit for TEGs. TEG attached on human skin has thermal resistance of $p - n$ legs R_{TEG} and three important parasitic resistances, e.g., (i) skin thermal resistance R_{skin} that is dependent on specific physiological parameters such as age, body fat, and gender; (ii) thermal contact resistance at TEG-skin interface $R_{interface}$, which has significant impact on TEG performance since the human skin is inherently rough surface; and (iii) heatsink thermal resistance R_{sink} at cold side of TEG to increase in convection efficiency [16].

Several theoretical models of the impact of thermal resistance have been proposed [16, 149–154]; and obtained results showed that lowering κ is more critical than increasing σ to optimize TEG's performance. Suarez et al. [16] developed quasi-3D model to study the impact of all critical design parameters, including heatsink and skin contact resistance, fill factor, geometry design, filler materials, and external spreader, on the performance of body-wearable TEGs.

The easiest way to reduce thermal resistance is to attach of metal heatsink or heat spreader to the cold side of TEG [16, 152, 155, 156] as heatsink can directly impact the temperature drop across TEG. The modification of heatsink shapes, especially, increasing fin height to increase in surface area, can minimize the thermal resistance of the cold side of TEG [152]. For wearable TEG, the use of external spreader can double the performance. Commercial TEGs on the market make use of aluminum heatsinks, which are bulky, rigid, and unsuitable to integrate in wearable applications. Alternatively, Kim et al. [148] developed a new type of polymer-based flexible heatsink where the polymer particles store a large amount of water and the water slowly evaporate over time. With this flexible heatsink, TEG generated output power density of 38 μW/cm^2 for the first 10 minutes compared to 8 μW/cm^2 generated by TEG with metal heatsink.

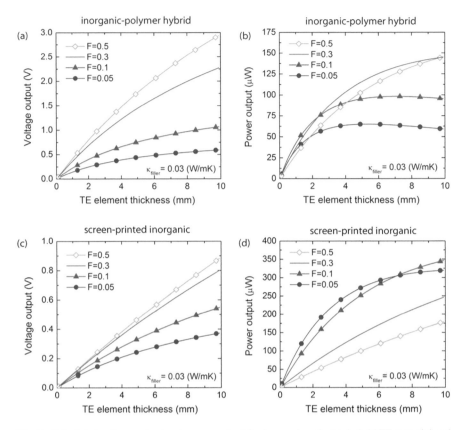

Fig. 9 Calculated voltage output and power output for inorganic polymer hybrid TE material and screen-printed inorganic TE material as a function of TE leg thickness for different fill factors, κ of gap filler is 0.03 W/(m×K) [104]

The fill factor, which is the fractional area coverage of TE legs to overall surface area of the whole device, and filler materials are two important parameters to indicate device performance [157]. The gap between TE legs can be filled with either air or good thermal-insulated polymer materials. Air has very low $\kappa \sim 0.025$ W/(m×K) at room temperature. The mostly used polymer filler for TEG devices is polydimethylsiloxane (PDMS) due to its good flexibility and low $\kappa \sim 0.15$ W/(m×K) [148, 158, 159]. In theory, Yazawa et al. [160] stated that using low fill factor could increase electrical power output per unit mass. Figure 9 presented the voltage and power output with various fill factor [104]. It can be seen that, with increasing fill factor, inorganic polymer and inorganic TE materials have same trend in voltage output but different trend in power output. For inorganic TEG, most of the heat flows through inorganic TE materials because κ of inorganic material is much larger than the filler. Therefore, the output power increases with the reduced fill factor. On the opposite side, κ of organic TE material is comparable to that of gap filler. If the fill factor is small, the parasitic heat conduction through the gap filler becomes significant and reduces the power output. Therefore, the higher output power can be generated with larger fill factor. Gordiz et al. [156] achieved large fill factor (~91%) for screen-printed organic TEG compared to typical commercial inorganic TEGs (~25%) by arranging $p - n$ legs in a closed-packed hexagonal layout and wiring those based on Hilbert space-filling curve pattern.

4 Hybrid Energy Harvesters

In many cases, energy scavenging simultaneously from multiple energy sources is preferred as (i) while the primary source powers electronic devices, the secondary transducer is able to power up auxiliary circuit in the ultralow-power management units [161]; and (ii) the combination can generate additional electrical energy and make up some limitations from single energy sources, e.g., solar cells have limitation on low-energy density, seasonal accessibility, and geographical dependency; piezoelectric and electromagnetic energy harvester generally generates relatively low output and possess complex fabrication processes; triboelectric harvester has very low current output due to the huge impedance of the insulation materials. However, it is not easy to combine as there are confrontations between different harvesting methods.

Photo-thermoelectric harvesters are the most widely investigated hybrid harvesters [162–172]. Concentrating photovoltaic (CPV) cell exhibits the highest conversion efficiency among solar cells, but the performance is affected by its high operating temperature due to higher internal recombination of charge carriers in PV cell [173]. The solar cell is usually placed on the hot side of TEG, while the bottom side of TEG is contacted to the heatsink. Utilizing dissipated thermal energy from solar cells to drive the temperature difference across TEG has significant potential to further improve conversion efficiency of the solar cells. However, low κ of TE materials blocked the heat dissipation from solar cell to heatsink as well. Kil et al. [168] developed high-performance hybrid CPV/TEG, as shown in Fig. 10, by replacing GaAs substrate with Si substrate as Si has higher κ and thus enhanced the thermal flow and reduced temperature of CPV. In addition, load resistance impedance matching is critical as TEG feedback of Peltier effects can further reduce CPV temperature, while maintaining high efficiency of TEG [168]. Within these two effects, hybrid CPV/TEG was able to achieve efficiency around 23%, which is ~3% higher than single CPV cell at the solar concentration of 50 suns. Flexible and lightweight photo-thermoelectric nanogenerator (PTENG) can be fabricated by [170] simply adhering molybdenum disulfide/polyurethane (MoS_2/PU) photothermal film onto TE Te nanowire/PEDOT layer with silver paste. By exposing this PTENG under outdoor sunlight with atmospheric temperature of 20 °C, it can deliver an output voltage of 1.48 mV.

TEGs have also been combined with pyroelectric energy harvesters [174]. Pyroelectric energy harvesting [6] converts temperature fluctuation into electrical energy, but the efficiency is dependent on the frequency of temperature changes. It

Fig. 10 Hybrid
photovoltaic/thermoelectric
generator [168]

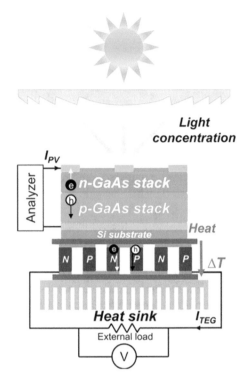

is not easy to collect human body heat using pyroelectric generators because the body temperature is stable and maintained when a person wears clothes. When a person sweats, the skin temperature would drop considerably as the sweat evaporates, and pyroelectric harvester would be much helpful than TEG. Circular mesh polyester knit fabric with p- and n-type TE materials $Bi_{0.5}Sb_{1.5}Te_3$ and $Bi_2Te_{2.7}Se_{0.3}$ was fabricated on the top of pyroelectric polyvinylidene fluoride (PVDF) film with circular-patterned electrodes. The hybrid harvester was then packaged by sewing a quick sweat-pick up/dry fabric. Although this is a hybrid harvester and it is a good demonstration for wearable applications, the level of generated energy was not sufficient to portable electronics [174].

Mechanical energy is common and abundant in environments featuring machinery, human activity, water, etc. Coupling mechanical and thermal energy harvesters is not easy because TEG is rigid and static, while mechanical energy harvester such as piezoelectric generators (PEG) is normally flexible and dynamic. In addition, TEGs are low resistive element and generate DC voltage signal, but PEG are capacitive dielectrics and generate AC signal [175]. If TEG and PEG connect electrically in series, then capacitor creates open circuit decreasing dramatically power output of TEG, if in parallel, then low resistance of TEG will discharge PEG capacitor, thereby causing only TEG to generate power [176]. Montgomery et al. [176] developed flexible thin film structure to overcome the coupling issue and allow TEG and

PEG to work orthogonally without reducing each other's performance. The hybrid generator TPEG consists of PEG based on piezoelectric film (PVDF / Poly(vinylidene) fluoride) and TEG made of alternating p-type CNT/PVDF films and n-type CNT/PVDF doped with PEI (Polyethyleneimine) films. This device was folded at $p - n$ junctions to allow for maximum thermal gradient across the thickness of the device and able to generate 89% of the maximum TE power and achieve 5.3 times more piezoelectric voltage when compared with traditional device. TEG can be integrated with triboelectric nanogenerator (TENG) as well. TENG [177] based on the coupling between the triboelectric effect and the electrostatic effect can generate large voltage but still present a challenge on the efficiency. Kim et al. [138] have developed a hybrid nanogenerator that integrates a sliding-type TENG and solid-state TEG by attaching a commercial TEG to the bottom of a polytetrafluoroethylene (PTFE) film with an adhesive phase change material layer, which ensures a good thermal connection. The friction heat results from the friction between sliding aluminum panel and PTFE layer (sliding velocity ~ 12 cm/s) and acts as thermal input for TEG. In this hybrid TENG/TEG, TENG and TEG can generate 21.9 V and 25.8 mV, respectively. Power density of hybrid harvester was 14.98 mW/m², and the charging rate is 13.3% higher than that of TENG alone.

5 Conclusions and Prospects

In this chapter, we have described thermoelectric materials and devices, with a particular emphasis on wearable applications and methods to improve power factor PF, figure of merit ZT, and overall efficiency of the conversion of thermal energy to electrical energy. The operational limits imposed by using body heat for wearable application in terms of maximum temperature gradients are discussed, along with the need for low weight, nontoxic, and flexible TE devices. Organic and inorganic TE materials have been described, including recent efforts to produce composite systems that use foams, particles, and fibers and the use of nanoscale structuring of TE material to control properties like Seebeck coefficient and balance the electrical and thermal conductivity. Organic and composite thermoelectrics provide potential for low cost, low density, and mechanical flexibility, but improvements are required to achieve properties of the more established inorganic materials. At the device level, a number of approaches have been considered to create TE elements for wearable applications that provide mechanical flexibility and durability and are able to produce sufficient power for wearable applications. In an effort to improve heat transfer for power generation, self-supporting devices have been developed along with the optimization of heatsinks for thermal matching to the human body. Electrical matching is also required. Since power levels are typically low due to the small area and low-temperature difference, the potential to combine TEG energy harvesting with other scavenging approaches is also being considered; these include photovoltaic, piezoelectric, and pyroelectric approaches. Nevertheless, the combination of improved materials and devices, combined with electronics which continue to require lower power levels, provides scope for the creation of self-powered wearable systems in the future.

Acknowledgment   Prof. Chris Bowen would like to acknowledge the funding from the European Research Council under the European Union's Seventh Framework Program (FP/2007–2013)/ ERC Grant Agreement no. 320963 on Novel Energy Materials, Engineering Science and Integrated Systems (NEMESIS). Prof. Sadao Kawamura and A/Prof. Shima Okada would like to acknowledge the funding from the Ritsumeikan Global Innovation Research Organization, Ritsumeikan University, on "Robotics Innovation Based on Advanced Materials".

References

1. A. Pantelopoulos, N.G. Bourbakis, A survey on wearable sensor-based systems for health monitoring and prognosis. IEEE Trans. Syst. Man Cybern. Part C Appl. Rev. **40**, 1–12 (2010)
2. A.P. Chandrakasan, N. Verma, D.C. Daly, Ultralow-power electronics for biomedical applications. Annu. Rev. Biomed. Eng. **10**, 247–274 (2008). https://doi.org/10.1146/annurev. bioeng.10.061807.160547
3. Y. Zang, F. Zhang, D. Huang, et al., Flexible suspended gate organic thin-film transistors for ultra-sensitive pressure detection. Nat. Commun. **6**, 6269 (2015). https://doi.org/10.1038/ ncomms7269
4. S. Priya, D.J. Inman, *Energy harvesting technologies* (Springer, New York, 2009)
5. G.J. Snyder, Thermoelectric energy harvesting, in *Energy Harvesting Technologies*, (Springer, New York, 2009), pp. 325–336
6. C.R. Bowen, J. Taylor, E. LeBoulbar, et al., Pyroelectric materials and devices for energy harvesting applications. Energy Environ. Sci. **7**, 3836–3856 (2014b). https://doi.org/10.1039/ C4EE01759E
7. D.M. Rowe, *Thermoelectrics Handbook: Macro to Nano* (CRC Press, Boca Raton, 2005)
8. G. Sebald, D. Guyomar, A. Agbossou, On thermoelectric and pyroelectric energy harvesting. Smart Mater. Struct. **18**, 125006 (2009). https://doi.org/10.1088/0964-1726/18/12/125006
9. X. Zhang, L.-D. Zhao, Thermoelectric materials: Energy conversion between heat and electricity. J. Mater. **1**, 92–105 (2015). https://doi.org/10.1016/j.jmat.2015.01.001
10. F.J. DiSalvo, Thermoelectric cooling and power generation. Science (80-) **285**, 703–706 (1999). https://doi.org/10.1126/science.285.5428.703
11. S.B. Riffat, X. Ma, Thermoelectrics: A review of present and potential applications. Appl. Therm. Eng. **23**, 913–935 (2003)
12. H.S. Kim, W. Liu, G. Chen, et al., Relationship between thermoelectric figure of merit and energy conversion efficiency. Proc. Natl. Acad. Sci. **112**, 8205–8210 (2015a). https://doi. org/10.1073/pnas.1510231112
13. A. Majumdar, Thermoelectricity in semiconductor nanostructures. Science (80-) **303**, 777–778 (2004)
14. P. Webb, Temperatures of skin, subcutaneous tissue, muscle and core in resting men in cold, comfortable and hot conditions. Eur. J. Appl. Physiol. Occup. Physiol. **64**, 471–476 (1992). https://doi.org/10.1007/BF00625070
15. H. Fang, J. Xia, K. Zhu, et al., Industrial waste heat utilization for low temperature district heating. Energy Policy **62**, 236–246 (2013). https://doi.org/10.1016/j.enpol.2013.06.104
16. F. Suarez, A. Nozariasbmarz, D. Vashaee, M.C. Öztürk, Designing thermoelectric generators for self-powered wearable electronics. Energy Environ. Sci. **9**, 2099–2113 (2016). https://doi. org/10.1039/c6ee00456c
17. G. Chen, W. Xu, D. Zhu, Recent advances in organic polymer thermoelectric composites. J. Mater. Chem. C **5**, 4350–4360 (2017)

18. H. Yao, Z. Fan, H. Cheng, et al., Recent development of thermoelectric polymers and composites. Macromol. Rapid Commun. **39**, 1700727 (2018)
19. J. Yang, H.-L. Yip, A.K.-Y. Jen, Rational design of advanced thermoelectric materials. Adv. Energy Mater. **3**, 549–565 (2013). https://doi.org/10.1002/aenm.201200514
20. T. Park, C. Park, B. Kim, et al., Flexible PEDOT electrodes with large thermoelectric power factors to generate electricity by the touch of fingertips. Energy Environ. Sci. **6**, 788–792 (2013). https://doi.org/10.1039/c3ee23729j
21. O. Bubnova, Z.U. Khan, H. Wang, et al., Semi-metallic polymers. Nat. Mater. **13**, 190–194 (2014). https://doi.org/10.1038/nmat3824
22. D. Lee, K. Cho, J. Choi, S. Kim, Effect of mesoscale grains on thermoelectric characteristics of aligned ZnO/PVP composite nanofibers. Mater. Lett. **142**, 250–252 (2015). https://doi.org/10.1016/j.matlet.2014.12.029
23. D.A. Mengistie, C. Chen, K.M. Boopathi, et al., Enhanced thermoelectric performance of PEDOT:PSS flexible bulky papers by treatment with secondary dopants. ACS Appl. Mater. Interfaces **7**, 94–100 (2015). https://doi.org/10.1021/am507032e
24. L. Persano, A. Camposeo, D. Pisignano, Active polymer nanofibers for photonics, electronics, energy generation and micromechanics. Prog. Polym. Sci. **43**, 48–95 (2015). https://doi.org/10.1016/j.progpolymsci.2014.10.001
25. Y.C. Sun, D. Terakita, A.C. Tseng, H.E. Naguib, Study on the thermoelectric properties of PVDF/MWCNT and PVDF/GNP composite foam. Smart Mater. Struct. **24**, 085034 (2015b). https://doi.org/10.1088/0964-1726/24/8/085034
26. M. Aghelinejad, S.N. Leung, Enhancement of thermoelectric conversion efficiency of polymer/carbon nanotube nanocomposites through foaming-induced microstructuring. J. Appl. Polym. Sci. **134**, 45073 (2017). https://doi.org/10.1002/app.45073
27. M. Eslamian, Inorganic and organic solution-processed thin film devices. Nano-Micro Lett. **9**, 3 (2017)
28. Z. Fan, P. Li, D. Du, J. Ouyang, Significantly enhanced thermoelectric properties of PEDOT:PSS films through sequential post-treatments with common acids and bases. Adv. Energy Mater. **7**, 1602116 (2017). https://doi.org/10.1002/aenm.201602116
29. O. Bubnova, X. Crispin, Towards polymer-based organic thermoelectric generators. Energy Environ. Sci. **5**, 9345–9362 (2012). https://doi.org/10.1039/c2ee22777k
30. B. Kamran, The Seebeck coefficient as a measure of entropy per carrier, in *Fundamentals of Thermoelectricity*, (Oxford University Press, Oxford, 2015)
31. M. Bharti, A. Singh, S. Samanta, D.K. Aswal, Conductive polymers for thermoelectric power generation. Prog. Mater. Sci. **93**, 270–310 (2018)
32. Z. Zhu, C. Liu, F. Jiang, et al., Effective treatment methods on PEDOT:PSS to enhance its thermoelectric performance. Synth. Met. **225**, 31–40 (2017c). https://doi.org/10.1016/j.synthmet.2016.11.011
33. B. Russ, M.J. Robb, F.G. Brunetti, et al., Power factor enhancement in solution-processed organic n-type thermoelectrics through molecular design. Adv. Mater. **26**, 3473–3477 (2014). https://doi.org/10.1002/adma.201306116
34. A.M. Glaudell, J.E. Cochran, S.N. Patel, M.L. Chabinyc, Impact of the doping method on conductivity and thermopower in semiconducting polythiophenes. Adv. Energy Mater. **5**, 149 (2015). https://doi.org/10.1002/aenm.201401072
35. A.G. MacDiarmid, A.J. Epstein, Secondary doping in polyaniline. Synth. Met. **69**, 85–92 (1995). https://doi.org/10.1016/0379-6779(94)02374-8
36. W. Ma, K. Shi, Y. Wu, et al., Enhanced molecular packing of a conjugated polymer with high organic thermoelectric power factor. ACS Appl. Mater. Interfaces **8**, 24737–24743 (2016). https://doi.org/10.1021/acsami.6b06899
37. A. Hamidi-Sakr, L. Biniek, J.L. Bantignies, et al., A versatile method to fabricate highly in-plane aligned conducting polymer films with anisotropic charge transport and thermoelectric properties: The key role of alkyl side chain layers on the doping mechanism. Adv. Funct. Mater. **27**, 1700173 (2017). https://doi.org/10.1002/adfm.201700173

38. J.H. Hsu, W. Choi, G. Yang, C. Yu, Origin of unusual thermoelectric transport behaviors in carbon nanotube filled polymer composites after solvent/acid treatments. Org. Electron. **45**, 182–189 (2017). https://doi.org/10.1016/j.orgel.2017.03.007

39. S. Liu, H. Deng, Y. Zhao, et al., The optimization of thermoelectric properties in a PEDOT:PSS thin film through post-treatment. RSC Adv. **5**, 1910–1917 (2015). https://doi.org/10.1039/c4ra09147g

40. R. Kroon, D.A. Mengistie, D. Kiefer, et al., Thermoelectric plastics: From design to synthesis, processing and structure-property relationships. Chem. Soc. Rev. **45**, 6147–6164 (2016)

41. B. O'Connor, R.J. Kline, B.R. Conrad, et al., Anisotropic structure and charge transport in highly strain-aligned regioregular poly(3-hexylthiophene). Adv. Funct. Mater. **21**, 3697–3705 (2011). https://doi.org/10.1002/adfm.201100904

42. L. Wang, Y. Liu, Z. Zhang, et al., Polymer composites-based thermoelectric materials and devices. Compos. Part B Eng. **122**, 145–155 (2017a)

43. H. Kaneko, T. Ishiguro, A. Takahashi, J. Tsukamoto, Magnetoresistance and thermoelectric power studies of metal-nonmetal transition in iodine-doped polyacetylene. Synth. Met. **57**, 4900–4905 (1993). https://doi.org/10.1016/0379-6779(93)90836-L

44. M. Culebras, B. Uriol, C.M. Gómez, A. Cantarero, Controlling the thermoelectric properties of polymers: Application to PEDOT and polypyrrole. Phys. Chem. Chem. Phys. **17**, 15140–15145 (2015). https://doi.org/10.1039/c5cp01940k

45. J. Li, X. Tang, H. Li, et al., Synthesis and thermoelectric properties of hydrochloric acid-doped polyaniline. Synth. Met. **160**, 1153–1158 (2010). https://doi.org/10.1016/j.synthmet.2010.03.001

46. B.T. McGrail, A. Sehirlioglu, E. Pentzer, Polymer composites for thermoelectric applications. Angew. Chem. Int. Ed. **54**, 1710–1723 (2015). https://doi.org/10.1002/anie.201408431

47. G.H. Kim, L. Shao, K. Zhang, K.P. Pipe, Engineered doping of organic semiconductors for enhanced thermoelectric efficiency. Nat. Mater. **12**, 719–723 (2013). https://doi.org/10.1038/nmat3635

48. Q. Zhang, Y. Sun, W. Xu, D. Zhu, Thermoelectric energy from flexible P3HT films doped with a ferric salt of triflimide anions. Energy Environ. Sci. **5**, 9639–9644 (2012). https://doi.org/10.1039/c2ee23006b

49. K. Shi, F. Zhang, C.A. Di, et al., Toward high performance n-type thermoelectric materials by rational modification of BDPPV backbones. J. Am. Chem. Soc. **137**, 6979–6982 (2015). https://doi.org/10.1021/jacs.5b00945

50. R.A. Schlitz, F.G. Brunetti, A.M. Glaudell, et al., Solubility-limited extrinsic n-type doping of a high electron mobility polymer for thermoelectric applications. Adv. Mater. **26**, 2825–2830 (2014). https://doi.org/10.1002/adma.201304866

51. Y. Wang, M. Nakano, T. Michinobu, et al., Naphthodithiophenediimide-Benzobisthiadiazole-based polymers: Versatile n-type materials for field-effect transistors and thermoelectric devices. Macromolecules **50**, 857–864 (2017c). https://doi.org/10.1021/acs.macromol.6b02313

52. S.J. Kim, J.H. We, B.J. Cho, A wearable thermoelectric generator fabricated on a glass fabric. Energy Environ. Sci. **7**, 1959–1965 (2014). https://doi.org/10.1039/c4ee00242c

53. J. Liu, H.Q. Yu, Thermoelectric enhancement in polyaniline composites with polypyrrole-functionalized multiwall carbon nanotubes. J. Electron. Mater. **43**, 1181–1187 (2014). https://doi.org/10.1007/s11664-013-2958-4

54. L. Liang, G. Chen, C.-Y. Guo, Polypyrrole nanostructures and their thermoelectric performance. Mater. Chem. Front. **1**, 380–386 (2017a). https://doi.org/10.1039/C6QM00061D

55. Y. Wang, J. Yang, L. Wang, et al., Polypyrrole/graphene/polyaniline ternary nanocomposite with high thermoelectric power factor. ACS Appl. Mater. Interfaces **9**, 20124–20131 (2017e). https://doi.org/10.1021/acsami.7b05357

56. J. Wu, Y. Sun, W.B. Pei, et al., Polypyrrole nanotube film for flexible thermoelectric application. Synth. Met. **196**, 173–177 (2014). https://doi.org/10.1016/j.synthmet.2014.08.001

57. Y.W. Park, Y.S. Lee, C. Park, et al., Thermopower and conductivity of metallic polyaniline. Solid State Commun. **63**, 1063–1066 (1987). https://doi.org/10.1016/0038-1098(87)90662-4

58. E. Dalas, S. Sakkopoulos, E. Vitoratos, Chemical preparation, direct-current conductivity and thermopower of polyaniline and polypyrrole composites. J. Mater. Sci. **29**, 4131–4133 (1994). https://doi.org/10.1007/BF00355982
59. K. Lee, S. Cho, H.P. Sung, et al., Metallic transport in polyaniline. Nature **441**, 65–68 (2006). https://doi.org/10.1038/nature04705
60. Q. Wang, Q. Yao, J. Chang, L. Chen, Enhanced thermoelectric properties of CNT/PANI composite nanofibers by highly orienting the arrangement of polymer chains. J. Mater. Chem. **22**, 17612–17618 (2012). https://doi.org/10.1039/c2jm32750c
61. D. Yoo, J.J. Lee, C. Park, et al., N-type organic thermoelectric materials based on polyaniline doped with the aprotic ionic liquid 1-ethyl-3-methylimidazolium ethyl sulfate. RSC Adv. **6**, 37130–37135 (2016). https://doi.org/10.1039/c6ra02334g
62. L. Wang, Q. Yao, H. Bi, et al., PANI/graphene nanocomposite films with high thermoelectric properties by enhanced molecular ordering. J. Mater. Chem. A **3**, 7086 (2015). https://doi.org/10.1039/c4ta06422d
63. L. Wang, Q. Yao, W. Shi, et al., Engineering carrier scattering at the interfaces in polyaniline based nanocomposites for high thermoelectric performances. Mater. Chem. Front. **1**, 741–748 (2017b). https://doi.org/10.1039/C6QM00188B
64. C. Nath, A. Kumar, Y.-K. Kuo, G.S. Okram, High thermoelectric figure of merit in nanocrystalline polyaniline at low temperatures. Appl. Phys. Lett. **105**, 13310 (2014). http://scitation.aip.org/docserver/fulltext/aip/journal/apl/105/13/1.4897146.pdf?expires=1428674603&id=id&accname=2102614&checksum=5C84CE30FAC03AF2551EC6461F77E525%5Cnhttp://scitation.aip.org/content/aip/journal/apl/105/13/10.1063/1.4897146
65. R.D. McCullough, The chemistry of conducting polythiophenes. Adv. Mater. **10**, 93–116 (1998). https://doi.org/10.1002/(SICI)1521-4095(199801)10:2<93::AID-ADMA93>3.0.CO;2-F
66. H. Shimotani, G. Diguet, Y. Iwasa, Direct comparison of field-effect and electrochemical doping in regioregular poly(3-hexylthiophene). Appl. Phys. Lett. **86**, 022104 (2005). https://doi.org/10.1063/1.1850614
67. M. He, J. Ge, Z. Lin, et al., Thermopower enhancement in conducting polymer nanocomposites via carrier energy scattering at the organic-inorganic semiconductor interface. Energy Environ. Sci. **5**, 8351–8358 (2012). https://doi.org/10.1039/c2ee21803h
68. O. Bubnova, Z.U. Khan, A. Malti, et al., Optimization of the thermoelectric figure of merit in the conducting polymer poly(3,4-ethylenedioxythiophene). Nat. Mater. **10**, 429–433 (2011). https://doi.org/10.1038/nmat3012
69. Q. Wei, M. Mukaida, K. Kirihara, et al., Recent progress on PEDOT-based thermoelectric materials. Materials (Basel) **8**, 732–750 (2015)
70. M. Culebras, C.M. Gómez, A. Cantarero, Enhanced thermoelectric performance of PEDOT with different counter-ions optimized by chemical reduction. J. Mater. Chem. A **2**, 10109–10115 (2014). https://doi.org/10.1039/c4ta01012d
71. J. Zhao, D. Tan, G. Chen, A strategy to improve the thermoelectric performance of conducting polymer nanostructures. J. Mater. Chem. C **5**, 47–53 (2017a). https://doi.org/10.1039/C6TC04613D
72. K. Sun, S. Zhang, P. Li, et al., Review on application of PEDOTs and PEDOT:PSS in energy conversion and storage devices. J. Mater. Sci. Mater. Electron. **26**, 4438–4462 (2015a)
73. L. Groenendaal, F. Jonas, D. Freitag, et al., Poly(3,4-ethylenedioxythiophene) and its derivatives: Past, present, and future. Adv. Mater. **12**, 481–494 (2000). https://doi.org/10.1002/(SICI)1521-4095(200004)12:7<481::AID-ADMA481>3.0.CO;2-C
74. P. Kar, *Doping in Conjugated Polymers* (Wiley, Hoboken, 2013)
75. X. Zhao, D. Madan, Y. Cheng, et al., High conductivity and electron-transfer validation in an n-type fluoride-anion-doped polymer for thermoelectrics in air. Adv. Mater. **29**, 1606928 (2017b). https://doi.org/10.1002/adma.201606928
76. S. Wang, H. Sun, U. Ail, et al., Thermoelectric properties of solution-processed n-doped ladder-type conducting polymers. Adv. Mater. **28**, 10764–10771 (2016). https://doi.org/10.1002/adma.201603731

77. N. Toshima, Recent progress of organic and hybrid thermoelectric materials. Synth. Met. **225**, 3–21 (2017). https://doi.org/10.1016/j.synthmet.2016.12.017

78. A.M. Marconnet, N. Yamamoto, M.A. Panzer, et al., Thermal conduction in aligned carbon nanotube-polymer nanocomposites with high packing density. ACS Nano **5**, 4818–4825 (2011). https://doi.org/10.1021/nn200847u

79. H. Deng, L. Lin, M. Ji, et al., Progress on the morphological control of conductive network in conductive polymer composites and the use as electroactive multifunctional materials. Prog. Polym. Sci. **39**, 627–655 (2014)

80. G.P. Moriarty, S. De, P.J. King, et al., Thermoelectric behavior of organic thin film nanocomposites. J. Polym. Sci. B **51**, 119–123 (2013). https://doi.org/10.1002/polb.23186

81. G. Mechrez, R.Y. Suckeveriene, E. Zelikman, et al., Highly-tunable polymer/carbon nanotubes systems: Preserving dispersion architecture in solid composites via rapid microfiltration. ACS Macro Lett. **1**, 848–852 (2012). https://doi.org/10.1021/mz300145a

82. K. Zhang, J. Qiu, S. Wang, Thermoelectric properties of PEDOT nanowire/PEDOT hybrids. Nanoscale **8**, 8033–8041 (2016). https://doi.org/10.1039/c5nr08421k

83. A.J. Minnich, M.S. Dresselhaus, Z. Ren, G. Chen, Bulk nanostructured thermoelectric materials: Current research and future prospects. Energy Environ. Sci. **2**, 466 (2009). https://doi.org/10.1039/b822664b

84. Z. Liang, M.J. Boland, K. Butrouna, et al., Increased power factors of organic-inorganic nanocomposite thermoelectric materials and the role of energy filtering. J. Mater. Chem. A **5**, 15891–15900 (2017b). https://doi.org/10.1039/c7ta02307c

85. J. Xiong, F. Jiang, H. Shi, et al., Liquid exfoliated graphene as dopant for improving the thermoelectric power factor of conductive PEDOT:PSS nanofilm with hydrazine treatment. ACS Appl. Mater. Interfaces **7**, 14917–14925 (2015). https://doi.org/10.1021/acsami.5b03692

86. L. Zhang, Y. Harima, I. Imae, Highly improved thermoelectric performances of PEDOT:PSS/SWCNT composites by solvent treatment. Org. Electron. **51**, 304–307 (2017a). https://doi.org/10.1016/j.orgel.2017.09.030

87. L. Yan, M. Shao, H. Wang, et al., High seebeck effects from hybrid metal/polymer/metal thin-film devices. Adv. Mater. **23**, 4120–4124 (2011). https://doi.org/10.1002/adma.201101634

88. D.G. Cahill, W.K. Ford, K.E. Goodson, et al., Nanoscale thermal transport. J. Appl. Phys. **93**, 793–818 (2003)

89. C. Dames, G. Chen, Thermal conductivity of nanostructured thermoelectric materials, in *Thermoelectrics Handbook Macro to Nano*, (Boca Raton, CRC Press, 2005), p. 1014

90. C. Yu, K. Choi, L. Yin, J.C. Grunlan, Light-weight flexible carbon nanotube based organic composites with large thermoelectric power factors. ACS Nano **5**, 7885–7892 (2011). https://doi.org/10.1021/nn202868a

91. K. Biswas, J. He, I.D. Blum, et al., High-performance bulk thermoelectrics with all-scale hierarchical architectures. Nature **489**, 414–418 (2012). https://doi.org/10.1038/nature11439

92. L.D. Zhao, V.P. Dravid, M.G. Kanatzidis, The panoscopic approach to high performance thermoelectrics. Energy Environ. Sci. **7**, 251–268 (2014c)

93. L. Wang, Q. Yao, H. Bi, et al., Large thermoelectric power factor in polyaniline/graphene nanocomposite films prepared by solution-assistant dispersing method. J. Mater. Chem. A **2**, 11107–11113 (2014). https://doi.org/10.1039/c4ta01541j

94. A. Bejan, Porous media, in *Heat Transfer Handbook*, ed. by A. Bejan, A. Kraus, (Wiley, Hoboken, 2003)

95. Q. Zhang, Y. Sun, W. Xu, D. Zhu, Organic thermoelectric materials: Emerging green energy materials converting heat to electricity directly and efficiently. Adv. Mater. **26**, 6829–6851 (2014b). https://doi.org/10.1002/adma.201305371

96. M. Antunes, J.I. Velasco, V. Realinho, et al., Heat transfer in polypropylene-based foams produced using different foaming processes. Adv. Eng. Mater. **11**, 811 (2009). https://doi.org/10.1002/adem.200900129

97. M. Antunes, J.I. Velasco, Multifunctional polymer foams with carbon nanoparticles. Prog. Polym. Sci. **39**, 486–509 (2014)

98. W. Yang, W. Zou, Z. Du, et al., Enhanced conductive polymer nanocomposite by foam structure and polyelectrolyte encapsulated on carbon nanotubes. Compos. Sci. Technol. **123**, 106–114 (2016b). https://doi.org/10.1016/j.compscitech.2015.12.009

99. M.P. Tran, C. Detrembleur, M. Alexandre, et al., The influence of foam morphology of multi-walled carbon nanotubes/poly(methyl methacrylate) nanocomposites on electrical conductivity. Polymer (United Kingdom) **54**, 3261–3270 (2013). https://doi.org/10.1016/j.polymer.2013.03.053

100. T. Zhu, Y. Liu, C. Fu, et al., Compromise and synergy in high-efficiency thermoelectric materials. Adv. Mater. **29**, 1605884 (2017a)

101. L.-D. Zhao, S.-H. Lo, Y. Zhang, et al., Ultralow thermal conductivity and high thermoelectric figure of merit in SnSe crystals. Nature **508**, 373–377 (2014b). https://doi.org/10.1038/nature13184

102. M. Saleemi, A. Ruditskiy, M.S. Toprak, et al., Evaluation of the structure and transport properties of nanostructured antimony telluride (Sb2Te3). J. Electron. Mater. **43**, 1927–1932 (2014). https://doi.org/10.1007/s11664-013-2911-6

103. X. Lu, D.T. Morelli, *Materials Aspect of Thermoelectricity* (CRC Press, Boca Raton, 2016)

104. J.H. Bahk, H. Fang, K. Yazawa, A. Shakouri, Flexible thermoelectric materials and device optimization for wearable energy harvesting. J. Mater. Chem. C **3**, 10362–10374 (2015)

105. M. Beekman, D.T. Morelli, G.S. Nolas, Better thermoelectrics through glass-like crystals. Nat. Mater. **14**, 1182–1185 (2015)

106. L.D. Hicks, M.S. Dresselhaus, Effect of quantum-well structures on the thermoelectric figure of merit. Phys. Rev. B **47**, 12727–12731 (1993). https://doi.org/10.1103/PhysRevB.47.12727

107. M.S. Dresselhaus, G. Chen, M.Y. Tang, et al., New directions for low-dimensional thermoelectric materials. Adv. Mater. **19**, 1043–1053 (2007). https://doi.org/10.1002/adma.200600527

108. B. Sothmann, R. Sánchez, A.N. Jordan, Thermoelectric energy harvesting with quantum dots. Nanotechnology **26**, 032001 (2015)

109. J. Mao, Z. Liu, Z. Ren, Size effect in thermoelectric materials. npj Quantum Mater. **1**, 16028 (2016). https://doi.org/10.1038/npjquantmats.2016.28

110. D.M. Rowe, *CRC Handbook of Thermoelectrics*, vol 16 (CRC Press, New York, 1995), pp. 1251–1256

111. T.M. Tritt, X. Tang, Q. Zhang, W. Xie, Solar thermoelectrics: Direct solar thermal energy conversion. MRS Bull. **34**, 366 (2011)

112. S.I. Kim, K.H. Lee, H.A. Mun, et al., Dense dislocation arrays embedded in grain boundaries for high-performance bulk thermoelectrics. Science (80-) **348**, 109–114 (2015b). https://doi.org/10.1126/science.aaa4166

113. R. Venkatasubramanian, E. Siivola, T. Colpitts, B. O'Quinn, Thin-film thermoelectric devices with high room-temperature figures of merit. Nature **413**, 597–602 (2001). https://doi.org/10.1038/35098012

114. L. Hu, H. Wu, T. Zhu, et al., Tuning multiscale microstructures to enhance thermoelectric performance of n-type bismuth-telluride-based solid solutions. Adv. Energy Mater. **5**, 1500411 (2015). https://doi.org/10.1002/aenm.201500411

115. H. Zhao, J. Sui, Z. Tang, et al., High thermoelectric performance of MgAgSb-based materials. Nano Energy **7**, 97–103 (2014a). https://doi.org/10.1016/j.nanoen.2014.04.012

116. Z. Liu, Y. Wang, J. Mao, et al., Lithium doping to enhance thermoelectric performance of MgAgSb with weak electron-phonon coupling. Adv. Energy Mater. **6**, 1502269 (2016). https://doi.org/10.1002/aenm.201502269

117. P. Ying, X. Li, Y. Wang, et al., Hierarchical chemical bonds contributing to the intrinsically low thermal conductivity in α-MgAgSb thermoelectric materials. Adv. Funct. Mater. **27**, 1604145 (2017). https://doi.org/10.1002/adfm.201604145

118. N. Yamada, R. Ino, Y. Ninomiya, Truly transparent p-type γ-CuI thin films with high hole mobility. Chem. Mater. **28**, 4971–4981 (2016). https://doi.org/10.1021/acs.chemmater.6b01358

119. C. Yang, M. Kneiβ, M. Lorenz, M. Grundmann, Room-temperature synthesized copper iodide thin film as degenerate p-type transparent conductor with a boosted figure of merit. Proc. Natl. Acad. Sci. **113**, 12929–12933 (2016a). https://doi.org/10.1073/pnas.1613643113

120. C. Yang, D. Souchay, M. Kneiß, et al., Transparent flexible thermoelectric material based on non-toxic earth-abundant p-type copper iodide thin film. Nat. Commun. **8**, 16076 (2017a). https://doi.org/10.1038/ncomms16076

121. K.C. See, J.J. Urban, R.A. Segalman, et al., Inorganic nanostructure-organic polymer hetero-structures useful for thermoelectric devices (2017)

122. C. Wan, X. Gu, F. Dang, et al., Flexible n-type thermoelectric materials by organic intercalation of layered transition metal dichalcogenide TiS 2. Nat. Mater. **14**, 622–627 (2015). https://doi.org/10.1038/nmat4251

123. Y. Du, K. Cai, S. Chen, et al., Facile preparation and thermoelectric properties of Bi2Te3 based alloy nanosheet/PEDOT: PSS composite films. ACS Appl. Mater. **6**, 5735–5743 (2014). https://doi.org/10.1021/am5002772

124. S. Yang, K. Cho, J. Yun, et al., Thermoelectric characteristics of γ-Ag2Te nanoparticle thin films on flexible substrates. Thin Solid Films **641**, 65–68 (2017b). https://doi.org/10.1016/j.tsf.2017.01.068

125. J.Y. Oh, J.H. Lee, S.W. Han, et al., Chemically exfoliated transition metal dichalcogenide nanosheet-based wearable thermoelectric generators. Energy Environ. Sci. **9**, 1696–1705 (2016). https://doi.org/10.1039/c5ee03813h

126. X. Meng, Z. Liu, B. Cui, et al., Grain boundary engineering for achieving high thermoelectric performance in n-type skutterudites. Adv. Energy Mater. **7**, 1602582 (2017). https://doi.org/10.1002/aenm.201602582

127. L. Yang, Z.G. Chen, M.S. Dargusch, J. Zou, High performance thermoelectric materials: Progress and their applications. Adv. Energy Mater. **8**, 1701797 (2018)

128. Y. Wang, Y. Shi, D. Mei, Z. Chen, Wearable thermoelectric generator for harvesting heat on the curved human wrist. Appl. Energy **205**, 710–719 (2017d). https://doi.org/10.1016/j.apenergy.2017.08.117

129. Y. Wang, Y. Shi, D. Mei, Z. Chen, Wearable thermoelectric generator to harvest body heat for powering a miniaturized accelerometer. Appl. Energy **215**, 690–698 (2018). https://doi.org/10.1016/j.apenergy.2018.02.062

130. Y. Shi, Y. Wang, D. Mei, et al., Design and fabrication of wearable thermoelectric generator device for heat harvesting. IEEE Robot Autom. Lett. **3**, 373–378 (2018). https://doi.org/10.1109/LRA.2017.2734241

131. P. Gokhale, B. Loganathan, J. Crowe, et al., Development of flexible thermoelectric cells and performance investigation of thermoelectric materials for power generation. Energy Procedia **110**, 281–285 (2017)

132. P. Fan, Z.H. Zheng, Y.Z. Li, et al., Low-cost flexible thin film thermoelectric generator on zinc based thermoelectric materials. Appl. Phys. Lett. **106**, 073901 (2015). https://doi.org/10.1063/1.4909531

133. J. Gao, L. Miao, C. Liu, et al., A novel glass-fiber-aided cold-press method for fabrication of n-type Ag2Te nanowires thermoelectric film on flexible copy-paper substrate. J. Mater. Chem. A **5**, 24740–24748 (2017). https://doi.org/10.1039/c7ta07601k

134. Z. Lu, M. Layani, X. Zhao, et al., Fabrication of flexible thermoelectric thin film devices by inkjet printing. Small **10**, 3551–3554 (2014)

135. D. Madan, Z. Wang, P.K. Wright, J.W. Evans, Printed flexible thermoelectric generators for use on low levels of waste heat. Appl. Energy **156**, 587–592 (2015). https://doi.org/10.1016/j.apenergy.2015.07.066

136. Z. Cao, E. Koukharenko, M.J. Tudor, et al., Flexible screen printed thermoelectric generator with enhanced processes and materials. Sens. Actuators A Phys. **238**, 196–206 (2016). https://doi.org/10.1016/j.sna.2015.12.016

137. Z. Cao, E. Koukharenko, M.J. Tudor, et al., Screen printed flexible Bi2Te3-Sb2Te3 based thermoelectric generator. J. Phys. Conf. Ser. **476**, 012031 (2013). https://doi.org/10.1088/1742-6596/476/1/012031

138. M.-K.K. Kim, M.-S. Kim, S.-E. Jo, Y.-J. Kim, Triboelectric–thermoelectric hybrid nanogenerator for harvesting frictional energy. Smart Mater. Struct. **25**, 125007 (2016)

139. J.H. We, S.J. Kim, B.J. Cho, Hybrid composite of screen-printed inorganic thermoelectric film and organic conducting polymer for flexible thermoelectric power generator. Energy **73**, 506–512 (2014). https://doi.org/10.1016/j.energy.2014.06.047

140. L. Zhang, S. Lin, T. Hua, et al., Fiber-based thermoelectric generators: Materials, device structures, fabrication, characterization, and applications. Adv. Energy Mater. **8**, 1700524 (2018)
141. Z. Lu, H. Zhang, C. Mao, C.M. Li, Silk fabric-based wearable thermoelectric generator for energy harvesting from the human body. Appl. Energy **164**, 57–63 (2016). https://doi.org/10.1016/j.apenergy.2015.11.038
142. A.R.M. Siddique, R. Rabari, S. Mahmud, B. Van Heyst, Thermal energy harvesting from the human body using flexible thermoelectric generator (FTEG) fabricated by a dispenser printing technique. Energy **115**, 1081–1091 (2016). https://doi.org/10.1016/j.energy.2016.09.087
143. J.A. Lee, A.E. Aliev, J.S. Bykova, et al., Woven-Yarn thermoelectric textiles. Adv. Mater. **28**, 5038–5044 (2016). https://doi.org/10.1002/adma.201600709
144. T. Zhang, K. Li, J. Zhang, et al., High-performance, flexible, and ultralong crystalline thermoelectric fibers. Nano Energy **41**, 35–42 (2017b). https://doi.org/10.1016/j.nanoen.2017.09.019
145. S.J. Kim, H.E. Lee, H. Choi, et al., High-performance flexible thermoelectric power generator using laser multiscanning lift-off process. ACS Nano **10**, 10851–10857 (2016). https://doi.org/10.1021/acsnano.6b05004
146. Y. Eom, D. Wijethunge, H. Park, et al., Flexible thermoelectric power generation system based on rigid inorganic bulk materials. Appl. Energy **206**, 649–656 (2017). https://doi.org/10.1016/j.apenergy.2017.08.231
147. E. Mu, G. Yang, X. Fu, et al., Fabrication and characterization of ultrathin thermoelectric device for energy conversion. J. Power Sources **394**, 17–25 (2018). https://doi.org/10.1016/j.jpowsour.2018.05.031
148. C.S. Kim, H.M. Yang, J. Lee, et al., Self-powered wearable electrocardiography using a wearable thermoelectric power generator. ACS Energy Lett. **3**, 501–507 (2018b). https://doi.org/10.1021/acsenergylett.7b01237
149. V. Leonov, P. Fiorini, Thermal matching of a thermoelectric energy scavenger with the ambience. Proceedings 5th European Conference on Thermoelectrics (ECT 07) (2007), pp. 129–133
150. M. Lossec, B. Multon, H. Ben Ahmed, Sizing optimization of a thermoelectric generator set with heatsink for harvesting human body heat. Energy Convers. Manag. **68**, 260–265 (2013). https://doi.org/10.1016/j.enconman.2013.01.021
151. A. Montecucco, J. Siviter, A.R. Knox, Constant heat characterisation and geometrical optimisation of thermoelectric generators. Appl. Energy **149**, 248–258 (2015). https://doi.org/10.1016/j.apenergy.2015.03.120
152. K. Pietrzyk, J. Soares, B. Ohara, H. Lee, Power generation modeling for a wearable thermoelectric energy harvester with practical limitations. Appl. Energy **183**, 218–228 (2016). https://doi.org/10.1016/j.apenergy.2016.08.186
153. R.A. Kishore, M. Sanghadasa, S. Priya, Optimization of segmented thermoelectric generator using Taguchi and ANOVA techniques. Sci. Rep. **7**, 16746 (2017). https://doi.org/10.1038/s41598-017-16372-8
154. C.S. Kim, G.S. Lee, H. Choi, et al., Structural design of a flexible thermoelectric power generator for wearable applications. Appl. Energy **214**, 131–138 (2018a). https://doi.org/10.1016/j.apenergy.2018.01.074
155. M. Wahbah, M. Alhawari, B. Mohammad, et al., Characterization of human body-based thermal and vibration energy harvesting for wearable devices. IEEE J. Emerg. Sel. Top. Circuits Syst. **4**, 354–363 (2014). https://doi.org/10.1109/JETCAS.2014.2337195
156. K. Gordiz, A.K. Menon, S.K. Yee, Interconnect patterns for printed organic thermoelectric devices with large fill factors. J. Appl. Phys. **122**, 124507 (2017). https://doi.org/10.1063/1.4989589
157. S. LeBlanc, Thermoelectric generators: Linking material properties and systems engineering for waste heat recovery applications. Sustain. Mater. Technol. **1**, 26–35 (2014). https://doi.org/10.1016/j.susmat.2014.11.002
158. L. Francioso, C. De Pascali, V. Sglavo, et al., Modelling, fabrication and experimental testing of an heat sink free wearable thermoelectric generator. Energy Convers. Manag. **145**, 204–213 (2017). https://doi.org/10.1016/j.enconman.2017.04.096

159. T. Nguyen Huu, T. Nguyen Van, O. Takahito, Flexible thermoelectric power generator with Y-type structure using electrochemical deposition process. Appl. Energy **210**, 467–476 (2018). https://doi.org/10.1016/j.apenergy.2017.05.005
160. K. Yazawa, A. Shakouri, Cost-efficiency trade-off and the design of thermoelectric power generators. Environ. Sci. Technol. **45**, 7548–7553 (2011). https://doi.org/10.1021/es2005418
161. J. Estrada-López, A. Abuellil, Z. Zeng, E. Sánchez-Sinencio, Multiple input energy harvesting systems for autonomous IoT end-nodes. J. Low Power Electron. Appl. **8**, 6 (2018). https://doi.org/10.3390/jlpea8010006
162. M. Hasan Nia, A. Abbas Nejad, A.M. Goudarzi, et al., Cogeneration solar system using thermoelectric module and fresnel lens. Energy Convers. Manag. **84**, 305–310 (2014). https://doi.org/10.1016/j.enconman.2014.04.041
163. J. Zhang, Y. Xuan, L. Yang, Performance estimation of photovoltaic-thermoelectric hybrid systems. Energy **78**, 895–903 (2014a). https://doi.org/10.1016/j.energy.2014.10.087
164. Y. Da, Y. Xuan, Q. Li, From light trapping to solar energy utilization: A novel photovoltaic-thermoelectric hybrid system to fully utilize solar spectrum. Energy **95**, 200–210 (2016). https://doi.org/10.1016/j.energy.2015.12.024
165. R. Lamba, S.C. Kaushik, Modeling and performance analysis of a concentrated photovoltaic–thermoelectric hybrid power generation system. Energy Convers. Manag. **115**, 288–298 (2016). https://doi.org/10.1016/j.enconman.2016.02.061
166. J. Zhang, Y. Xuan, Investigation on the effect of thermal resistances on a highly concentrated photovoltaic-thermoelectric hybrid system. Energy Convers. Manag. **129**, 1–10 (2016). https://doi.org/10.1016/j.enconman.2016.10.006
167. W. Zhu, Y. Deng, L. Cao, Light-concentrated solar generator and sensor based on flexible thin-film thermoelectric device. Nano Energy **34**, 463–471 (2017b). https://doi.org/10.1016/j.nanoen.2017.03.020
168. T.H. Kil, S. Kim, D.H. Jeong, et al., A highly-efficient, concentrating-photovoltaic/thermoelectric hybrid generator. Nano Energy **37**, 242–247 (2017). https://doi.org/10.1016/j.nanoen.2017.05.023
169. A. Rezania, L.A. Rosendahl, Feasibility and parametric evaluation of hybrid concentrated photovoltaic-thermoelectric system. Appl. Energy **187**, 380–389 (2017). https://doi.org/10.1016/j.apenergy.2016.11.064
170. M. He, Y.J. Lin, C.M. Chiu, et al., A flexible photo-thermoelectric nanogenerator based on MoS2/PU photothermal layer for infrared light harvesting. Nano Energy **49**, 588–595 (2018). https://doi.org/10.1016/j.nanoen.2018.04.072
171. E. Yin, Q. Li, Y. Xuan, Optimal design method for concentrating photovoltaic-thermoelectric hybrid system. Appl. Energy **226**, 320–329 (2018). https://doi.org/10.1016/j.apenergy.2018.05.127
172. Y.P. Zhou, M.J. Li, W.W. Yang, Y.L. He, The effect of the full-spectrum characteristics of nanostructure on the PV-TE hybrid system performances within multi-physics coupling process. Appl. Energy **213**, 169–178 (2018). https://doi.org/10.1016/j.apenergy.2018.01.027
173. M.A. Green, General temperature dependence of solar cell performance and implications for device modelling. Prog. Photovolt. Res. Appl. **11**, 333–340 (2003). https://doi.org/10.1002/pip.496
174. M.S. Kim, M.K. Kim, K. Kim, Y.J. Kim, Design of wearable hybrid generator for harvesting heat energy from human body depending on physiological activity. Smart Mater. Struct. **26**, 095046 (2017). https://doi.org/10.1088/1361-665X/aa82d5
175. C.R. Bowen, H.A. Kim, P.M. Weaver, S. Dunn, Piezoelectric and ferroelectric materials and structures for energy harvesting applications. Energy Environ. Sci. **7**, 25–44 (2014a). https://doi.org/10.1039/C3EE42454E
176. D.S. Montgomery, C.A. Hewitt, D.L. Carroll, Hybrid thermoelectric piezoelectric generator. Appl. Phys. Lett. **108**, 263901 (2016). https://doi.org/10.1063/1.4954770
177. F.R. Fan, Z.Q. Tian, Z. Lin Wang, Flexible triboelectric generator. Nano Energy **1**, 328–334 (2012). https://doi.org/10.1016/j.nanoen.2012.01.004

Novel Materials and Device Design for Wearable Energy Harvesters

Masakazu Nakamura, Mitsuhiro Ito, Naofumi Okamoto, and Ichiro Yamashita

1 Introduction

Thermoelectric generators (TEGs) are expected to be used for energy harvesting of low-temperature waste heat in applications such as small-scale distributed power sources for wearable electronics [1, 2]. Therefore, wide-area, low-cost, and mechanically flexible TEGs are strongly desired to cost-effectively collect low-density heat flux [3–5]. Researchers who study TE materials/devices use an index named dimensionless figure of merit, $ZT = S^2\sigma T/\kappa$ (S, σ, κ, and T are Seebeck coefficient, electrical conductivity, thermal conductivity, and absolute temperature, respectively) because it is known to be directly related to energy conversion efficiency. However, is it the most important for the wearable applications?

The work introduced in this chapter was carried out to answer this question. To satisfy various requirements and limitations in practical situations of energy harvesting from the human body, different strategies for material development and different device design from conventional ones are required. We hope the concepts proposed in this work can function as a guideline in the development of the materials and TEGs for wearable energy harvesters.

M. Nakamura (✉) · M. Ito · N. Okamoto
Division of Materials Science, Nara Institute of Science and Technology, Ikoma, Nara, Japan
e-mail: mnakamura@ms.naist.jp

I. Yamashita
Graduate School of Engineering, Osaka University, Suita, Osaka, Japan

© Springer Nature Switzerland AG 2021
S. Skipidarov, M. Nikitin (eds.), *Thin Film and Flexible Thermoelectric Generators, Devices and Sensors*, https://doi.org/10.1007/978-3-030-45862-1_2

2 Requirements for Wearable Thermoelectric Generators

First of all, we have to emphasize the importance of comfortability, which increases chance and motivation to use TEGs for practical wearable applications. One may easily imagine that lightweight and sufficient mechanical flexibility are necessary to use TEGs without stimulating unpleasant feelings to the users. Therefore, one of the goals of wearable TEGs is to be similar to regular clothing, such as T-shirts, headbands, wristbands, and caps. The degree of the heat conductance must be also tuned to be similar to such clothing to avoid unwanted dissipation of body heat.

The thermal design is also important to maximize the output power. If, for example, TE material is coated on supporting film substrate, non-negligible temperature drop appears in the substrate and results in loss of energy conversion efficiency. And if, for example, active TE material is inserted in supporting substrate to directly connect cold and hot sides, the output power is still reduced by the volume fraction of the inactive supporting part. Therefore, very high and very low κ are required for the supporting material in the former and the latter cases, respectively.

Furthermore, when heat energy from the human body or living environment is to be harvested, the low-temperature side would be, in most cases, the air, which is far from the ideal cold bath. Besides, the contact between human body and TEG could not be good enough when comfortableness is the highest priority requirement. Accordingly, the thermal resistance against a heat source and/or the air will determine the temperature difference applied to TEG.

Here, we assume that skeleton-type TEG is tightly attached to a human body with temperature $T_{body} = 310$ K and that TEG surface is naturally cooled by air $T_{room} = 295$ K; the temperature of the device surface $T_{surface}$ satisfies the following equation:

$$\frac{\kappa}{d}\left(T_{body} - T_{surface}\right) = J\left(T_{surface} - T_{room}\right) + \left(T_{surface}^{4} - T_{room}^{4}\right)F\sigma_{SB}, \tag{1}$$

where d, F, J, and σ_{SB} are thickness, emissivity, convective heat transfer coefficient, and Stefan-Boltzmann constant, respectively. With this calculated temperature difference ($\Delta T = T_{body} - T_{surface}$) applied actually to TEG, power conversion efficiency is calculated by:

$$\eta = \frac{\left(S\Delta T\right)^{2}}{2RW_{in}}, \tag{2}$$

where R and W_{in} are electrical resistance of TEG and the heat energy input from human body, respectively. The results are normalized by the maximum efficiency among all of the calculations η/η_{max} and plotted in Fig. 1. Here, we fixed the product of κ and R (κR) to fix ZT value, and κ was varied from 0.1 W/(m×K), which is relatively small value for organic TE materials, to 6.4 W/(m×K), which is relatively large value among various modern inorganic TE materials. Considering the

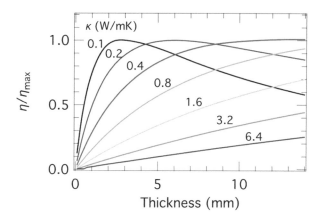

Fig. 1 Dependence of η/η_{max} on device thickness and κ. The efficiency η was calculated by assuming skeleton-type TEG, negligible temperature drops due to the electrodes, $T_{body} = 310$ K (37 °C), and $T_{air} = 295$ K (22 °C). The calculated data were normalized by the maximum value η_{max}

bottlenecking heat resistance between TEG and ambient air, one may understand that κ as low as 0.1 W/(m×K) and relatively large device thickness, over 3 mm, are required to achieve sufficient temperature difference. To effectively generate power with less than 1 mm thick TEGs, κ much lower than 0.1 W/(m×K) is desirable.

Decreasing κ to 0.1 W/(m×K) is feasible if we use organic or hybrid TE materials. Recently, flexible TEGs using organic or hybrid materials have been studied extensively. However, most of TEGs use lateral temperature gradient. There are two important issues: difficulty in using conventional deposition techniques used for thin-film devices and the necessity of three-dimensional structural designs to maintain mechanical flexibility. To solve these problems, simply following conventional design of TEGs does not work, and total design from materials to device structure is strongly desired.

3 Device Design and Demonstration of Prototype Module

To fulfill the difficult requirements of millimeter thicknesses and sufficient flexibility, we have proposed using CNT/polymer composite yarn as TE component in the fabrication of thickness-controllable "thermoelectric fabrics" [11]. CNTs are expected to have high σ and desirable mechanical properties for flexible devices [6, 12–14], and extremely high aspect ratio is favorable for forming yarn. The name, thermoelectric fabric, has been already used in other works [7, 8]. However, TEGs fabricated in other works operate with in-plane temperature gradient. It is obvious that such lateral structure is not suitable for wearable energy harvesters because most clothing has temperature gradient in its thickness direction.

(a)   (b)

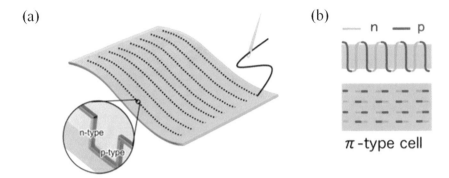

Fig. 2 Schematic drawings of (**a**) the structure and fabrication process of "thermoelectric fabric" and (**b**) its sectional and planar structure with *p*/*n*-striped doping composing *π*-type cells

Figure 2 depicts the basic structure of proposed TE fabric using a stripe-doping-patterned CNT/polymer composite yarn. After forming CNT-based TE material into thin flexible yarn, it is periodically doped to make *p*/*n*-striped pattern. By sewing a soft and thick fabric used as a substrate with CNT/polymer composite yarn of which doping pitch is predesigned to fit to the thickness of fabric, a series connection of *π*-type cells is densely formed, as indicated in Fig. 2. Although we introduce only a method using stripe-doped yarn, one can also form a series connection of *π*-type cells with uniformly *p*- and *n*-doped yarns by knitting those on a fabric.

This device design and fabrication method has several advantages:

(1) Thermally insulating cloths, such as felt or fleece, can be used as a substrate.
(2) Thickness of the module is easily tuned so as to satisfy the required output power, flexibility, and thermal conductance.
(3) The area, shape, and density of TE material are tunable depending on the heat source and the performance-vs-cost design.
(4) Electrode formation to interconnect *p*- and *n*-blocks is omittable.
(5) Durable against bending, stretching, and twisting.

Point (1) contributes to low thermal conductance of TEGs. Points (2)–(4) provide designability to optimize device performance, cost, and comfortableness. Point (5) is, no doubt, the most important for wearable applications. Point (5) comes from the basic structure of proposed device: TE yarn is sewn into, but not tightly bonded to the substrate. Therefore, active component is nearly free from the stress caused by the motion of the substrate. Although details of device structure and materials to satisfy this basic concept would have various candidates, flexible TEG demonstrated using CNT/polymer composite yarn fabricated through the wet-spinning method [11] is explained in this chapter as an example.

Single-walled CNTs synthesized by the enhanced direct injection pyrolytic synthesis (eDIPS) method [15] were used as TE material. CNTs were used without any intentional purification, but the process used in this work possibly removes amorphous carbon impurities. CNT material was first mixed with ionic liquid,

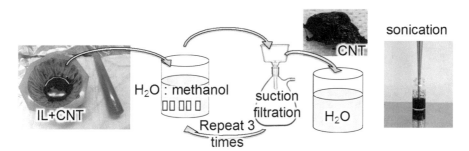

Fig. 3 Preparation method of CNT dispersion in water

1-butyl-3-methylimidazolium hexafluorophosphate ([BMIM]PF$_6$), and ground using mortar and pestle. The strong van der Waals force between CNTs is shielded by the ionic liquid, and CNTs are easily disentangled [16]. By using this primary dispersion step with ionic liquid, we can minimize the total amount of ultrasonic-wave application, which increases defects in CNTs [17]. After the primary dispersion step, [BMIM]PF$_6$ was removed by repeating rinse in water-methanol mixture, and CNT was finally dispersed in sodium dodecyl sulfate (SDS, 4 wt. %) aqueous solution by sonication for short time (400 W, 2 min) as shown in Fig. 3. The concentration of CNTs in dispersant was 0.15 wt. %.

Polyethylene glycols (PEG) were then added to be 0, 0.01, 0.05, and 0.10 wt. % for reinforcement of the yarn [18]. The dispersion was then injected into methanol, which works as a coagulant, in rotating vessel as shown in Fig. 4a, b through polytetrafluoroethylene tube (inner diameter: 0.5 mm). The line speeds of CNT dispersion and methanol at the injection point were set to 764 and 452 m/hour, respectively. After leaving it in methanol for 24 hours, gel-like yarn was pulled up and dried in air (Fig. 4c, d). The diameter of obtained CNT/PEG composite yarn was 40–50 μm (Fig. 4e). The maximum length of the thread obtained in our experiments was over 3 m.

Figure 5 shows TE performance of CNT/PEG composite yarn as a function of PEG concentration. All samples exhibited positive S with values almost independent on PEG concentration. On the other hand, σ decreases with increasing PEG concentration because PEG disturbs charge carriers transport between CNTs, and κ decreases also by adding PEG. As a result, ZT value of CNT/PEG composite yarn is the largest when PEG concentration is 0.01–0.05 wt. %. In the following experiments, we used CNT/PEG composite yarn made using the dispersion with 0.01 wt. % PEG.

As seen in Fig. 5a, as-fabricated CNT/PEG composite yarn exhibits p-type TE property. Therefore, n-type doping is necessary to fabricate π-type cells. The same ionic liquid as primary dispersion step was used as n-type doping agent. CNT/PEG threads were partially immersed in [BMIM]PF$_6$ containing 10 wt. % of dimethyl sulfoxide (DMSO), which is used to decrease viscosity, for ca. 24 hours (Fig. 6a), and the agent was wiped off with cotton cloths. Although the electron-doping mechanism has not been fully clarified yet, immersion in [BMIM]PF$_6$ was confirmed to convert CNT/PEG threads into stable n-type semiconductors, and high viscosity of

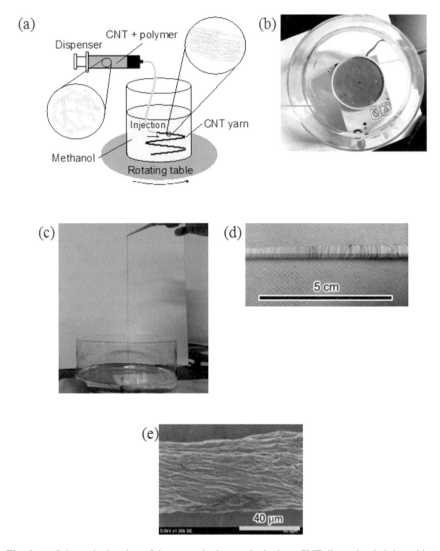

Fig. 4 (**a**) Schematic drawing of the wet-spinning method where CNT dispersion is injected into a coagulant in rotating vessel. (**b**) Photo during the formation of gel-like yarn in rotating vessel. (**c**) Gel-like yarn pulled out from the coagulant. (**d**) Photo of CNT/PEG yarn after drying in air, where 3-m-long yarn is wrapped around wooden chopstick. (**e**) Scanning electron microscope (SEM) image of CNT/PEG yarn

[BMIM]PF$_6$ is suitable to avoid the doping agent infiltrated to the neighboring sections. TE voltage of doped ΔV_L and undoped ΔV_R parts and the sum ΔV_{total} were measured separately as depicted in the inset of Fig. 6b. Before the partial doping, both ΔV_L and ΔV_R exhibited p-type S, and the sum was nearly zero because they cancelled each other in this configuration. After the partial doping with optimum condition, ΔV_L exhibited n-type with $S = -49.1$ μV/K, while ΔV_R continued to

Fig. 5 (**a**) Electrical conductivity σ, Seebeck coefficient S, (**b**) thermal conductivity κ, and ZT value of CNT/PEG composite yarn formed by changing PEG concentration in the dispersion

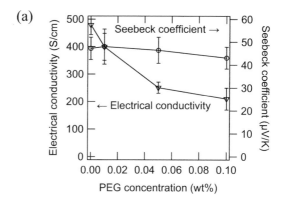

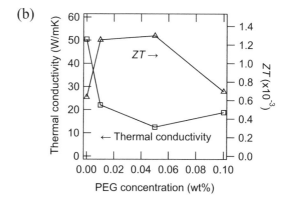

exhibit p-type with $S = 47.8$ μV/K, and $\Delta V_{total} = 101.5$ μV was obtained between both ends (Fig. 6b). S of CNT/PEG yarn similarly n-doped by this method was tested in air and confirmed to be stable for at least 10 days without any passivation coating.

To complete fabric-type TEG, numbers of π-type cells must be connected in series. Therefore, doping method was extended to form p/n-striped pattern on CNT/PEG yarn. After winding a yarn around long and narrow plastic plate, the doping agent was dropped onto only one side and wiped off (Fig. 7). The width of the plate determines p/n pitch and can be varied to fit to the thickness of fabric substrate. Boundaries between p- and n-sections were marked with conductive silver paste so as to make p/n boundary visible. The stripe-doping-patterned yarn was, then, sewn into a piece of felt fabric (3 mm in thickness and low $\kappa < 0.05$ W/(m$\times$K) with a sewing needle such that n-type sections and p-type sections traversed the fabric in downward and upward directions, respectively, as seen in Fig. 7.

Figure 8a is a photo of prototype TE fabric on which eight units of π-type structures are formed. The stability of the device characteristics was tested by bending TE fabric in air. Figure 8b shows the resistance of the module during the bending stress test. The resistance was stable even when the module was bent to make the inner

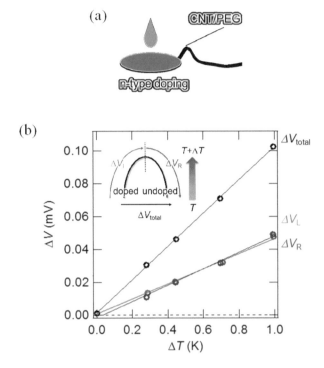

Fig. 6 (**a**) Schematic illustration showing partial n-type doping. (**b**) Thermoelectromotive force of doped part ΔV_L, undoped part ΔV_R, and the sum of them ΔV_{total} after partial n-type doping with optimum conditions

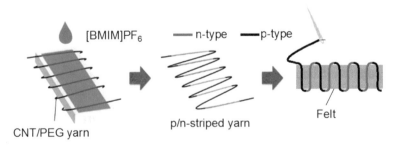

Fig. 7 Schematic drawing of the doping method to form p/n-striped pattern on CNT/PEG yarn and cross-sectional view of a part of the final module structure

radius of curvature almost zero, which means that the module was folded, as shown in Fig. 8b. Within 160 cycles of the bending-stretching test, there is no obvious change (less than 2%) in the resistance. The stability of this module is remarkably higher than that of other flexible TEGs [5, 19]. This is because sewn CNT/PEG yarn, which is not tightly fixed to the fabric substrate, is thought not to be subjected to stress from substrate bending as pointed out in the previous section. This is a notable advantage for wearable applications.

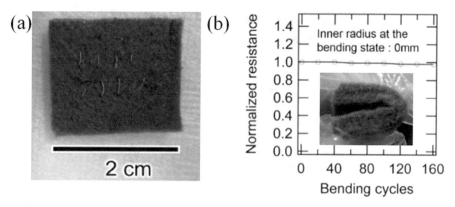

Fig. 8 (a) Photo of prototype TE fabric with eight units of π-type cells. (b) Results of bending (folding) test by monitoring sample resistance

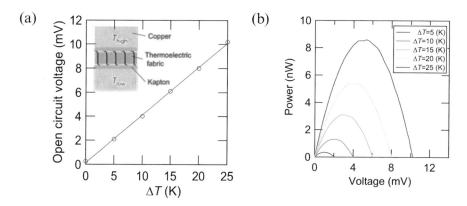

Fig. 9 (a) $V_{OC} - \Delta T$ and (b) $P_{out} - V_{OC}$ characteristics of prototype TE fabric composed of eight units of π-type structures

Figure 9 shows the output characteristics of TE fabric. Open circuit voltage V_{OC} is exactly proportional to the temperature difference, and dependences of output power P_{out} on voltage are parabolic, as expected. From this result, mean S per cell is estimated as ca. 60 μV/K, which is smaller than that obtained in Fig. 6. This is presumably because the thermal resistance between the heating and cooling blocks and TE fabric where thin Kapton films are inserted for electrical insulation (inset of Fig. 9) at zero contact pressure is still too large to avoid a temperature drop.

Operation of TE fabric in more realistic situation is demonstrated in Fig. 10. One side of the device is gently touched by a finger, and the other side is naturally cooled by air under windless condition. The output voltage (i.e., the temperature difference between the inside and outside of the device) becomes stable within 4 seconds, and steady-state TE voltage of approximately 2.3 mV is obtained. From this output voltage, the temperature difference in the device is estimated to be slightly less than

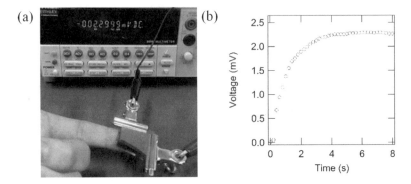

Fig. 10 Demonstration of prototype TE fabric with a human body: (**a**) a photograph showing the power generation by finger touch vs natural air cooling and (**b**) evolution of output voltage after finger touch

5 K, which is smaller than that between the finger (306 K) and air (297 K). However, this is much larger than the temperature difference, less than 0.6 K, that appeared in conventional inorganic TEG (*Thermal Electronics Corp.*, TEC1-03104, 31 couples) tested under the same condition. It indicates that the thermal conductance of conventional TEG modules is too large for the wearable applications explained in Sect. 2. Considering the order of magnitude, larger temperature difference, and Eq. (2), the effective power factor of TE fabric could be as low as almost 1/100 of the conventional TEGs to generate the same power.

4 Further Material Design

Among many conductive materials, CNT is promising for flexible TE applications not only because of its mechanical strength, lightness, and richness in element resources but also its high σ [8, 10, 19]. The demonstration of the novel device design in Sect. 3 utilizes this character of CNT by using CNT/PEG composite yarn as a basic component of flexible TEGs. High σ could be advantageous to obtain high ZT. However, CNT is also known as high κ material, which is disadvantageous to obtain high ZT. As explained in Sect. 2, very low κ is desired especially for wearable applications. Furthermore, it is inevitable to use metal/semiconductor mixture of CNTs to save the material cost. Relatively low S of such mixed state is also a disadvantage. Although several efforts have been reported to control S by chemical doping CNTs [20] or κ by the formation of porous composite with polymer [21], it is still difficult to control all these parameters to obtain higher ZT, because S and σ are interconnected by Mott relation [22]; moreover, σ and κ drop simultaneously by decreasing composition ratio of CNTs in the module. Therefore, innovative method is highly desired. Here, we introduce another material design concept which improves TE performance of CNT/polymer composites by controlling both thermal and charge carriers transport at CNT/CNT junctions using bio-based molecules [9].

Figure 11 depicts schematically proposed junction structure. A particle or molecule composed of semiconductor core and electrically insulating soft shell is placed at CNT/CNT junction, where the quantum tunneling of charge carriers is controlled by selecting semiconductor core material and shell thickness. Under heat flow through the junction, a steep temperature gradient across the junction appears due to the phonon scattering at CNT/soft-shell interface. The steep temperature gradient generates asymmetric Fermi-Dirac distribution of electrons between both sides of the junction, and, as a result, electrons (or holes) flow unidirectionally through the conduction (or valence) band of the core, namely, Seebeck effect at molecular junction. Seebeck coefficient can be tuned by the density-of-state (DOS) function of the core material against Fermi energy of CNTs. By connecting such junctions in series within composite material, enhancement of TE performance can be expected via suppressed thermal transport and enhanced Seebeck effect.

To realize the conceptual model in Fig. 11 with high probability of the series connection of CNT/molecule/CNT junctions, we used cage-shaped protein named *Listeria innocua* Dps (DNA-binding protein from starved cells) [23, 24]. Dps is 9.5 nm in outer diameter and 4.5 nm in inner diameter. It can naturally mineralize the iron oxide core in the cavity [25], and several types of nanoparticle cores can be artificially synthesized in the cavity [26–28]. To provide the ability to adsorb on CNTs, we genetically added a peptide aptamer with an affinity for carbonaceous materials (NHBP-1 [29] to the subunit of Dps [30]). Twelve mutant Dps subunits self-assemble into cage-shaped protein shell, and the aptamer peptide is on the outer surface of each subunit; this is herein referred to as C-Dps. Figure 12a shows a schematic drawing of C-Dps molecule with semiconductor core. The thickness of the soft shell is presumed to decrease at CNT/C-Dps junction point owing to large attractive force between those, which makes the quantum tunneling of charge carriers easier. Via the aptamer peptides, C-Dps attaches to CNT and covers the whole surface of CNT [30]. Therefore, in CNT/C-Dps composites, CNTs are inevitably connected through CNT/C-Dps/CNT junctions (Fig. 12b).

Single-walled CNT material (Aldrich, 50–70% purity) with an outer diameter of 1.2–1.4 nm was used in this work. As core materials, cadmium sulfide, cadmium

Fig. 11 Schematic drawings of Seebeck effect at junction with core-shell particle

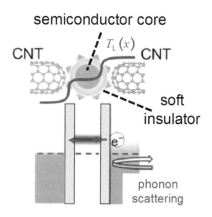

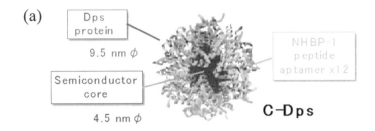

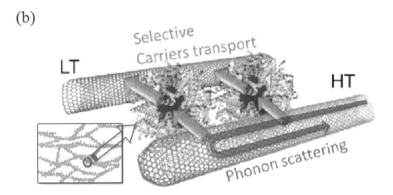

Fig. 12 Schematic drawings of Seebeck effect at biomolecular junction; (**a**) cage-shaped C-Dps molecule, of which diameter is ca. 9 nm, with modified peptides and semiconductor core, diameter is ca. 4.5 nm, and (**b**) two C-Dps molecules bridging two CNTs

selenide, ferrihydrite, or cobalt oxide were introduced to the inner space of C-Dps via biomineralization function of ferritin families [25, 27]. Figure 13 shows X-ray diffraction patterns of C-Dps powders with these core materials. By good agreement with the standard powder diffraction patterns, we consider that the core materials are included as CdS, CdSe, ferrihydrite $Fe_2O_3 \cdot nH_2O$, and Co_3O_4 forms. The molecules are hereinafter referred to as C-Dps(CdS), C-Dps(CdSe), C-Dps(Fe), or C-Dps(Co), respectively. According to Williamson-Hall analysis [31] for these diffraction peaks, the peak widths of the ferrihydrite are determined by small crystallite size, 1.3 nm, and those of CdSe by random strain of lattice. These results imply that semiconductor cores are nanocrystalline and contain many defects.

CNT dispersion (0.2 mg/ml) was mixed with C-Dps solution (0.3 mg/ml) by sonication. CNTs not attached to C-Dps molecules were precipitated and removed by centrifugation at 8500 rpm. C-Dps-adsorbed CNTs (CNT/C-Dps) were obtained as a precipitation by further centrifugation at 80,000 rpm. C-Dps/CNT was then dispersed in water by sonication. Figure 14 shows TEM image of CNT/C-Dps(Fe). Protein shells of C-Dps molecules appear as white circles and ferrihydrite cores as dark dots. The areal coverage of attached C-Dps molecules to CNT surface is estimated to be ca. 53% by analyzing many fibers in the similar TEM images. From this surface coverage, we can expect that C-Dps molecules are inserted at every CNT/

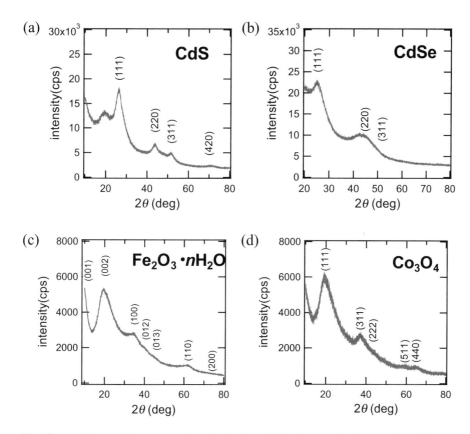

Fig. 13 $\theta - 2\theta$ X-ray diffraction profiles of (**a**) C-Dps(CdS), (**b**) C-Dps(CdSe), (**c**) C-Dps(Fe), and (**d**) C-Dps(Co). CuKα is used as X-ray source. Diffraction planes taken from the standard data are indicated in the figure

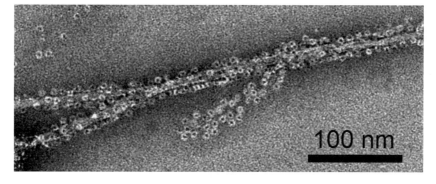

Fig. 14 TEM image of CNT/C-Dps(Fe). In image, protein shells were stained with phosphotungstic acid and appear as white circles. Two CNT bundles are closing at the center of image

CNT junction with high probability and work in parallel as "phonon blocking and electron (hole) transmitting" junctions (Fig. 12b). More recently, another purification method without using centrifugation has been developed, which is more suitable for mass production of C-Dps/CNT dispersion.

The dispersion was dropped on UV/O$_3$-treated glass substrate and dried for approximately 24 hours under ambient condition. Values of σ and S in in-plane direction were measured at 300 K in vacuum by laboratory-made apparatus [32]. For κ measurements, thicker films (ca. 30 μm) were prepared by pressure filtration. Thermal diffusivity toward the thickness direction was measured by *ai-Phase Mobile 2* (*ai-Phase*), employing the temperature wave analysis method [33].

Figure 15 summarizes κ, σ, and S of pristine CNT (blue marks) and CNT/C-Dps(Fe) (red marks) thin films. By inserting the molecular junctions, κ dropped from 17.2 to 0.13 W/(m×K) although the density increased from 0.80 to 1.76 g/cm^3. In a more recent study, the minimum κ of 0.06 W/(m×K) has been obtained with CNT/C-Dps films of 1.5 g/cm^3 density by optimizing the adsorption density of C-Dps on CNT. These results suggest that the thermal boundary resistance with protein shell is very effective for suppressing the heat transport.

CNT material itself used in this work exhibits *p*-type semiconductor nature, and ferrihydrite core, which also behaves as *p*-type one, enhanced S by adding Seebeck effect at molecular junction. On the contrary, when CdS or CdSe, which works as *n*-type semiconductor, is used as the core material, Seebeck coefficient was confirmed to decrease. These results indicate that the current is primarily flowing through the molecular junctions and the core DOS contributes to TE property of the

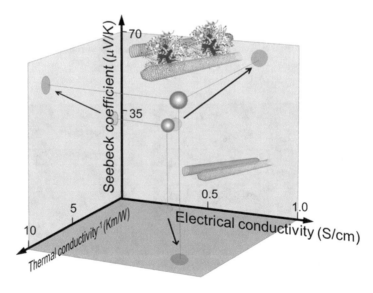

Fig. 15 Three-dimensional graph showing κ, σ, and S of pristine CNT (blue marks) and CNT/C Dps(Fe) (red marks) thin films. Note that only κ is expressed by the reciprocal number. Farther from the origin, therefore, indicates higher performance on any axes

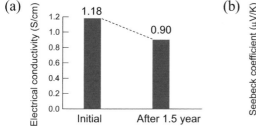
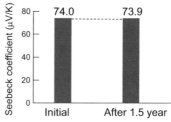

Fig. 16 Variation of TE properties after the storage of CNT/C-Dps film in dry air for 1.5 years: (**a**) σ and (**b**) S

composite material. Additionally, the conductivity of the composite material often increased by the addition of C-Dps with p-type cores, which is presumed to be due to the charge transfer between CNTs and core semiconductors.

Figure 16 shows results of long-term stability test of CNT/C-Dps film without any passivation coating. TE properties are almost unchanged even after the storage in dry air for 1.5 years. The small decrease in σ is often observed not only with CNT/C-Dps films but also with pristine CNT films. The result indicates that the electric material containing protein as a functional element is compatible for long-term use and strict gas-barrier layer is not required as in, e.g., organic electroluminescent devices.

5 Summary

For wearable applications of flexible TEGs, heat transport from TEGs to the air is, in many cases, a bottleneck in the device operation. Not only having high ZT but also low κ is, therefore, necessary to obtain high performance. Very low κ of 0.1 W/(m×K) or less and millimeter thickness are required to effectively generate electricity through sufficient temperature difference between front (cold) and back (hot) sides of TEG. To satisfy these requirements while maintaining mechanical flexibility, total design from materials to device structures is important.

As an example of such attempts, a novel device structure and fabrication method using CNT/PEG yarn with p/n-striped doping and fabric substrate was introduced. By totally developing the materials, components, and device structures, thickness-controllable and thermally insulating TE fabric has been realized without interconnecting many p- and n-blocks by electrodes. Another remarkable advantage of TE fabric is its high durability against bending or stretching, which is due to the mechanical isolation of the continuous TE components from the flexible substrate.

Another material design suppressing κ in condensed phase of CNTs, the utilization of core-shell-type biomolecule, was also demonstrated. Thermal conductivity is dramatically suppressed by phonon scattering at the biomolecular junction, which can be a big advantage for wearable TEGs. As a result of the blocking of heat transport by the biomolecule, the macroscopic temperature gradient is transformed into the local temperature difference at the junction, and thus Seebeck effect at the junction is effectively added to the macroscopic TE property. As a next step, a study to combine these device and material designs is under progress.

References

1. V. Leonov, R.J.M. Vullers, Wearable electronics self-powered by using human body heat: The state of the art and the perspective. J. Renew. Sustain. Energy **1**, 062701 (2009)
2. G. Xu, Y. Yang, Y. Zhou, J. Liu, Wearable thermal energy harvester powered by human foot. Front. Energy **7**, 26 (2013)
3. O. Bubnova, X. Crispin, Towards polymer-based organic thermoelectric generators. Energy Environ. Sci. **5**, 9345 (2012)
4. T. Park, C. Park, B. Kim, H. Shin, E. Kim, Flexible PEDOT electrodes with large thermoelectric power factors to generate electricity by the touch of fingertips. Energy Environ. Sci. **6**, 788 (2013)
5. S.J. Kim, J.H. We, B.J. Cho, A wearable thermoelectric generator fabricated on a glass fabric. Energy Environ. Sci. **7**, 1959 (2014)
6. N. Toshima, K. Oshima, H. Anno, T. Nishinaka, S. Ichikawa, A. Iwata, Y. Shiraishi, Novel hybrid organic thermoelectric materials: three-component hybrid films consisting of a nanoparticle polymer complex, carbon nanotubes, and vinyl polymer. Adv. Mater. **27**, 2246 (2015)
7. K. Suemori, Y. Watanabe, S. Hoshino, Carbon nanotube bundles/polystyrene composites as high-performance flexible thermoelectric materials. Appl. Phys. Lett. **106**, 113902 (2015)
8. Y. Nonoguchi, K. Ohashi, R. Kanazawa, K. Ashibara, K. Hata, T. Nakagawa, C. Adachi, T. Tanase, T. Kawai, Systematic conversion of single walled carbon nanotubes into n-type thermoelectric materials by molecular dopants. Sci. Rep. **3**, 3344 (2013)
9. M. Ito, N. Okamoto, R. Abe, H. Kojima, R. Matsubara, I. Yamashita, M. Nakamura, Enhancement of thermoelectric properties of carbon nanotube composites by inserting biomolecules at nanotube junctions. Appl. Phys. Express **7**, 065102 (2014)
10. C.A. Hewwit, A.B. Kaiser, S. Roth, M. Craps, R. Czerw, D.L. Carroll, Multilayered carbon nanotube/polymer composite based thermoelectric fabrics. Nano Lett. **12**, 1307 (2012)
11. M. Ito, T. Koizumi, H. Kojima, T. Saito, M. Nakamura, From materials to device design of a thermoelectric fabric for wearable energy harvesters. J. Mater. Chem. A **5**, 12068 (2017)
12. W. Glatz, S. Muntwyler, C. Hierold, Optimization and fabrication of thick flexible polymer based micro thermoelectric generator. Sens. Actuators A **132**, 337 (2006)
13. V. Leonov, Human Machine and Thermoelectric Energy Scavenging for Wearable Devices. ISRN Renewable Energy **2011**, 785380 (2011)
14. Y. Kurazumi, L. Rezgals, A.K. Melikov, Convective heat transfer coefficients of the human body under forced convection from ceiling. J. Ergon. **1**, 1000126 (2014)
15. T. Saito, S. Ohshima, T. Okazaki, S. Ohmori, M. Yumura, S. Iijima, Selective diameter control of single-walled carbon nanotubes in the gas-phase synthesis. J. Nanosci. Nanotechnol. **8**, 6153 (2008)
16. J. Wang, H. Chu, Y. Li, Why single-walled carbon nanotubes can be dispersed in imidazolium-based ionic liquids. ACS Nano **2**, 2540 (2008)
17. P. Vichchulada, M.A. Cauble, E.A. Abdi, E.I. Obi, Q. Zhang, M.D. Lay, Sonication power for length control of single-walled carbon nNanotubes in aqueous suspensions used for 2-dimensional network formation. J. Phys. Chem. **114**, 12490 (2010)
18. B. Vigolo, A. Pénicaud, C. Coulon, C. Sauder, R. Pailler, C. Journet, P. Bernier, P. Poulin, Macroscopic fibers and ribbons of oriented carbon nanotubes. Science **290**, 1331 (2000)

19. K. Suemori, S. Hoshino, T. Kamata, Flexible and lightweight thermoelectric generators composed of carbon nanotube–polystyrene composites printed on film substrate. Appl. Phys. Lett. **103**, 153902 (2013)
20. M. Piao, M.R. Alam, G. Kim, U. Dettlaff-Weglikowska, S. Roth, Effect of chemical treatment on the thermoelectric properties of single walled carbon nanotube networks. Phys. Status Solidi B **249**, 2353 (2012)
21. K. Zhang, M. Davis, J. Qiu, L.H. Weeks, S. Wang, Thermoelectric properties of porous multi-walled carbon nanotube/polyaniline core/shell nanocomposites. Nanotechnology **23**, 385701 (2012)
22. M. Cutler, N.F. Mott, Observation of Anderson localization in an electron gas. Phys. Rev. **181**, 1336 (1969)
23. S.G. Wolf, D. Frenkiel, T. Arad, S.E. Finkel, R. Kolter, A. Minsky, DNA protection by stress-induced biocrystallization. Nature **400**, 83 (1999)
24. M. Bozzi, G. Mignogna, S. Stefanini, D. Barra, C. Longhi, P. Valenti, E. Chiancone, A novel non-heme iron-binding ferritin related to the DNA-binding proteins of the Dps family in Listeria innocua. J. Biol. Chem. **272**, 3259 (1997)
25. P. Ceci, G.D. Cecca, M. Falconi, F. Oteri, C. Zamparelli, E. Chiancone, Effect of the charge distribution along the ferritin-like pores of Dps proteins on the iron incorporation process. J. Biol. Inorg. Chem. **16**, 869 (2011)
26. K. Iwahori, T. Enomoto, H. Furusho, A. Miura, K. Nishio, Y. Mishima, I. Yamashita, Cadmium sulfide nanoparticle synthesis in Dps protein from Listeria innocua. Chem. Mater. **19**, 3105 (2007)
27. M. Okuda, Y. Suzumoto, K. Iwahori, S. Kang, M. Uchida, T. Douglas, I. Yamashita, Bio-templated CdSenanoparticle synthesis in a cage shaped protein, Listeria-Dps, and their two dimensional ordered array self-assembly. Chem. Commun. **46**, 8797 (2010)
28. M. Allen, D. Willits, M. Young, T. Douglas, Constrained synthesis of cobalt oxide nanomaterials in the 12-subunit protein cage from Listeria innocua, Inorg. Chem. **42**, 6300 (2003)
29. D. Kase, J.L. Kulp, M. Yudasaka, J.S. Evans, S. Iijima, K. Shiba, Affinity selection of peptide phage libraries against single-wall carbon nanohorns identifies a peptide aptamer with conformational variability. Langmuir **20**, 8939 (2004)
30. M. Kobayashi, S. Kumagai, B. Zheng, Y. Uraoka, T. Douglas, I. Yamashita, A water-soluble carbon nanotube network conjugated by nanoparticles with defined nanometre gaps. Chem. Commun. **47**, 3475 (2011)
31. G.K. Williamson, W.H. Hall, X-ray line broadening from filed aluminium and wolfram. Acta Metall. **1**, 22 (1953)
32. M. Nakamura, A. Hoshi, M. Sakai, K. Kudo, Evaluation of thermopower of organic materials toward flexible thermoelectric power generators, MRS Proc. **1197**, 1197-D09-07 (2009)
33. J. Morikawa, T. Hashimoto, Thermal diffusivity of aromatic polyimide thin films by temperature wave analysis. J. Appl. Phys. **105**, 113506 (2009)

Solution-Processed Metal Chalcogenide Thermoelectric Thin Films

Seung Hwae Heo, Seungki Jo, Soyoung Cho, and Jae Sung Son

1 Introduction

Recently, small thermoelectric (TE) power generation systems as a power source for wearable and smart devices are emerging as a new alternative. To this end, it is necessary to secure the functionality-flexibility and processability of the material for minimizing thermal resistance to low-potential heat such as body heat as well as the performance of TE material. Conventional well-known TE material groups like Bi_2Te_3, BiSbTe, PbTe, Cu_2Se, SnSe, etc. are mostly inorganic materials, which can achieve high efficiency of TE energy conversion but are difficult to attach to heat source with curved surface due to low flexibility. Accordingly, researches on organic-based TE materials are being conducted worldwide for wearable TE power generation systems. However, despite significant efforts, TE properties of such materials remain lower on 10–30% of existing inorganic materials.

One possible solution for this challenge is using film-based inorganic thermoelectric materials that are thin enough to be flexible. Moreover, ink-based processing can further provide benefits to the processability of materials in large-scale and cost-effective fabrication of thin films. For example, organic-free inorganic solutions have been of great interest as TE inks for thin film fabrication, due to intrinsically high TE performance and potential to ink processing. Examples of these include soluble chalcogenidometallates of Sb_2Te_3 and Cu_2Se and nanoparticles capped with these ionic compounds, which create excellent inks for building pure inorganic-phase materials through thermal decomposition. Here, we review recent advances in the development of inorganic metal chalcogenide TE solutions for fabrication of thin films. We discuss the basic chemistry to synthesize soluble inorganic

S. H. Heo · S. Jo · S. Cho · J. S. Son (✉)
School of Materials Science and Engineering, Ulsan National Institute of Science
and Technology (UNIST), Ulsan, Republic of Korea
e-mail: jsson@unist.ac.kr

© Springer Nature Switzerland AG 2021
S. Skipidarov, M. Nikitin (eds.), *Thin Film and Flexible Thermoelectric
Generators, Devices and Sensors*, https://doi.org/10.1007/978-3-030-45862-1_3

ions of metal chalcogenides and present solution-processed TE thin films. Finally, we discuss the prospects of inorganic TE inks.

2 Basic Chemistry

2.1 Dimensional Reduction

The synthesis of soluble precursors for solution processing of inorganic materials stands on the basis of the concept of "dimensional reduction" which represents breaking M-X-M (M, metal; X, anion) framework of MX_a compound when reacting with reducing agent of A_bX (A, cation) to produce $A_{nb}MX_{a+n}$ compound [1]. Upon reaction, X anions not forming a strong covalent bonding with highly electropositive A cation isolate a three-dimensional (3D) corner-sharing MX_6 octahedron to lower-dimensional (2D, 1D or 0D) oligomeric or monomeric components which can be dissolved in desired solvents and are available for solution-based fabrication of thin films (Fig. 1a) [1]. Typically, thin film fabrication includes synthesis of molecular precursors and deposition of soluble molecules using solution-based processing such as spin coating, spray coating, doctor blading or dip coating, and subsequent heat treatment for decomposition of A cations to obtain desired crystalline MX_a product.

2.2 Hydrazine-Based Synthetic Route

Hydrazine has been considered as a solvent to synthesize inorganic molecular precursor using a dimensional reduction. Upon reaction, excess chalcogens (X: S, Se, Te) are generally added to improve reducing ability for metal chalcogenides, where nucleophilic chalcogenide anions X_n^{2-} reduced from extra chalcogens assist in dismantling the metal chalcogenide framework to produce low-dimensional soluble chalcogenidometallate anions [4]. For example, tin chalcogenide (SnX_2) molecular precursor can be synthesized via the following reaction:

$$2SnX_2 + 2X + 5N_2H_4 \rightarrow Sn_2X_6^{4-} + 4N_2H_5^+ + N_2 \,(gas). \tag{1}$$

These chalcogenidometallate anions are highly soluble in polar solvents such as dimethylformamide (DMF), dimethyl sulfoxide (DMSO), and N-methylformamide (NMF) and, additionally, can be easily decomposed and recovered to crystalline phase of metal chalcogenide upon heating (~400 °C) through the following reaction:

$$(N_2H_5)_4 Sn_2X_6 \rightarrow 2H_2X + 4N_2H_4 + 2SnX_2. \tag{2}$$

Thus, there have been significant efforts to facilitate chalcogenidometallate anion solutions as inorganic ink solutions to fabricate semiconductor thin films for

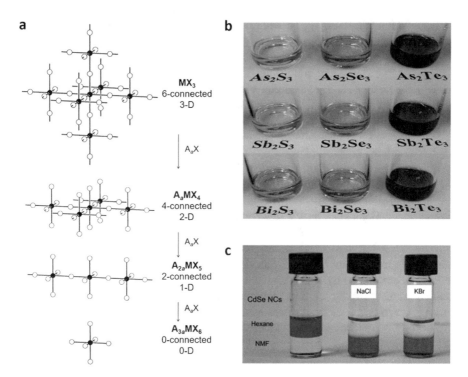

Fig. 1 (**a**) Schematic illustration showing dimensional reduction of metal chalcogenides. (Reprinted permission from [1]. Copyright 2001, American Chemical Society). Photographs of (**b**) solutions of metal chalcogenides in a solvent mixture of amine and thiol and (**c**) colloidal solutions of CdSe nanoparticles undergoing ligand exchange reaction with NaCl and KBr. (Reprinted permission from [2, 3]. Copyright 2013, American Chemical Society. Copyright 2014, American Chemical Society)

electronic and optoelectronic applications. Mitzi et al. opened this field of hydrazine-based synthetic route starting with the synthesis of $SnS_{2-x}Se_x$ solution for fabrication of uniform and continuous thin film applied to active channel layer of field effect transistor (FET) [5]. Furthermore, they have demonstrated successful examples of the applications for FET and photovoltaic (PV) devices including In_2Se_3 [6], $GeSe_2$ [7], Cu_2S [8], and $Cu(In, Ga)Se_{2-x}S_x$ [9–11]. However, despite the facile synthesis and thin film fabrication process, hydrazine-based synthetic route has not been widely deployed because of its high toxicity and explosiveness fatal to the environment.

2.3   Cosolvent Approach

Thanks to the continuous efforts to discover a less toxic solvent with a comparable solvation power to hydrazine, in the year 2013, Webber and Brutchey developed "alkahest" using a cosolvent system of 1,2-ethylenediamine (en) and 1,2-ethanedithiol

(EDT), which can dissolve the wide nomenclature of chalcogen (S, Se, Te) [12], metal chalcogenides (As_2S_3, As_2Se_3, Sb_2S_3, Sb_2Se_3, and Sb_2Te_3) [2], and oxide materials (PbO, Ag_2O, CdO, Sb_2O_3, Bi_2O_3, ZnO, and SnO) [13] with high solubility (10–30 wt. %) at ambient conditions (Fig. 1b). Fully dissolved solution provides a clear solution containing the solutes less than 1 nm in size which is evidenced by dynamic light scattering (DLS) analysis, indicating that bulk metal chalcogenides reduced to molecule level and any nano- or microscale structures were not created. They suggested that these dissolved species might be in the form of chacogenido-metallate anions counterbalanced by enH^+ [12]. Similar to hydrazine-based route, synthesized molecular solutions are easily recovered to crystalline phase after mild heat treatment (~400 °C); therefore, there have been studies to exploit this alkahest for solution process of metal chalcogenide thin film for electronic and optoelectronic applications such as PV devices [14], photodetectors [15], electrocatalysts [16], and thermoelectrics [17].

The developed chemistry was inspired from the previous report that proposes the dissolution mechanism for bulk sulfur in solvent mixture of amine and thiol, where thiol is deprotonated by amine giving a nucleophilic environment to produce polysulfide anions [18]. In the first report on cosolvent approach, they suggested that strong solvation power to covalent-bonded inorganic semiconductors is originated from N–H⋯S hydrogen bond which stabilizes $(enH^+)_2(EDT^{2-})(en)$ crystal [12]. Also, a series of experiments revealed that both 1,2-chelating amine and thiol are required for high solvent power, demonstrated by poor solubility in solvent mixture of monoamine and thiol or 1,3-chelating thiol, e.g., 1,3-propanedithiol or n-propanethiol in en [12].

2.4 All-Inorganic Nanoparticles as Soluble Precursor

Chemically synthesized colloidal nanoparticles are used as a precursor or building block for fabrication of inorganic thin films, as those can be dispersed in desired solvents for solution processing. Also, electronic and optical properties can be precisely controlled by varying size and shape of the nanoparticles, which give a degree of freedom to tailor the properties of resulting film materials [19–22]. Similar to other precursor solutions, colloidal nanoparticles also can be deposited on diverse substrates by solution process and produce the uniform thin film after heat treatment. Another advantage of using nanoparticles as precursors for solution process is to facilitate the decrease in melting point with the decrease in size of materials, which enables one to sinter the nanoparticles at low temperature (~300 °C) [21, 22]. However, long-chain hydrocarbon surfactants or bulky organometallic molecules capping the nanoparticles for stabilization in solvents act as barriers for charge transport. Furthermore, these large species are not fully decomposed sometimes and remain as inhibitors for crystallization and grain growth [23].

To handle this issue, molecular inorganics have been used as surface ligands for nanoparticle (Fig. 1c) [3, 24]. Organic ligands on nanoparticle surfaces exchange via phase transfer of nanoparticles from nonpolar solvent to polar solvent that dissolves the molecular inorganics. Metal chalcogenide complex (MCC) ligands provide colloidal stability in polar solvent as well as better electronic connection among nanoparticles after deposition of thin films, making more suitable for electronic and optoelectronic applications [23, 24]. Furthermore, designing the combination between nanoparticles and inorganic ligands allows to control easily the compositions, providing the potential to adjust electronic properties of diverse composite materials.

3 Thermoelectric Thin Film

Even possessing excellent TE properties, inorganic metal chalcogenides had often been disregarded for solution processing due to strong covalent bonds between metal and chalcogen atoms, which are ironically also an origin of its superb TE properties [5]. Huge efforts have been devoted to synthesis of metal chalcogenide-based TE inks, but most researches were limited in inorganic-organic hybrid system, exploiting solution processability of organic materials [25–28]. However, despite organic materials having great dispersibility and could assist inorganic TE powders to form viscous TE ink, the organic residue after film fabrication deteriorates electrical properties [29]. Thanks to the recent development of solution chemistry for inorganic TE materials, researches on TE thin films entered a new stage. Below, we will discuss numerous studies on solution-processed TE thin films.

3.1 V-VI Semiconductors

Arguably, working horse materials for near-room temperature TE generator (TEG) are Bi_2Te_3 (n-type) and Sb_2Te_3 (p-type) [30]. Both materials are considered as strong candidates for TE applications since the 1950s until the recent date which physical properties are well-established thanks to thorough investigations over several decades [31–33]. Because Bi_2Te_3 and Sb_2Te_3 show best performance around room temperature, both are regarded as the most suitable materials for wearable devices, harvesting body heat to operate portable or wearable electronic devices. Naturally, it is no surprise that numerous researches attempt to fabricate thin or thick films consisted of Bi_2Te_3 and Sb_2Te_3 in accordance with advances of organic binders for inorganic powders, which provide full thermal conduct between rough and round-shaped surfaces of heat sources and TE materials and flexibility to tolerate strain [34, 35].

Consequently, when the concept of dimensional reduction was introduced, it attracted huge attention from TE society since organic additive-free solution processing could provide a novel approach for TE thin film fabrication. David B. Mitzi, the pioneer of solution process for inorganic materials, also suggested Sb_2X_3 (X = S,

Se or Te) as the next candidate for solution-processable inorganic materials for TE application in his work [5].

Solution processing of both Bi_2Te_3 and Sb_2Te_3 was accomplished in the late 2000s, using an analogous process of preliminary results. For Bi(III) chalcogenide ionic salts materials, bismuth sulfide (Bi_2S_3) precursor was synthesized by reacting bulk Bi_2S_3 with excess sulfur in the medium of distillated hydrazine [36]. The precursor forms a black viscous liquid after sufficient stirring, which composition of chalcogen atoms could be flexibly controlled through simply adding an appropriate amount of selenium or tellurium. The authors demonstrate Bi_2S_3 precursor can be transformed into any composition of $Bi_2Te_{3-x}Se_x$ and verified it through fabricating four different thin films, Bi_2Se_3, Bi_2TeSe_2, Bi_2Te_2Se, and Bi_2Te_3. As indicated in Fig. 2a and b, SEM images and XRD patterns of all four films show well-developed crystalline structure, without any by-product. Remarkably, even without delicate optimizing concentration of charge carriers and alloying, Bi_2Se_3, Bi_2TeSe_2, and Bi_2Te_2Se thin films exhibit good TE properties (Fig. 2c–f) and record ZT values of 0.40, 0.39, and 0.22, respectively, which are the highest ZT values among fully solution-processed TE materials at that time. This astonishing result supports strongly the effectiveness and applicability of newly developed solution chemistry for inorganic TE materials.

However, Bi(III) chalcogenide could not form stable molecular-sized precursors with chalcogen elements (S, Se, and Te) in hydrazine [37], limiting its further development and applications to other chemistry. Whereas, as introduced in the previous section, Sb_2Te_3 solution could form chalcogenidometallates and is proven as effective inorganic ligand for nanocrystals [24], numerous researches tried to establish the solution process of Sb_2Te_3 and exploit ligand characteristic of Sb_2Te_3 chalcogenidometallate-capped nanocrystals to synthesize raw materials of nanostructured Sb_2Te_3 TE thin films. In the year 2010, Kovalenko et al. reported nanostructured TE thin films using solution synthesized by reacting Bi_2S_3 nanorod with Sb_2Te_3 chalcogenidometallate in hydrazine [37]. As-synthesized Sb_2Te_3 solution is not suitable for film fabrication since large excess Te was detected through XRD analysis of powder, which could not be removed without harsh heat treatment. However, as previous research proved, Se and Te tend to react with Bi_2S_3 and could completely substitute S [36]. By using this chemical reaction, excess Te could be effectively collected by adding Bi_2S_3 nanocrystal in Sb_2Te_3 solution, and because S evaporates at a much lower temperature compared to Te, (Bi, Sb)$_2Te_3$ thin films could be handled at relatively milder heating condition. The detailed chemical reaction for the formation of (Bi, Sb)$_2Te_3$ phase was suggested by the following equations:

$$Sb_2Te_3 - MCC \rightarrow Sb_2Te_3 + Te + N_2H_4, \tag{3}$$

$$Bi_2S_3 \ nanorods + 3Te \rightarrow Bi_2Te_3 + 3S \uparrow, \tag{4}$$

$$xSb_2Te_3 + (2-x)Bi_2Te_3 \rightarrow 2Bi_{2-x}Sb_xTe_3, \tag{5}$$

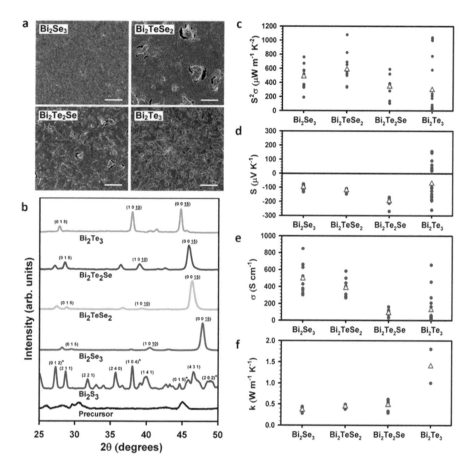

Fig. 2 Analyses of Bi_2S_3, Bi_2TeSe_2, Bi_2Te_2Se, and Bi_2Te_3 thin films fabricated via hydrazine chemistry. (**a**) SEM images and (**b**) XRD patterns of the films. The scale bar in all images is 5 μm. The graphs on the right side show (**c**) power factors, (**d**) Seebeck coefficients, (**e**) electrical conductivities, and (**f**) thermal conductivities of thin films. (Reprinted permission from [36]. Copyright 2010, American Chemical Society)

Both p-type $(Bi, Sb)_2Te_3$ and n-type $(Bi, Sb)_2(Te_{0.9}Se_{0.1})_3$ films produced through this process exhibit Seebeck coefficients of ±200–250 μV/K and conductivities up to 450 S/cm, which are surprisingly high value for thin film and even close to the values required for commercialization. Because it is well known that thermoelectrical properties of both Bi_2Te_3 and Sb_2Te_3 could be improved by alloying between those [30], the establishment of flexible control of atomic compositions with solution process could be regarded as a firm foothold for further advancement.

Moreover, nanostructuring techniques, including energy filtering or phonon scattering, could be realized through systematically designed nanocrystal ligand chemistry. Considering that nanostructuring is such a powerful method to enhancing the efficiency of TE materials [38–40], applying nanostructuring with simple solution process is a highly important achievement for not only TE thin films but the whole thermoelectric society. The first study on synthesizing nanocomposite consisting of metal nanocrystal and metal chalcogenide precursor was reported in the year 2011 by Ko et al., and authors inserted successfully Pt nanoparticles in hydrazine-based Sb_2Te_3 solution without any organic additives [41]. TEM images and selective area electron diffractions (SAED) of as-deposited Pt-Sb_2Te_3 nanocomposite solution indicate Pt nanoparticles are capped by Sb_2Te_3 chalcogenidometallate (Fig. 3a, b). Also, XRD pattern of the films after proper heat treatment shows clearly crystalline Sb_2Te_3 peaks and superimposed patterns of Pt, signifying that inserted Pt nanoparticles kept its nano-size structure even after heat treatment (Fig. 3c). Theoretical speculation of band diagram for Pt-Sb_2Te_3 nanocomposite predicted that low-energy charge carriers should scatter at the interfaces of Sb_2Te_3 and Pt nanoparticles as shown in Fig. 3d. The prediction was proven by the results of Seebeck coefficient measurement which exhibit significant enhancement of Seebeck coefficient by 34%, compared to thin film fabricated with Sb_2Te_3 solution only (Fig. 4). The other important finding of this study is developing n-trioctylphosphine (TOP) treatment, controlling the composition of thin films by dissolving excess Te from thin films after mild heat treatment.

Sb_2Te_3 could also be applied to amine-thiol cosolvent strategy, different from Bi(III) chalcogenides, which solubility to amine-thiol cosolvent is insufficient for practical applications [2]. However, even though its solubility to amine-thiol cosolvent was developed in the year 2013, the establishment of fundamental chemical principles for Sb_2Te_3 precursor and reliable methods for fabricating Sb_2Te_3 thin films were delayed to the year 2019 due to several obstacles [42]. Unexpectedly, when Sb_2Te_3 was dissolved in amine-thiol cosolvent, it does not immediately form molecular size anions but rather form tens of nm-size polymeric Sb-Te structure. Due to large size of solute, thin films fabricated with as-synthesized solution show numerous voids that could influence severely on transport of charge carriers across thin films (Fig. 5a). This phenomenon might originate from insufficient reducing power of amine-cosolvent system, so it could be successfully handled through providing a stronger reducing environment, specifically adding superhydride in Sb_2Te_3 solution. A clear change in the size of solute after superhydride treatment is depicted in Fig. 5b. However, there is also another side effect, that is, Sb_2Te_3 solution transformed into $Sb_2Te_7^{4-}$ cluster after superhydride treatment. When thin films heated to temperature enough to adjusting the composition of thin films, excess Te evaporated and left pores and voids in the microstructure of thin films (Fig. 5c). Thankfully, same as previous studies using hydrazine chemistry, TOP treatment could effectively control the amount of Te contained in the solution by removing excess Te. Finally, high-quality Sb_2Te_3 thin films could be fabricated with pretreated solution through simple spin coating. Its significant improvement in microstructure could be confirmed through SEM image (Fig. 5d).

Thanks to its excellent microstructure, the thin film exhibits power factor of $0.5~mW \times m^{-1} \times K^{-2}$ at room temperature to $0.85~mW \times m^{-1} \times K^{-2}$ at 423 K, one of the best results for Sb_2Te_3 thin films. In addition, because the size of Sb_2Te_3 precursor is less than 1 nm, it can be directly applied to ligand chemistry for nanocrystal-embedded TE precursor synthesis. Nanostructuring for Sb_2Te_3 thin films was achieved by introducing FePt nanocrystals, and significant decrease in lattice

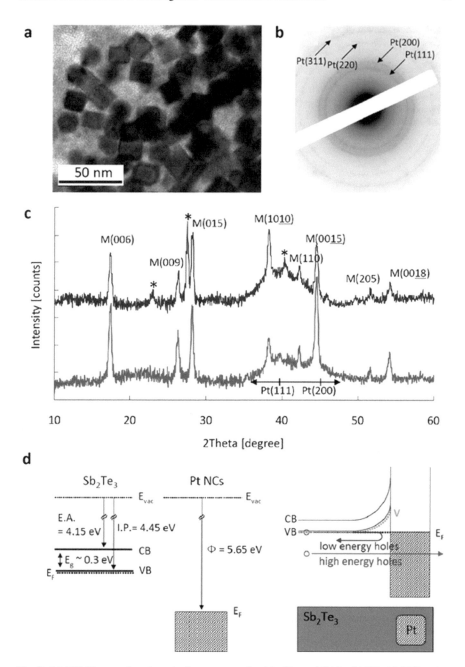

Fig. 3 (**a**) TEM image of as-deposited nanocomposite thin film and (**b**) its SAED. (**c**) Diffraction pattern of thin film powder shows broad Pt-related peaks, indicating nanocrystal maintains its crystalline structure. Diffraction peaks arising from Sb_2Te_3 matrix are denoted as M. (**d**) Calculated band alignment in Pt-Sb_2Te_3 nanocomposite. (Reprinted permission from [41]. Copyright 2011, American Chemical Society)

Fig. 4 Enhancement of
Seebeck coefficient by
introducing
nanostructuring to Sb_2Te_3
thin films. (Reprinted
permission from [41].
Copyright 2011, American
Chemical Society)

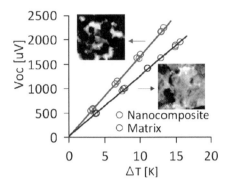

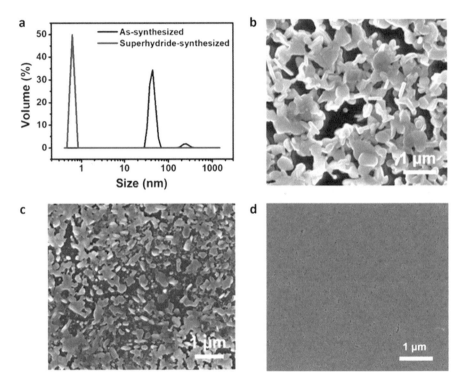

Fig. 5 (**a**) DLS analysis showing size distribution of the solute. SEM images of thin films fabricated by spin coating of (**b**) polymeric Sb_2Te_3, (**c**) Sb_2Te_7, and (**d**) molecular size Sb_2Te_3 solution. All films are heat treated at 300 °C. (Reprinted permission from [42]. Copyright 2019, American Chemical Society)

thermal conductivity from 0.573 W × m^{-1} × K^{-1} to 0.473 W × m^{-1} × K^{-1} at 423 K was observed, which is induced by phonon scattering effect (Fig. 6a–d). Lastly, Jo et al. [42] verified also the possibility of producing flexible devices out of Sb_2Te_3 solution (Fig. 6e). Surprisingly, Sb_2Te_3 thin film deposited on the flexible polyimide substrate shows nearly negligible change in resistance for several hundred times of

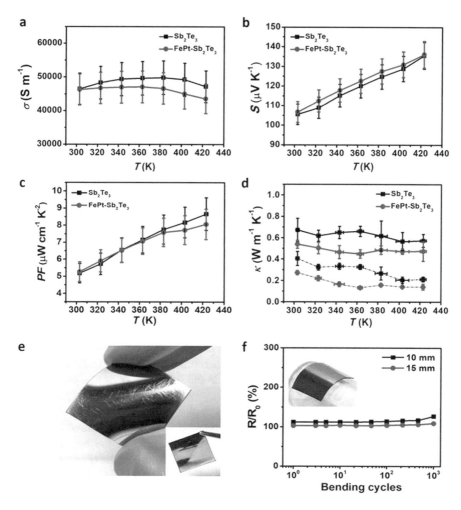

Fig. 6 Temperature dependence of (**a**) electrical conductivities, (**b**) Seebeck coefficients, (**c**) power factors, and (**d**) thermal conductivities of Sb_2Te_3 thin films. (**e**) Photographs showing flexible Sb_2Te_3 thin film deposited on polyimide substrate. (**f**) Relative change in resistance of Sb_2Te_3 thin film on polyimide substrate with the number of bending cycles at bending radius of 10 and 15 mm. (Reprinted permission from [42]. Copyright 2019, American Chemical Society)

bending test, and increase in resistance was less than 9% and 28% after 1000 times of bending at bending radius of 15 and 10 mm, respectively (Fig. 6f). This result suggests strongly applicability of the solution chemistry for inorganic TE materials.

3.2 Copper Selenide

The development of solution processing for Cu chalcogenide semiconducting materials was started at the early stage of chalcogenidometallate chemistry. The purpose of the researches, however, mainly focused on usage of p-type transistor or PV devices by forming compounds with other chalcogenidometallate clusters, including $CuInSe_2$ and $CuInTe_2$ [9, 10]. However, excellent TE properties of Cu_2Se were revealed in the year 2012 [43], and solution processing for Cu_2Se also received significant attention. At the same time, hydrazine-free, amine-thiol cosolvent process was developed around the year 2013. As discussed in the previous section, the cosolvent approach is much more suitable strategies in concern of mass production and commercialization. Taking benefits of advancing solution process chemistry, the behavior of Cu_2X (X = S, Se) within cosolvent environment was analyzed systematically in the year 2015 [44], and Cu_2X thin films could be produced out of cosolvent solution with simple spin-coating or drop-casting method, followed by heat treatment step. Several analyses, including XRD, EDS, XPS, and UV-Vis spectrum, are conducted to confirm high crystallinity and stoichiometry of the films (Fig. 7a–f), and SEM images show remarkable film uniformity (Fig. 7g, h). Electrical characterization of Cu_2S and Cu_2Se thin films revealed electrical conductivity of 127 S/cm and 1168 S/cm, respectively, which are the highest values among Cu_2X thin films thanks to remarkably high mobility of charge carriers of 2.4 cm$^2 \times$ V$^{-1} \times$ s^{-1} for Cu_2S and 3.9 cm$^2 \times$ V$^{-1} \times$ s^{-1}, for Cu_2Se. More detailed investigations over TE properties of Cu_2Se thin films are managed in the follow-up research, published in the year 2017 [17]. As suggested in preliminary research, uniform, high-quality thin films could be achieved and four different samples annealed at different temperatures of 573 K, 623 K, 703 K, and 773 K, respectively. To avoid unwanted annealing effect during measurement, each sample is measured up to the corresponding annealing temperature. It is well known that Cu vacancies are the origin of p-type conduction behavior of Cu_2Se and determine the concentration of charge carriers, indicating that electrical properties of Cu_2Se thin films could be handled through composition control via controlling annealing temperature. Corresponding to the theoretical prediction, as annealing temperature of the films increase, electrical conductivity increases, and Seebeck coefficient decreases. Optimal temperature for electrical properties was turned out to be 703 K, and highest power factor was 0.62 mW $\times$ m$^{-1} \times$ K^{-2} at 684 K, which is significantly higher value compared to other Cu_2Se thin films, generally exhibiting power factor smaller than 0.1 mW $\times$ m$^{-1} \times$ K^{-2}. Furthermore, Cu_2Se thin film deposited on flexible substrate shows also power factor of 0.46 mW $\times$ m$^{-1} \times$ K^{-2}, at 664 K, which is slightly lower than that of Cu_2Se thin film on rigid Al_2O_3 substrate but still significantly high value. This flexible Cu_2Se TE thin film shows only 8% change of resistance and negligible change of thermal conductivity after 1000 bending cycles, indicating great feasibility of TE thin films for practical applications.

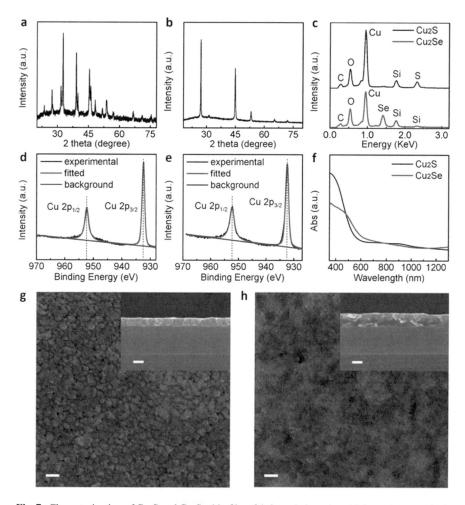

Fig. 7 Characterization of Cu_2S and Cu_2Se thin films fabricated via amine-thiol cosolvent method. XRD pattern of powder of (**a**) Cu_2S and (**b**) Cu_2Se thin films, respectively. (**c**) EDS for thin films. X-ray photoelectron microscopy (XPS) of (**d**) Cu_2S and (**e**) Cu_2Se thin films on SiO_2/Si substrate. (**f**) UV-vis spectra for Cu_2S and Cu_2Se thin films on glass substrate. SEM images of (**g**) Cu_2S and (**h**) Cu_2Se thin films. Scale bar is 100 nm for all images. (Reprinted permission from [44]. Copyright 2015, American Chemical Society)

3.3 Tin Selenide

Tin selenide (SnSe) is one of the most highlighted TE materials since the year 2014 after SnSe set a new record for ZT value at its very first publication by Zhao et al. [45]. Excellent TE efficiency of SnSe is originated from its extremely low thermal

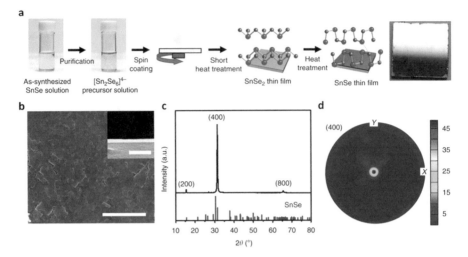

Fig. 8 (**a**) Schematic illustration of the fabrication process of SnSe thin films. (**b**) SEM image of SnSe thin film fabricated with purified precursor solution shows full coverage of SnSe on the substrate. Scale bar, 5 μm. Cross-sectional SEM image (inset) shows uniform thickness of 85 nm. Scale bar, 500 nm. (**c**) XRD pattern of SnSe thin film fabricated with purified precursor solution. Vertical lines indicate the pattern for orthorhombic SnSe reference (JCPDS 32–1382). (**d**) Pole figure of (400) plane in SnSe thin film fabricated with purified precursor solution. (Reprinted permission from [56]. Copyright 2019, Springer Nature)

conductivity, which is derived from unique crystal structure. However, at the same time, the crystal structure of SnSe is one of the main obstacles for high-performance polycrystalline SnSe, because of unfavorable electrical properties along *a*-axis. Huge efforts have been devoted to controlling crystal structure or electrical properties of SnSe through hot pressing and doping strategies, but numerous results failed to reach enough efficiency [46–54]. Nevertheless, considering layered structure is vulnerable against cleaving, which makes SnSe inadequate for mass production and practical usage [55], finding a reliable method for producing high-quality polycrystalline SnSe is essential for further applications.

In the year 2019, to deal with the problems, research on the fabrication of highly textured SnSe thin films with a solution process was conducted [56]. Figure 8a shows thin film fabrication method, the film's unique plate-like morphology, strongly textured structure verified with XRD pattern, and the pole figure (Fig. 8b–d). Different from Cu_2Se, which could be deposited on a substrate without pretreatment for a solution, removing thiol from SnSe precursor solution is essential because it is well known that Sn and S from thiol tend to react very actively, forming SnS [57]. In the presence of thiol, S easily substitutes Se during heat treatment, and the crystalline structure of thin film becomes SnS_xSe_{1-x}, not SnSe. To minimize the effect of thiol, the purification step was implemented, and SnSe precursor dissolved in ethylenediamine solely after the purification. Using several analyses, including absorption and Raman spectrum analyses, it turned out that purified SnSe precursor solution contained well-known molecular anion $Sn_2Se_6^{4-}$, which could be coordinated with ethylenediammonium. The purification step is beneficial for not only removing the

Fig. 9 (a) XRD patterns of SnSe thin films upon heating. At 300 °C, only (00l) peaks of SnSe$_2$ were observed in XRD pattern. At higher temperatures, these peaks progressively disappeared, and (h00) peaks of SnSe were exclusively detected at 400 °C. (Reprinted permission from [56]. Copyright 2019, Springer Nature)

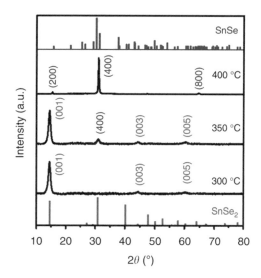

unwanted side reaction but also controlling atomic composition, which leads to exceptional level of texturing to SnSe thin film. Because SnSe precursor initially contains much higher Se compared to Sn, SnSe$_2$ is a more favorable structure for the deposited thin films. It is noteworthy that SnSe$_2$ is one of the famous 2D materials and has strong tendency to form unique layered structure. It was not an exception for solution-based thin films, and XRD pattern of thin film fabricated by spin coating followed by heat treatment at 300 °C show (001), (003), and (005) peaks of SnSe$_2$. When heat treatment temperature is increased to 350 °C, (400) peak of SnSe starts to emerge, and lastly, at 400 °C, all of SnSe$_2$ related peaks completely disappear, and only (200), (400), and (800) peaks of SnSe are observed (Fig. 9). The estimated orientation factor of (h00) plane is 0.89, which is surprisingly high value for thin films and even comparable to that of bulk SnSe sample produced by zone-melting method. Same results are observed for thin film produced by spray-coating method, in which orientation factor of (h00) plane was 0.75, indicating SnSe precursor solution could be applied to general thin film-fabricating methods.

These results signify that layered structure of SnSe$_2$ thin film could be maintained during evaporation of Se and change of crystal structure, as explained above, because *b-c* plane of SnSe has much superior TE properties compared to *a*-axis; it can be expected that highly textured SnSe thin films have great TE properties. Even more, composition of SnSe thin films could be systematically controlled with heat treatment. It is widely accepted that *p*-type nature of SnSe is originated by intrinsically existing Sn vacancies, which act as hole dopant [58, 59]. In other words, concentration of charge carriers in SnSe thin films could be tailored by simple heat treatment process. As a result, both structural and electrical properties could be managed with simple and cost-effective solution process for optimum TE efficiency. The maximum TE power factor of thin film records 0.43 mW $\times$ m^{-1} $\times$ K^{-2} at 550 K, which is by far the best performance among SnSe thin films at the moment [60, 61] and even higher than that of SnSe single crystal at the same temperature (Fig. 10).

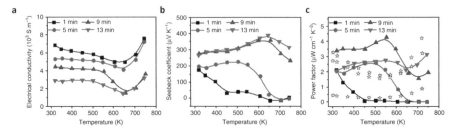

Fig. 10 Temperature dependences of (**a**) electrical conductivities, (**b**) Seebeck coefficients, and (**c**) power factors for SnSe thin films annealed at 400 °C for 1, 5, 9, and 13 min. Stars in (**c**) indicate data points obtained with SnSe single crystals in *a* (black), *b* (red), and *c* (blue) axes. (Reprinted permission from [56]. Copyright 2019, Springer Nature)

4 Outlook

In summary, we reviewed the recent advances in soluble inorganic metal chalcogenides applied to TE thin films, highlighting several benchmark examples. So far, solution-processed inorganic thin films have been developed in limited material classes of metal chalcogenides due to limitation of the synthesis. Although metal chalcogenides are the most dependable materials in TE community, the different classes of TE materials such as clathrates, skutterudites, and half-Heusler alloys are actively studied too. These classes of TE materials are metal pnictide-based materials in general. However, soluble pnictide inorganic compounds have rarely been studied due to the lack of the synthesis. This limitation in the synthesis may originate from the underlying challenge to understand the basic chemistry of metal pnictides in a solution. In other words, there is a huge room for studying this chemistry to extend different classes of TE materials beyond metal chalcogenides.

The biggest payoff of soluble TE inorganics is the potential for manufacturing highly efficient TE thin films in a cost-effective way. However, the current status of this research is still at the early stage to understand the chemistry and the properties of thin films. For pursuing industrialization, inorganic TE solutions should apply to printing techniques for patterning TE layers and electrodes without loss of ZT values, which require researchers' further efforts for the realization of thermoelectric inorganic solution technology to be viable.

References

1. E.G. Tulsky, J.R. Long, Dimensional reduction: A practical formalism for manipulating solid structures. Chem. Mater. **13**, 1149–1166 (2001)
2. D.H. Webber, R.L. Brutchey, Alkahest for V2VI3 chalcogenides: Dissolution of nine bulk semiconductors in a diamine-dithiol solvent mixture. J. Am. Chem. Soc. **135**, 15722–15725 (2013)
3. H. Zhang, J. Jang, W. Liu, D.V. Talapin, Colloidal nanocrystals with inorganic halide, pseudo-halide, and halometallate ligands. ACS Nano **8**, 7359–7369 (2014)

4. D.B. Mitzi, Solution processing of chalcogenide semiconductors via dimensional reduction. Adv. Mater. **21**, 3141–3158 (2009)
5. D.B. Mitzi, L.L. Kosbar, C.E. Murray, M. Copel, A. Afzali, High-mobility ultrathin semiconducting films prepared by spin coating. Nature **428**, 299–303 (2004)
6. D.B. Mitzi, M. Copel, S.J. Chey, Low-voltage transistor employing a high-mobility spin-coated chalcogenide semiconductor. Adv. Mater. **17**, 1285–1289 (2005)
7. D.B. Mitzi, Synthesis, structure, and thermal properties of soluble hydrazinium germanium(IV) and tin(IV) selenide salts. Inorg. Chem. **44**, 3755–3761 (2005)
8. D.B. Mitzi, N4H9Cu7S4: A hydrazinium-based salt with a layered Cu7S4-framework. Inorg. Chem. **46**, 926–931 (2007)
9. D.J. Milliron, D.B. Mitzi, M. Copel, C.E. Murray, Solution-processed metal chalcogenide films for p-type transistors. Chem. Mater. **18**, 587–590 (2006)
10. D.B. Mitzi, M. Copel, C.E. Murray, High-mobility p-type transistor based on a spin-coated metal telluride semiconductor. Adv. Mater. **18**, 2448–2452 (2006)
11. D.B. Mitzi, M. Yuan, W. Liu, A.J. Kellock, S.J. Chey, V. Deline, A.G. Schrott, A high-efficiency solution-deposited thin-film photovoltaic device. Adv. Mater. **20**, 3657–3662 (2008)
12. D.H. Webber, J.J. Buckley, P.D. Antunez, R.L. Brutchey, Facile dissolution of selenium and tellurium in a thiol–amine solvent mixture under ambient conditions. Chem. Sci. **5**, 2498–2502 (2014)
13. C.L. McCarthy, D.H. Webber, E.C. Schueller, R.L. Brutchey, Solution-phase conversion of bulk metal oxides to metal chalcogenides using a simple thiol-amine solvent mixture. Angew. Chem. Int. Ed. **54**, 8378–8381 (2015)
14. Q. Tian, Y. Cui, G. Wang, D. Pan, A robust and low-cost strategy to prepare Cu2ZnSnS4 precursor solution and its application in Cu2ZnSn(S,Se)4 solar cells. RSC Adv. **5**, 4184–4190 (2015)
15. M.R. Hasan, E.S. Arinze, A.K. Singh, S.M. Thon, R. Debnath, An antimony selenide molecular ink for flexible broadband photodetectors. Adv. Electron. Mater. **2**, 1600182 (2016)
16. C.L. McCarthy, C.A. Downes, E.C. Schueller, K. Abuyen, R.L. Brutchey, Method for the solution deposition of phase-pure CoSe2 as an efficient hydrogen evolution reaction electrocatalyst. ACS Energy Lett. **1**, 607–611 (2016)
17. Z. Lin, C. Hollar, J.S. Kang, A. Yin, Y. Wang, H.Y. Shiu, Y. Huang, Y. Hu, Y. Zhang, X. Duan, A solution processable high-performance thermoelectric copper selenide thin film. Adv. Mater. **29**, 1606662 (2017)
18. B.D. Vineyard, Versatility and the mechanism of the n-butyl-amine-catalyzed reaction of thiols with sulfur. J. Org. Chem. **32**, 3833–3836 (1967)
19. V.L. Colvin, M.C. Schlamp, A.P. Alivisatos, Light-emitting diodes made from cadmium selenide nanocrystals and a semiconducting polymer. Nature **370**, 354–357 (1994)
20. X. Peng, L. Manna, W. Yang, J. Wickham, E. Scher, A. Kadavanich, A.P. Alivisatos, Shape control of CdSe nanocrystals. Nature **404**, 59–61 (2000)
21. P. Buffat, J.P. Borel, Size effect on the melting temperature of gold particles. Phys. Rev. A **13**, 2287–2298 (1976)
22. A.N. Goldstein, C.M. Echer, A.P. Alivisatos, Melting in semiconductor nanocrystals. Science **256**, 1425–1427 (1992)
23. A. Nag, H. Zhang, E. Janke, D.V. Talapin, Inorganic surface ligands for colloidal nanomaterials. Z. Phys. Chem. **229**, 85–107 (2015)
24. M.V. Kovalenko, M. Scheele, D.V. Talapin, Colloidal nanocrystals with molecular metal chalcogenide surface ligands. Science **324**, 1417–1420 (2009)
25. B. Zhang, J. Sun, H.E. Katz, F. Fang, R.L. Opila, Promising thermoelectric properties of commercial PEDOT:PSS materials and their bi2Te3 powder composites. ACS Appl. Mater. Interfaces **2**, 3170–3178 (2010)
26. N. Toshima, M. Imai, S. Ichikawa, Organic–inorganic nanohybrids as novel thermoelectric materials: Hybrids of polyaniline and bismuth(III) telluride nanoparticles. J. Electron. Mater. **40**, 898–902 (2010)

27. K. Kato, H. Hagino, K. Miyazaki, Fabrication of bismuth telluride thermoelectric films containing conductive polymers using a printing method. J. Electron. Mater. **42**, 1313–1318 (2013)
28. C. Ou, A.L. Sangle, A. Datta, Q. Jing, T. Busolo, T. Chalklen, V. Narayan, S. Kar-Narayan, Fully printed organic-inorganic nanocomposites for flexible thermoelectric applications. ACS Appl. Mater. Interfaces **10**, 19580–19587 (2018)
29. S. Jo, S. Choo, F. Kim, S.H. Heo, J.S. Son, Ink processing for thermoelectric materials and power-generating devices. Adv. Mater. **31**, 1804930 (2019)
30. G.J. Snyder, E.S. Toberer, Complex thermoelectric materials. Nat. Mater. **7**, 105–114 (2008)
31. H.J. Goldsmid, R.W. Douglas, The use of semiconductors in thermoelectric refrigeration. Br. J. Appl. Phys. **5**, 386–390 (1954)
32. T.C. Harman, B. Paris, S.E. Miller, H.L. Goering, Preparation and some physical properties of Bi2Te3, Sb2Te3, and As2Te3. J. Phys. Chem. Solids **2**, 181–190 (1957)
33. I.J. Ohsugi, T. Kojima, I.A. Nishida, Orientation analysis of the anisotropic galvanomagnetism of sintered Bi2Te3. J. Appl. Phys. **68**, 5692–5695 (1990)
34. S.J. Kim, J.H. We, J.S. Kim, G.S. Kim, B.J. Cho, Thermoelectric properties of P-type Sb2Te3 thick film processed by a screen-printing technique and a subsequent annealing process. J. Alloys Compd. **582**, 177–180 (2014)
35. S.J. Kim, J.H. We, B.J. Cho, A wearable thermoelectric generator fabricated on a glass fabric. Energy Environ. Sci. **7**, 1959–1965 (2014)
36. R.Y. Wang, J.P. Feser, X. Gu, K.M. Yu, R.A. Segalman, A. Majumdar, D.J. Milliron, J.J. Urban, Universal and solution-processable precursor to bismuth chalcogenide thermoelectrics. Chem. Mater. **22**, 1943–1945 (2010)
37. M.V. Kovalenko, B. Spokoyny, J. Lee, M. Scheele, A. Weber, S. Perera, D. Landry, D.V. Talapin, Semiconductor nanocrystals functionalized with antimony telluride zintl ions for nanostructured thermoelectrics. J. Am. Chem. Soc. **132**, 6686–6695 (2010)
38. A. Majumdar, Thermoelectricity in semiconductor nanostructures. Science **303**, 777–778 (2004)
39. J.P. Heremans, V. Jovovic, E.S. Toberer, A. Saramat, K. Kurosaki, A. Charoenphakdee, S. Yamanaka, G.J. Snyder, Enhancement of thermoelectric efficiency in PbTe by distortion of the electronic density of states. Science **321**, 554–557 (2008)
40. Y. Pei, H. Wang, G.J. Snyder, Band engineering of thermoelectric materials. Adv. Mater. **24**, 6125–6135 (2012)
41. D.K. Ko, Y. Kang, C.B. Murray, Enhanced thermopower via carrier energy filtering in solution-processable Pt-Sb2Te3 nanocomposites. Nano Lett. **11**, 2841–2844 (2011)
42. S. Jo, S.H. Park, H. Shin, I. Oh, S.H. Heo, H.W. Ban, H.W. Jeong, F. Kim, S. Choo, D.H. Gu, S. Baek, S. Cho, J.S. Kim, B. Kim, J.E. Lee, S. Song, J. Yoo, J.Y. Song, J.S. Son, Soluble telluride-based molecular precursor for solution-processed high-performance thermoelectrics. ACS Appl. Energy Mater. **2**, 4582–4589 (2019)
43. H. Liu, X. Shi, F. Xu, L. Zhang, W. Zhang, L. Chen, Q. Li, C. Uher, T. Day, G.J. Snyder, Copper ion liquid-like thermoelectrics. Nat. Mater. **11**, 422–425 (2012)
44. Z. Lin, Q. He, A. Yin, Y. Xu, C. Wang, M. Ding, H. Cheng, B. Papandrea, Y. Huang, X. Duan, Cosolvent approach for solution-processable electronic thin films. ACS Nano **9**, 4398–4405 (2015)
45. L.D. Zhao, S.H. Lo, Y. Zhang, H. Sun, G. Tan, C. Uher, C. Wolverton, V.P. Dravid, M.G. Kanatzidis, Ultralow thermal conductivity and high thermoelectric figure of merit in SnSe crystals. Nature **508**, 373–377 (2014)
46. D. Feng, Z.H. Ge, D. Wu, Y.X. Chen, T. Wu, J. Li, J. He, Enhanced thermoelectric properties of SnSe polycrystals via texture control. Phys. Chem. Chem. Phys. **18**, 31821–31827 (2016)
47. Y. Li, F. Li, J. Dong, Z. Ge, F. Kang, J. He, H. Du, B. Li, J.F. Li, Enhanced mid-temperature thermoelectric performance of textured SnSe polycrystals made of solvothermally synthesized powders. J. Mater. Chem. C **4**, 2047–2055 (2016)

48. Y. Fu, J. Xu, G.Q. Liu, J. Yang, X. Tan, Z. Liu, H. Qin, H. Shao, H. Jiang, B. Liang, J. Jiang, Enhanced thermoelectric performance in p-type polycrystalline SnSe benefiting from texture modulation. J. Mater. Chem. C **4**, 1201–1207 (2016)
49. G. Han, S.R. Popuri, H.F. Greer, J.G. Bos, D.H. Gregory, Facile surfactant-free synthesis of p-type SnSe nanoplates with exceptional thermoelectric power factors. Angew. Chem. Int. Ed. Engl. **55**, 6433–6437 (2016)
50. G. Han, S.R. Popuri, H.F. Greer, L.F. Llin, D.H. Gregory, Chlorine-enabled electron doping in solution-synthesized SnSe thermoelectric nanomaterials. Adv. Energy Mater. **7**, 1602328 (2017)
51. N.K. Singh, S. Bathula, B. Gahtori, K. Tyagi, D. Haranath, A. Dhar, The effect of doping on thermoelectric performance of p-type SnSe: Promising thermoelectric material. J. Alloys Compd. **668**, 152–158 (2016)
52. X. Wang, J. Xu, J. Liu, Y. Fu, Z. Liu, X. Tan, H. Shao, H. Jiang, T. Tan, J. Jiang, Optimization of thermoelectric properties in n-type SnSe doped with BiCl3. Appl. Phys. Lett. **108**, 083902 (2016)
53. M. Gharsallah, F. Elhalouani, J.A. Alonso, Giant Seebeck effect in Ge-doped SnSe. Sci. Rep. **6**, 26774 (2016)
54. C.L. Chen, H. Wang, Y.Y. Chen, T. Day, G.J. Snyder, Thermoelectric properties of p-type polycrystalline SnSe doped with Ag. J. Mater. Chem. A **2**, 11171–11176 (2014)
55. V.Q. Nguyen, T.H. Nguyen, V.T. Duong, J.E. Lee, S.D. Park, J.Y. Song, H.M. Park, A.T. Duong, S. Cho, Thermoelectric properties of hot-pressed bi-doped n-type polycrystalline SnSe. Nanoscale Res. Lett. **13**, 200 (2018)
56. S.H. Heo, S. Jo, H.S. Kim, G. Choi, J.Y. Song, J.Y. Kang, N.J. Park, H.W. Ban, F. Kim, H. Jeong, J. Jung, J. Jang, W.B. Lee, H. Shin, J.S. Son, Composition change-driven texturing and doping in solution-processed SnSe thermoelectric thin films. Nat. Commun. **10**, 864 (2019)
57. J.J. Buckley, C.L. McCarthy, J.D. Pilar-Albaladejo, G. Rasul, R.L. Brutchey, Dissolution of Sn, SnO, and SnS in a thiol-amine solvent mixture: insights into the identity of the molecular solutes for solution-processed SnS. Inorg. Chem. **55**, 3175–3180 (2016)
58. G. Duvjir, T. Min, T.T. Ly, T. Kim, A.T. Duong, S. Cho, S.H. Rhim, J. Lee, J. Kim, Origin of p-type characteristics in a SnSe single crystal. Appl. Phys. Lett. **110**, 262106 (2017)
59. A.T. Duong, V.Q. Nguyen, G. Duvjir, V.T. Duong, S. Kwon, J.Y. Song, J.K. Lee, J.E. Lee, S. Park, T. Min, J. Lee, J. Kim, S. Cho, Achieving ZT=2.2 with bi-doped n-type SnSe single crystals. Nat. Commun. **7**, 13713 (2016)
60. P.K. Nair, A.K. Martinez, A.R.G. Angelmo, E.B. Salgado, M.T.S. Nair, Thermoelectric prospects of chemically deposited PbSe and SnSe thin films. Semicond. Sci. Technol. **33**, 035004 (2018)
61. M.R. Burton, T. Liu, J. McGettrick, S. Mehraban, J. Baker, A. Pockett, T. Watson, O. Fenwick, M.J. Carnie, Thin film tin selenide (SnSe) thermoelectric generators exhibiting ultralow thermal conductivity. Adv. Mater. **30**, 1801357 (2018)

Recent Advances in Functional Thermoelectric Materials for Printed Electronics

A. L. Pires, J. A. Silva, M. M. Maia, S. Silva, A. M. L. Lopes, J. Fonseca, M. Ribeiro, C. Pereira, and André M. Pereira

Abbreviations

BST	$Bi_{0.5}Sb_{1.5}Te_3$
BTS	$Bi_2Te_{2.7}Se_{0.3}$
CNTs	Carbon nanotubes
CVD	Chemical vapor deposition
CSA	Camphor sulfonic acid
DCB	Dichlorobenzene
DEG	Diethylene glycol
DMSO	Dimethyl sulfoxide
EG	Ethylene glycol
F8BT	Poly(9,9-di-n-octylfluorene-alt-benzothiadiazole)
FGA	Forming gas annealing
κ	Thermal conductivity
κ_e	Electronic thermal conductivity
κ_l	Lattice thermal conductivity
MWCNTs	Multi-walled carbon nanotubes
NMP	N-Methyl-2-pyrrolidone
P3HT	Poly(3-hexylthiophene)
PA	Polyacetylene
PANi	Polyaniline

A. L. Pires · M. M. Maia · A. M. L. Lopes · A. M. Pereira (✉)
IFIMUP - Institute of Physics for Advanced Materials, Nanotechnology and Photonics, Physics and Astronomy Department, University of Porto, Porto, Portugal
e-mail: ampereira@fc.up.pt

J. A. Silva · S. Silva · J. Fonseca · M. Ribeiro
CeNTI – Centre for Nanotechnology and Smart Materials, Vila Nova de Famalicão, Portugal

C. Pereira
REQUIMTE/LAQV, Chemistry and Biochemistry Department, Faculty of Sciences, University of Porto, Porto, Portugal

© Springer Nature Switzerland AG 2021
S. Skipidarov, M. Nikitin (eds.), *Thin Film and Flexible Thermoelectric Generators, Devices and Sensors*, https://doi.org/10.1007/978-3-030-45862-1_4

PC	Polycarbonate
PEDOT	Poly(3,4-ethylenedioxythiophene)
PEI	Polyethyleneimine
PET	Polyethylene terephthalate
PF	Power factor
PG	Propylene glycol
PMMA	Poly(methyl methacrylate)
PPy	Polypyrrole
PS	Polystyrene
PSS	Poly(styrenesulfonate)
PTH	Polythiophene
PVA	Poly(vinyl alcohol)
PVAc	Polyvinyl acetate
PVDF	Polyvinylidene fluoride
PVP	Polyvinylpyrrolidone
R2R	Roll-to-roll
RT	Room temperature
S	Seebeck coefficient
σ	Electrical conductivity
SEM	Scanning electron microscopy
SLM	Selective laser melting
SWCNTs	Single-walled carbon nanotubes
T	Absolute temperature
TE	Thermoelectric
TEG	Thermoelectric generator
Tos	Tosylate
ZT	Figure of merit
ΔT	Temperature difference

1 Introduction

Flexible energy harvesting devices have been considered as very promising for the emerging market needs of flexible and wearable electronics. Currently, available energy harvesting devices are rigid and bulky, which is incompatible with the next generation of flexible electronics. Therefore, the development of high-performance and reliable power sources that are light, thin, and flexible becomes critical. Besides, these energy harvesters need to be functional under various mechanical requirements and cost-effective, since it is a key point to succeed in the market.

Printed flexible electronics seems to be the solution to overcome these challenges, since it holds lightness and flexibility of the devices and can be produced by low-cost and scalable processing techniques. As a reference, *IDTechEx* estimated that the market would grow from $29 billion to $73 billion between 2017 and 2027

[1] demonstrating high potential of this technology. For centuries, printing techniques have been widely used by the publishing industry (books, journals, magazines, etc.), packaging, ceramics, and glassware industry. Over time, these techniques have naturally evolved, expanding application to different industrial sectors. In the last years, triggered by the development of functional inks and technology miniaturization, printing techniques have reached new areas such as micro- and flexible electronics [2]. Thus, several printed electronic devices have been envisioned for many applications including smart packaging [3], flexible solar cells [4], therapeutic and medical diagnostic devices [5], batteries [6], energy harvesting and storage [7], as well as lighting systems [8].

Naturally, energy harvesting emerges as a solution to growing energy demands, promoting sustainable carbon-free energy consumption. Since the beginning of the decade, research in energy harvesting field increased as depicted in Fig. 1a, which represents the number of publications per year in this area since 2010.

Figure 2 shows several energy sources belonging to energy harvesting field such as radio frequency, vibration, light, and thermal with respective power outputs [9].

In the segment of thermal energy harvesting, TE materials and devices are considered as practical solution. Notably, TE devices have gained substantial attention over the years as can be observed in Fig. 1b, which shows the number of publications per year since 1980. This technology has the capability to convert directly of waste thermal energy into electricity. A pictorial view of the effect is depicted in Fig. 2b. It is estimated that in EU around 800 TWh of thermal energy are lost every year to the environment as waste heat from the industrial processes [10], which represents a strong opportunity for this technology.

TE effect was discovered in 1821 by Thomas Seebeck, and since then, an increasing number of studies in inorganic and organic TE materials can be found in the literature. This increase is revealed in the number of report reviews over the years [11–16]. These reports aim at understanding the properties of TE materials to further increase in performance of TEGs.

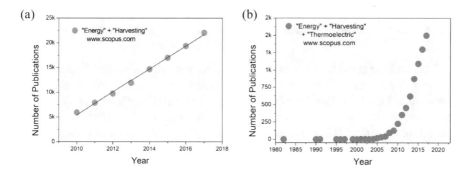

Fig. 1 (**a**) Number of publications per year from 2010 to 2018 using the keywords "Energy" and "Harvesting" and (**b**) publication number per year since 1980 using the keywords "Energy" and "Harvesting" and "Thermoelectric". (Source: www.scopus.com) (Color figure online)

(a)   (b)

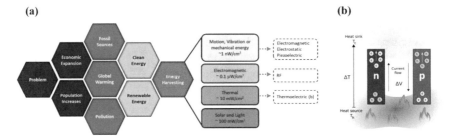

Fig. 2 (**a**) Taxonomy of energy harvesting concerning the issues that led to arising of this type of technology with the different individual sources. (**b**) Illustration of Seebeck effect where temperature gradient is converted into electricity using two different TE materials, n-type material with electron conductivity and p-type material with hole conductivity (Color figure online)

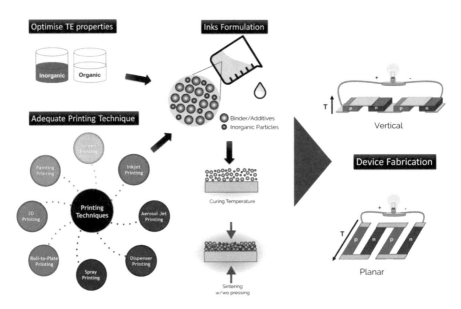

Fig. 3 Graphical abstract of the main contents of the chapter (Color figure online)

This chapter provides an overview of the current state-of-the-art in printed TE devices. A schematic illustration summarizing the main contents of the chapter is depicted in Fig. 3. Section 1 outlines the most relevant TE properties of TE materials from inorganic to organic materials, and the importance of combining these two classes of materials. Then, in Sect. 2, a description of the leading printing techniques is provided and will be an asset to Sect. 3 where the state-of-the-art TE inks will be addressed. In that section, special attention will be devoted to currently used inorganic TE materials and those combinations with organic materials such as binders and additives. In the last section, the most relevant TEGs will be presented followed by future prospects in the field.

2 Thermoelectric Materials

2.1 Thermoelectric Fundaments

The discovery of Seebeck effect, at the end of the eighteenth century, unleashed extensive research in the field of TE materials and devices that can now be applied in our daily lives. At the beginning of the nineteenth century, the concept of figure of merit was introduced by Altenkirch, who revealed that good TE material should have large S and high σ to minimize Joule heating. Moreover, he unveiled that κ should be low to preserve large temperature gradient at the material's junctions [17]. The synergic combination of these three components makes the research in TE materials very challenging. Ioffe, in 1957, defined the figure of merit as quantification of the efficiency of TE materials, and nowadays, the dimensionless thermoelectric figure of merit ZT is presented as follows:

$$ZT = \frac{S^2 \sigma T}{\kappa}, \tag{1}$$

where S is Seebeck coefficient, σ is electrical conductivity, κ is thermal conductivity, and T is absolute temperature [17, 18].

Herein, S and σ for simple 3D geometry can be expressed by the following equations:

$$S = \frac{8\pi^2 k_B^2}{3eh^2} m^* T \left(\frac{\pi}{3n}\right)^{2/3}, \tag{2}$$

and

$$\sigma = e\mu n(p), \tag{3}$$

where e is electron charge, $n(p)$ is concentration of charge carriers, μ is mobility of charge carriers, k_B is Boltzmann constant, h is Planck constant, and m^* is effective mass of charge carriers [17, 19]. Moreover, κ in Eq. (1) results from the sum of electronic (by charge carriers) κ_e and lattice (by phonons) κ_l thermal conductivity contributions. κ_e can be expressed through Wiedemann-Franz law:

$$\kappa_e = L\sigma T, \tag{4}$$

where Lorentz number $L = 2.44 \times 10^{-8}$ W × Ohm×K^{-2}.

κ_l can be written as:

$$\kappa_l = \frac{n_v v \lambda_p C_v}{3N_A}, \tag{5}$$

where n_v is number of particles per unit volume, $\langle v \rangle$ is mean particle speed, λ_p is phonon mean free path, C_v is molar heat capacity, and N_A is Avogadro number. Notice that in the absence of κ_l, Eq. (1) is simplified due to Eq. (4) and ZT will only depend on the square of S. Nevertheless, this is not effectively observed, and κ_l has an important role on κ. Indeed, tuning κ_l is one of the major challenges on the new generation of high-performance TE materials.

In inorganic TE materials, the main question relies on obtaining materials with low κ since it directly affects ZT and consequently the performance of produced device. Phonon engineering [20, 21] and band engineering [22–24] have been pointed at two promising strategies to enhance ZT through reduction in κ_l, and increase in power factor PF, where $PF = S^2\sigma$.

Nanostructuring arises as a tuning materials engineering to reach high performance of TEGs since nanoconfinement and nanoconstrain allow the tailoring of lattice vibration. This approach can also lead to the enhancement of the density of states (DOS) near Fermi level and, therefore, increase in thermopower, which provides a way to decouple thermopower and σ [25]. Moreover, the mean free path of electrons is much smaller than that of phonons in heavily doped semiconductors. The nanostructuring will carry a large density of interfaces in which phonons can be scattered more effectively over a large mean free path range. This will lead to reduction in κ_l while preserving charge carriers' mobility and electronic conduction [25]. Finally, other strategies to improve TE properties have effectively been applied, namely, by point defect engineering [26, 27], hierarchical structuring [28], and embedding dense dislocation arrays in grain boundaries [21]. These topics will be covered throughout the chapter.

2.2 State-of-the-Art of Thermoelectric Materials

2.2.1 Inorganic Materials: Chalcogenides and Carbon-Based Materials

Until the beginning of the 2000s, it was believed that ZT value of most common commercial TE materials was limited to values close to unit. Until that time, TE materials were mainly based on Bi_2Te_3, $PbTe$, and $SiGe$ alloys from which Bi_2Te_3 was the most widely used for manufacturing TE materials, namely, in temperature control systems used as Peltier devices.

Bi_2Te_3 belongs to the tetradymite family, presenting an outstanding TE performance at room temperature. The rhombohedral crystal structure formed by D_{3d}^5 ($R\bar{3}m$) space group with five atoms in the trigonal unit cell is responsible for excellent TE properties [29, 30]. Moreover, the remarkable anisotropic behavior resulting from high c-axis length ratio of the rhombohedral structure is behind its high TE performance [31]. Other materials have been identified as promising TE materials, namely, skutterudites, clathrates, metal chalcogenides, and oxides materials. These materials presented high ZT and operate in different temperature ranges (further information about these materials can be found in [32]).

Bi_2Te_3 based materials are widely investigated and can be produced with different synthesis methods. The most common growth methods of TE materials are, among others, solvothermal or hydrothermal [33–35], spark plasma sintering [36], arc melting [37], Bridgman [38, 39], and mechanical alloying [40].

On the other hand, with the advent of nanotechnology and with the unquestionable versatility of nanostructuring of the materials, two main approaches are being pursued toward this goal: top-down and bottom-up approaches. In top-down approaches, larger sets of materials are processed to obtain smaller and/or structurally contained materials, allowing fabrication to be faster than the counterpart. On the other hand, bottom-up processes allow the fine tuning of material's shape and size. Regarding the top-down approach, a breakthrough was achieved by Venkatasubramanian et al. [41] that reported high $ZT \sim 2.4$ for p-type Bi_2Te_3/Sb_2Te_3 superlattices which consist of alternate nanometer-thick layers of each material. The nanostructure allows electrons to flow within thin film while suppressing phonon transport. Although being an attractive result, this fabrication process is very expensive and time-consuming restraining its use for large-scale production. Concerning the bottom-up approach, Gou et al. [34] have used solvothermal method to produce hierarchical Bi_2Te_3 nanoflowers assembled by 2D thin nanosheets with defects that shows ZT of 0.68 at 475 K. Regarding traditional spherical nanoparticles, Scheele et al. [42] were able to produce, with high quality, sub-10 nm Bi_2Te_3 nanoparticles prepared by spark plasma sintering after removing the particle ligands by hydrazine hydrate based etching process, with κ as low as 0.8 W $\times$ m^{-1} $\times$ K^{-1} (at 200 K). $ZT \approx 0.2$ at 300 K was achieved with this methodology. The main efforts in this area are being devoted toward increasing in ZT of such materials.

Traditional inorganic TE materials have higher ZT at RT, as verified for chalcogenide-based material. However, disadvantages like toxicity, high cost, and brittleness have limited those technological application [43]. Consequently, in the last years, a new class of carbon-based nanomaterials have been explored due to low cost and ease of process, becoming attractive for TE applications. Carbon nanotubes (CNTs) or graphene materials are being envisaged for TE applications. These carbon-based materials present good σ, high mechanical strength and modulus, large aspect ratio, lightness and low-cost, when compared with Te-based materials. However, ZT of the pristine carbon-based materials shows usually lower S, around 3–80 μV/K, and high κ that can reach up to 53 W $\times$ m^{-1} $\times$ K^{-1} [44, 45]. Among CNTs, SWCNTs in pristine form present higher ZT than MWCNTs, due to the unique electronic structure (one layer) arising from the shape of DOS near Fermi energy [46].

Nevertheless, the ability of all carbon-based materials to be easily manipulated on charge carriers' type (n-type and p-type), to tailor the atomic structure by chemical functionalization and/or through the mechanical process leading to atomic structure changes, is an asset for TE applications. One possible approach to decrease high κ of these materials is through the fine tuning of phonon thermal transport by increasing in degree of disorder [47]. For example, Xiao et al. [48] showed that the use of oxygen plasma treatment in a few layers of graphene allow the increase in S

from 80 to 700 μV/K but no significant change in σ is observed. This higher S is due to the generation of disorders in graphene which opens up $\pi - \pi^*$ gap. Chemical functionalization using strong acids can create defects on the sidewalls and reduce σ owing to disruption of extended π conjugation, for instance, in CNTs [46]. Other strategies toward the enhancement of TE properties have been widely reported, e.g., doping and stabilization of CNTs. Zhou et al. [49] achieved high PF of 2482 μW $\times$ m^{-1} $\times$ K^{-2} at RT for as-synthesized SWCNT buckypaper through floating catalyst CVD with O_2 doping. This achievement represents approximately half to PF reported for Bi_2Te_3 (~ 5062 μW $\times$ m^{-1} $\times$ K^{-2} at $T = 320$ K) [50].

On the other hand, CNTs have much lower density than Bi_2Te_3, i.e., 1.1 to 7.9 g/cm^3, respectively. Despite the enormous improvements made in the last decade, more research are still necessary to reach competitive values. Nevertheless, promising TEGs may benefit from the use of these materials.

2.2.2 Organic Materials: Polymer Based

In the last century, organic/polymer materials have aroused interest in thermoelectricity due to low κ, good flexibility, ease of fabrication, low cost, and low toxicity when compared with inorganic materials [51]. Poor or negligible σ and poor air stability, mainly for n-type materials, are the disadvantages [43]. The efficiency of these materials remains very low, and research works to fulfill the demands of practical application are still at the initial stage.

The most widely studied organic polymers which exhibit adjustable σ due to doping level are PEDOT:PSS [52], PANi [53], P3HT [54], PA [55], PPy [56], PEI [57], and others. Unlike the inorganic materials, the doping in organic ones requires the addition of electron donor (n-type) or electron acceptor (p-type) molecular species to induce mobile charge carriers along the backbone polymer [58]. PEDOT:PSS-based materials have shown high transparency and good solubility. These properties make this polymer an alternative TE material to be processed using printing techniques [59].

As specific examples, Kim et al. [60] reached ZT of 0.42 for DMSO-mixed PEDOT:PSS at RT and 0.28 for EG-mixed PEDOT:PSS. They discussed the importance of reducing dopant volume to optimize concentration of charge carriers while maximizing ZT of the organic semiconductors. High ZT was also measured in combination of n-type poly[Nax(Ni-ett)] and p-type poly[Cux(Cu-ett)] (ZT of 0.1 at 400 K) [61]. However, poor solubility can be a problem for printing processes. Other polymers can be found with ZT values of 1.1×10^{-2} at 423 K for PANi, 2.9×10^{-2} at 250 K for PTH, 0.25 at RT for PEDOT:tosylate (Tos), and 3×10^{-2} at 423 K for PPy [62]. However, low stability and humidity sensitivity in ambient conditions could limit practical application.

2.2.3 Hybrid Materials in Modern Thermoelectric Composites

Hybrid TE composites are becoming the mainstream materials for flexible TEGs due to ability to combine in flexibility of organic materials and high TE performance of inorganic component. Moreover, this strategy allows taking advantage of nanomaterials properties arising from the nanoconfinement that have shown to improve ZT effectively. In this case, polymer matrix has often acted as binding element of the nanomaterials. Guo et al. [63] suggest that three key factors enhance the performance of organic-inorganic TE composites: (1) inorganic component should possess high σ and S; (2) it needs to be well dispersed in polymer matrix; and (3) one has to ensure the formation of a large number of strong interfaces between inorganic element and polymer matrix leading to improvement in mechanical properties combined by an increase in S and σ.

Figure 4 displays ZT values that can be found in the literature for several hybrid composites, where PANi and PEDOT are clearly the most popular polymers to be used showing very large improvement of ZT when combined with different inorganic materials.

Notice that all reported ZT values are lower than unity at RT. For example, Zhang et al. [85] reported that by combining PEDOT:PSS with $Bi_{0.5}Sb_{1.5}Te_3$ (BST), the resulting composite presented poor mechanical properties such as low flexibility, low tensile strength, and poor fracture strain and found that using extra addition of PVA could improve mechanical properties. It was concluded that nanocomposite film with 36 wt. % BST, 39 wt. % PEDOT:PSS, and 25 wt. % PVA using drop casting method has ZT outcome of 0.05, tensile strength of 79.3 MPa, and fracture strain

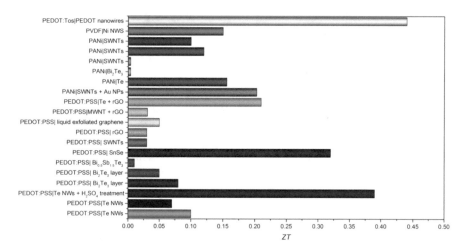

Fig. 4 Resume of typical ZT value of TE composites reported in the literature at room temperature. (▢[64] ▣ [65] ■ [66] ▣ [67] ▢ [68] ▢ [69] ■ [70] ▣ [71] ▢ [72] ▢ [73] ▢ [74] ▢ [75] ■ [76] ■ [77] ▢ [78] ■ [79] ■ [80] ■ [81] ■ [82] ▢ [83]). (Adapted from [84]) (Color figure online)

of 32.4% which enables this composite to be used in low-power flexible/wearable electronic devices.

Chalcogenide-based organic nanocomposites are one of the most studied systems. Fabrication process and chemical composition are the main degrees of freedom toward high-performance TE material. For example, Wang et al. [86] fabricated *PbTe*-polyaniline nanocomposite in situ by interfacial polymerization at RT and cold-pressed composite achieving *PF* of 0.713 $\mu W \times m^{-1} \times K^{-2}$. Chatterjee et al. [69] produced Bi_2Te_3-polyaniline nanocomposite using simultaneous electrochemical reaction and deposition method. With a composition of 30 wt. % Bi_2Te_3, they achieved low κ of 0.11 $W \times m^{-1} \times K^{-1}$ and *ZT* of 0.0043 at RT. More recently, Mitra et al. [87] reported the use of conducting polymer-chalcogenide composite fabricated via combining in situ oxidative polymerization and solvothermal method. For 30 wt. % of Bi_2Se_3 nanoplates in PANi matrix, low κ of 0.19 $W \times m^{-1} \times K^{-1}$ and *ZT* of 0.046 at RT were achieved, which is almost 20 times higher than that achieved for PANi polymer. Also, Guo et al. [63] reported a novel inorganic/polymer TE composite fabricated by CSA-doped PANi and BST using intercalation/exfoliation method. The dispersion was achieved after CSA:PANi matrix was treated by cryogenic grinding forming a large number of hybrid organic-inorganic interfaces. As a result, a remarkable enhancement of TE performance in these composites is observed. This achievement associated to energy filtering of charge carriers on interfaces between the nanoplates and CSA:PANi. It was also observed that *PF* increased from 16.5×10^{-8} to 84.4×10^{-4} $\mu W \times m^{-1} \times K^{-2}$ as the concentration of nanoplates increased from 5 to 20 wt. %.

In situ interfacial polymerization has been reported as excellent technique to disperse homogeneously inorganic material in polymeric matrix, and prevent the oxidation of materials during this process, reducing interface resistance [62].

Besides chalcogenides, carbon nanomaterials are also used as inorganic component for hybrid TE materials where Fermi level control is a parameter that remains crucial due to requiring the balance between *S* and σ. Doping of inorganic materials has been demonstrated as highly effective way to increase in TE properties of such materials and to prevent the oxidation of CNTs that will result in *p*-type material [88, 89]. Several reports indicate that doping of composite with chemical reagents may improve TE performance. Recently, Zhou et al. [90] reported on doping of composites F8BT/SWCNTs with ferric chloride. A maximum *PF* of 1.7 $\mu W \times m^{-1} \times K^{-2}$ and *ZT* of 7.1×10^{-4} at 303 K were achieved for doped composite with 60% of SWCNTs, which was around 1.5 times higher than in composite with undoped F8BT. The polymer, in composite fabrication, promotes the increase in phonon scattering at inter-tube interfaces while decreasing in κ of CNTs in several orders of magnitude [91, 92]. Choi et al. [93] produced a robust carbon nanotube yarn with excellent σ (~ 3147 S/cm) due to the increase in longitudinal mobility of charge carriers resulting from highly aligned structure. Excellent *PF* of 2387 $\mu W \times m^{-1} \times K^{-2}$ for *p*-type doped with Fe_3Cl_3 ethanol solution and 2456 $\mu W \times m^{-1} \times K^{-2}$ for *n*-type doped with PEI ethanol solution were achieved. Yu et al. [57] reported on nontoxic polymer composite, fabricated with 20 wt. % of CNT in PVAc that exhibits high σ of 4800 S/m, low κ of 0.34 $W \times m^{-1} \times K^{-1}$, and

ZT of 0.006 at RT. This improvement in σ is associated to CNT network in the composite that tune TE properties, i.e., due to the junctions between CNT networks which are electrically connected but thermally disconnected.

Piao et al. [94] envisaged the incorporation effects of SWCNT networks, so-called buckypaper, in neutral polymers and impact on *S*, σ. Neutral polymers like PVDF, PMMA, PVA, PS, and PC were used with the aim to hinder phonon transport through CNT networks, increasing *S* up to 50 µV/K [94]. Also, CNTs were functionalized with PEI and incorporated in PVA. The chemically modified surfaces result in an enhancement of σ as high as 1500 S/m and *S* of -100 µV/K [95].

3   Printing Techniques

The use of conventional printing techniques for manufacturing of electronic devices, known as printed electronics, enables more cost-efficient mass production of electronics and reduction in material waste compared with more traditional technologies [96, 97]. In general, printing processes are roll-to-roll (R2R), being particularly suitable to produce lightweight, highly flexible, and rollable devices [96–98]. However, there are several challenges in the adaptation of printing technologies to print electronic devices, e.g., non-uniformities and poor control of thickness, voids (holes), and other defects that need to be avoided because it can have an adverse impact on the performance of electronic devices [99]. Another relevant aspect is change in the properties of functional materials when processed by printing technologies: e.g., σ of printed films tends to be lower than films produced by other techniques. Regarding printed TE devices, more specific issues have been pointed out. For example, common substrates used in printed electronics have moderate κ reducing the efficiency of TE devices [97]. The referred constraints have limited the choice of printing technologies to manufacture electronic devices. The most common printing technologies are screen, inkjet, gravure and flexographic printing, slot-die coating, and offset (e.g., lithography). An overview of the key parameters of each technique is summarized in Table 1.

Recently, multiple research groups have reported the use of printing techniques to produce printed TE devices, mainly using screen and inkjet printing technologies. Therefore, these two techniques will be mainly focused herein.

3.1   Screen Printing

Figure 5 illustrates typical screen printing process when the ink is transferred to the surface to be printed by forcing it through porous mesh image carrier, or screen, in which printing area is open, and non-printing area is blocked (stencil) [97, 98, 100].

Table 1 Main characteristics of printing techniques

Parameter	Units	Screen printing	Inkjet	Offset	Slot-die	Flexography	Gravure
Print resolution	μm	30–100	15–100	20–50	200	30–80	50–200
Layer thickness	μm	3–30	0.01–0.5	0.6–2	0.15–60	0.17–8	0.02–12
Printing speed	m/min	0.6–100	0.02–5	0.6–15	0.6–5	5–180	8–100
Ink viscosity	Pa × s	0.5–5	0.001–0.10	5–2	0.002–5	0.01–0.5	0.01–1.1
Solution surface tension	mN/m	38–47	15–25	–	65–70	13.9–23	41–44

Adapted from [99]

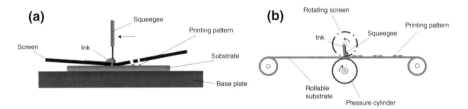

Fig. 5 Screen printing technique with (**a**) flat-bed and (**b**) rotary screen (Color figure online)

The screen printing is commonly designated as flat-bed printing (Fig. 5a), where the screen is flat, enabling printing on rigid and flexible substrates (including paper, plastics, metal, glass, textile, among many others). Usually, the screens are made of polyester or stainless steel. The mesh (number of wires per unit length) and the wire diameter are the most relevant screen properties, controlling film thickness and having a substantial impact on printing resolution and speed. The smaller the wire diameter and the higher the mesh, the thicker the printed art will be. With the increase in printed pattern thickness, a decrease in resolution is also expected [101]. Other parameters affect the printed pattern properties such as screen tension, snap-off distance, squeegee pressure, angle and velocity, squeegee shape and hardness, and stencil thickness. The choice of the squeegee shape and hardness is mostly dictated by the substrate roughness and porosity [99]. It is possible to achieve smaller printed patterns using sharper and harder squeegees, but performance on rough/porous substrates decreases.

The screen can also be cylindrical, known as rotary screen, Fig. 5b. Its working principle is similar to the flat-bed technology. The rotary screen allows a continuous process in R2R mode and, therefore, only compatible with flexible substrates. In the rotary screen process, the squeegee is inside the screen, at a fixed position. The screen rotates at a constant rate, matching the speed of substrate movement. The rotary screen technique allows a higher production rate and is usually integrated into R2R systems that include complementary techniques such as lamination and plasma treatment [100].

This technique is particularly versatile and straightforward, being compatible with a wide range of substrates, from rigid to flexible, allowing to print a great variety of inks, including conductive, electroluminescent, electroactive, UV curable, adhesive inks, among others.

One of the most critical factors to obtain reliable and reproducible printing results is the ink formulation. Since ink must be pushed through the screens, the printing quality is strongly dependent on the ink's rheological properties. For instance, when applied shear stress, the ink viscosity needs to decrease and recover when the application was interrupted.

The main distinguishing characteristic of the screen printing technique is the possibility of printing thicker printed pattern compared with other printing techniques, which can be relevant in the manufacturing of functional structures for electronic devices. The deposition of relatively thick layers of ink is possible; however, it must have enough viscosity to maintain its shape, which could require longer curing times [98]. Besides, inks composed of volatile solvents can reduce the resolution of the printed pattern. Another challenge is the control of the pattern morphology with thick ink layers, which are prone to cracking, especially in flexible devices [97].

Several research groups have reported on the development of TE devices manufactured by screen printing. Typically, used meshes were around 200 to 325 (thread/inch), commonly made of steel screen [102–104]. A more detailed description of the work of these research groups on printed TE devices can be found in Sect. 4.

3.2 Inkjet Printing

Another strategy to produce TE devices besides screen printing technique is the inkjet printing. This technique is a computer-assisted printing technique which allows the deposition of tiny droplets (2–12 pL) on a large variety of substrates (plastic, metal, glass, paper, etc.) with high precision, resolution, and reproducibility [96]. Inkjet printing may be performed with different inkjet printing head technologies: piezoelectric, thermal, acoustic, and electrostatic. Piezoelectric and thermal are the most common [105] and are schematically represented in Fig. 6. In piezoelectric printheads, the ink ejection is controlled using piezoelectric materials actuated by an electric field. The thermal technology is based on ink emission resulting from a vapor bubble that is produced on the surface of a hot resistor located inside each nozzle. As the temperature rises, the vapor bubble grows, increasing in pressure and forcing the ink to be expelled. Comparing both thermal and piezoelectric inkjet technologies, the first has become very popular mainly due to its lower cost. However, since the inks in thermal inkjet must be vaporized during printing process, the solvent choice for the inks is severely limited by specificities of printing head itself, making piezoelectric technology the most preferred for functional ink printing [97].

The viscosity of the ink prepared for inkjet printing when compared to screen printing should be much lower, as can be seen in Table 1. This low viscosity could

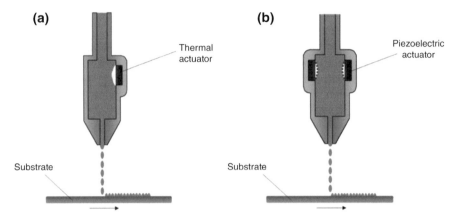

Fig. 6 Scheme of inkjet printing process with (**a**) thermal and (**b**) piezoelectric actuators (Color figure online)

be an essential asset in TE printed materials because it allows the production of films with lower additive concentration and, therefore, preserving the functional materials properties. Additionally, no physical contact between printhead and substrate enables inkjet printing technique on substrates sensitive to mechanical contact. On the other hand, both technologies, thermal and piezoelectric, require a very fine tune of ink's physical properties including viscosity, density, and surface tension [98, 106]. Likewise, ink must be free from large particles, meaning that size should be much smaller than the nozzle diameter, in order to prevent the blocking narrow channels and nozzles. Therefore, in this technique, the filtration process is a crucial step. Notice that this may constitute a disadvantage for manufacturing TE devices, since TE inks are, in general, composed of particles suspended in a liquid [97, 98], thus, nanomaterials suggest a solution to overcome this limitation. Additionally, in the case of thermal printing, inks must possess a very high thermal stability because inks must be vaporized in a short period of time when ejected. A more detailed description of the work of these research groups on printed TE devices can be found in the next section.

4 Inks Processing and Post-treatments

Printed electronics adapts printing methods to reduce production costs, and to simplify device fabrication. In TE industry, the production costs and low TE performance are the main limitations to revolutionize the market. A possible approach to overcome this hurdle is to reduce production cost of TE devices, resorting to printing techniques to produce printed TE films with high ZT. To this end, the development of useful and efficient inks for the printing techniques, mentioned above, is crucial and will be addressed in the following section.

In ink formulation, it is crucial that the resulting composite has the same performance as TE material itself. For this, an exhaustive and appropriate choice of organic additives should be made with the purpose of combining with TE material as described in Sect. 2.2.3. Specific requirements to ink formulations include rheology of the paste, namely, viscosity or drying kinetics which should correspond to non-Newtonian liquid. These factors contribute to successful transfer of the pattern onto the substrate. In the case of screen or dispenser printing, the typical values of viscosity are of the magnitude of several thousand mPa × s, whereas for inkjet printing in the range of 10 mPa × s. This specific viscosity value relies on the requirement of screen printing inks or pastes to have both pseudoplastic and thixotropic behaviors [2, 97], i.e., the viscosity decreases as the shear rate increases (pseudoplastic), and when subject to a constant shear rate, those viscosity also decreases (thixotropic) [2]. In the case of inkjet printing, the surface tension and viscosity of inkjet printing inks determine the formation of droplets on the substrate, including the specific size and shape of the droplet [107].

Regarding TE inks, the presence of organic components to provide adequate viscosity can be a main drawback due to the negative impact on ZT of the final composite. Undoubtedly, effective removal of the additives/organic compounds from the cured inks is a crucial step to improve TE properties and needs to be performed without promoting the oxidation, and high porosity on final TE-printed films. Thermal treatment is the most common process to remove organic additives. However, this process limits the use of some polymers and/or substrates due to the heat tolerance of these materials. High degree of mechanical stability and flexibility is always a requirement for printed TEG and needs to be considered in inks' formulation and processing [103].

Before proceeding to a more detailed description of the current state of the scientific literature on TE inks, it is necessary to acquaint with the concept of the *post*-printing process, since it will be used subtly in several examples that we will give for printed techniques.

4.1 Post-printing Process: Thermal Treatments

The physical properties have an active role in the principles of the ink formulation. The process requires a homogeneous printed layer, where the agglomeration must be reduced or even eliminated, allowing to achieve high and homogeneous green density. Thus, powder consolidation and microstructure of the green sample followed by a sintering process need to be considered.

Post-printing processes such as curing temperature and sintering reveal to be a common step in TE film fabrication to improve TE properties of the final printed films [108]. These two additional steps have distinct results. The thermal curing of the printed ink results in solvent evaporation, and it can promote chemical reactions to enhance adhesion and strength of the film, as well as it can improve the conductance between particles. Moreover, the annealing (sintering) step, which occurs at high temperatures (above the curing temperature but below the melting point of the inorganic material), allows for complete degradation of any organic compound present in the film. Thus, the outcome is the bonding of particles and densification of the film. The sintering process depends on several parameters such as temperature, time, environment, particle size, initial density, and applied stress [108, 109]. In this process, three overlapping stages can be identified: (1) *initial stage*, bonding adjacent particles with formation of the necks but with limited densification; (2) *intermediate stage*, solid and porous phases connect, and considerable densification occurs; and (3) *final stage*, pores are isolated, and the grain coarsening is controlled by interaction between pores and grain boundaries [109]. Otherwise, in coarsening, in some cases, average grain size increases due to the rise of large grains and dissolution of small ones in a matrix, so-called Ostwald ripening [108, 110, 111]. The driving force for material transporting results from the difference in chemical potential of atoms under curved surfaces as represented in Fig. 7. Thus, redistribution of the material on particle surface via material transporting from particle surface to neck surface without densification, and then material transport from the grain boundary induces the densification.

Concerning the diffusion process, i.e., the rate of material transport from material source (grain boundary or particle surface) to the neck surface can be expressed by the following equation:

$$\frac{dV}{dt} = JAV_m = \left(-\frac{D}{RT} \nabla P \right) AV_m, \tag{6}$$

where V is volume of material transported to the neck, t is sintering time, J is material flux, A is diffusional area, V_m is molar volume, D is diffusion constant, R is gas constant, T is absolute temperature, and ∇P is stress (pressure) gradient. Therefore, σ can be improved through the formation of large necks, which decreases in constriction resistance and ultimately forms a metallic crystal structure with a small number of grain boundaries. In the last decade, ink formulation, Ostwald ripening

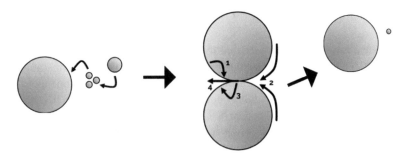

Fig. 7 Schematic illustration of several atomic diffusion paths between two contacting particles and Ostwald ripening process. 1. Lattice diffusion (no densification). 2. Surface diffusion (no densification). 3. Through-lattice diffusion (densification). 4. Grain boundary diffusion (densification). (Adapted from [108])

effect, and sintering have gained interest due to the requirement of early curing and sintering of the particles at temperatures well below the melting point of the bulk materials.

Another effect that occurs during the densification process is the increase in packing factor and minimization of pore surface due to the capillary pressure. These rearrangements correspond to a viscous flow process during liquid d-phase and are represented by the following expression [112]:

$$\frac{\Delta L_s}{L_0} \sim t^{1+y}, \tag{7}$$

where L_s is linear dimension of the sample and t denotes sintering time, respectively. The exponent $1 + y$ is considerably larger than unity, since pore size decreases and driving force increases.

Endeavoring low-cost production, both temperature and time required for sintering should be clearly reduced, and this has been the focus of many works over the last few years. As an example, Wu et al. [113] used non-contact dispenser printing combined with selective laser melting (SLM) as a single step of sintering. SLM is a technique designed to use a high-density laser to melt and fuse metallic powders together [114]. Combining printed techniques with SLM allowed achievement of n-type materials with excellent TE performance and a compact structure without porosity as shown in Fig. 8a, b. n-type $Bi_2Te_{2.7}Se_{0.3}$ bulk sample was fabricated by stacking 75 layers on top of each other and shows 96% relative density [113]. Additionally, incorporation of other components such as Sb_2Te_3 and Te during the sintering step is used to improve the densification of the final printed inks. Figure 8c represents the microstructure evolution diagram for undoped and doped materials.

Despite the evidence that annealing process improves TE properties, some precautions should be considered. In the case of $BiTe$ and $SbTe$ alloys, high-temperature annealing processes may cause the formation of anti-site defects of Bi and Sb in Te sites (Bi_{Te} and Sb_{Te}) and vacancy defects at Bi sites (V_{Bi}) due to Te evaporation (V_{Te}) [115, 116]. Formed defects suppress growth of electron concentration but increase in hole concentration. Additionally, this complex approach will increase in cost of the process, reducing reproducibility appearing to be inefficient to control concentration of charge carriers, despite high ZT reached. Other approaches can be tested to sinter the printed films such as plasma sintering, photonic curing, and pulse electric current sintering. The effect on transport and rheological properties should be investigated for these techniques due to the formation of harmful cracks in the final printed TE films.

Another difficulty to overcome is temperature limitation of plastic substrates, such as PET or PC, because of low glass transition temperature, T_g. Another approach is the use of nanoparticles instead of microparticles to reduce the sintering temperature. These developments are relevant to the fabrication of flexible TE films.

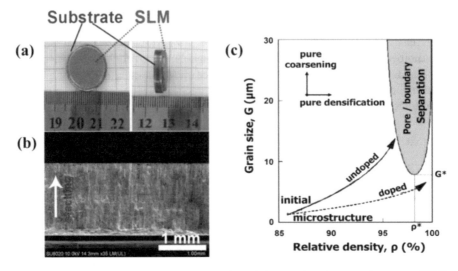

Fig. 8 (**a**) Real images of SLM prepared sample and (**b**) scanning electron microscopy (SEM) images of cross section of SLM bulk sample showing the stacking direction is textured along (110) direction (*ab* plane of Bi_2Te_3); and (**c**) diagram of the microstructure evolution, i.e., the grain size as a function of the relative density for undoped and doped materials. (Reprinted with permission from [109, 113] copyright 2017, John Wiley and Sons, Inc.)

4.2 Screen-Printed Thermoelectric Films

The highest fraction of publications in the field of flexible TE-printed films is related to screen printing method. To formulate an adequate ink for screen printing, some aspects must be taken into account, such as viscosity, curing temperature, curing time, and sintering as mentioned above. Further, we discuss features of TE thin films fabricated by screen printing method toward optimization.

4.2.1 Curing Temperature

Yuan et al. [103, 117] have studied the influence of curing temperature on the printed films. TE inks have been composed by *p*-type $Bi_{0.5}Sb_{1.5}Te_3$, Sb_2Te_3, and *n*-type $Bi_2Te_{2.7}Se_{0.3}$, sieved through 325 mesh screen, with a polymer binder formulated with PG, diglycidyl ether, epoxy resin, and methylhexahydrophthalic anhydride in a ration epoxy/hardener of 1:0.85 [103]. Further, to reduce in viscosity, butyl acetate (nonreactive diluent) was added into resin blend. The volume fraction between inorganic and organic materials was 45–50 vol. %. In the printing process, successively curing at 90 °C and printing were performed for 2 h. Then, different curing temperatures of square printed samples were studied, i.e., from 150 to 350 °C for 6 h in N_2 atmosphere [103]. An increase in σ was observed for *n*-type $Bi_2Te_{2.7}Se_{0.3}$, *p*-type

$Bi_{0.5}Sb_{1.5}Te_3$, and Sb_2Te_3 being in the range of 24.57–165.8 S/cm and it is observed S of −176.6, 223.3, and 139.7 μV/K, respectively, measured at RT.

One of the main drawbacks of the curing at higher temperature is binder residue in printed films, which results in low ZT outcome compared with ZT obtained after usual thermal annealing. If the thermal treatment is not used properly, it will result in densification reduction and promoting the porosity that is responsible for reducing σ in TE printed film.

4.2.2   Porosity

Kato et al. [118] studied the influence of the porosity on σ using combination of n- and p-type Bi_2Te_3 nanoparticles and polyimide in ionic liquid. The printed films were heated at 300–450 °C for 60 min in the air with H_2 to enhance TE performance. The authors claimed that σ, around 140–160 S/cm, of produced n- and p-type films, respectively, was enhanced due to the use of conductive polymers, focusing discussion around the importance of filling random pores with conductive polymer to improve the percolation effect. The achievement of impressive ZT of 0.87 and 0.5 for p- and n-type materials was an important milestone. This effect arises from extremely low $\kappa = 0.23–0.33$ W $\times$ m^{-1} $\times$ K^{-1}. Despite these results, Choi et al. [116] explored other approaches, namely, additional *post*-annealing followed by mechanical pressure. This methodology aimed to reduce the porosity and/or promote the densification of the materials, to overcome low ZT resulted from printed processes. Screen printing inks were prepared by p-type $Bi_{0.5}Sb_{1.5}Te_3$ and n-type $Bi_2Te_{2.7}Se_{0.3}$ materials with ethyl cellule and butyl carbitol acetate in 87:13 wt. % ratio, respectively. TE ink was printed several times on Al_2O_3 substrate and dried on hot plate at 110 °C for 10 min to remove the solvent. Then, two crystallizing steps were adopted: first, annealing in vacuum at 230 °C for 30 min to remove organic binder, and, second, recrystallization at 500 °C to promote grain growth. Finally, additional annealing with mechanical pressure was performed at 500 °C for 20 min to densify of printing ink. The final films showed 650 μm of thickness and achieved high ZT values (0.89 for p-type and 0.57 for n-type screen printing films at RT). The applied pressure during recrystallization step is responsible for increase in mobility of charge carriers by promoting better contact decreasing in inter-grain barrier which significantly contributes to the increase in σ.

Later, the same group [119] reported an simplified process to produce screen printed TE materials without conventional large furnace required and without counterbalancing the loss of TE property during thermal process based on the same inorganic materials. Notice that this loss is responsible for inducing transport parameter deviations and creating pores. Excess of Te powder has been incorporated directly on TE paste during preparation step, which allows the enhancement of screen-printed TE film. The authors support that it is necessary to add enough TE powder to the paste to minimize formation of vacancies, and to adjust concentration of charge carriers by optimizing crystallization conditions, i.e., annealing time and temperature. With this approach, an increase in 5% for *p*-type (*ZT* ~ 0.93) and 12% for *n*-type materials (*ZT* ~ 0.64) at RT was verified when compared with non-simplified approach [116]. However, in the case of *n*-type material, also an extra added step of glass frites was performed to help recrystallization of TE materials during sequential annealing process, i.e., nucleation centers in recrystallization process were promoted [102]. It was concluded that *Te* drives electron transfer channels, thus lowering the barrier energy between the grains of inorganic TE materials.

Han et al. [120] used *Te* powder in production of *PbTe* − *SrTe* based TE screen printed films. The excess of *Te* serves as a sintering aid, and the samples without excess of *Te* present typically a very coarse microstructure with many voids between the grains, i.e., excess of *Te* melting and induced liquid phase sintering to fill pores densely between the grains [121, 122]. Menon et al. [123] reported also on other hybrid materials for screen printing like *p*-type PEDOT/PSS with *Te* nanowires and *n*-type poly[Kx(Ni-ett)] blended with PVDF/DMSO.

4.2.3 Artificial Densification

Incorporation of other elements in the matrix is also reported as an approach to increase in TE properties by inducing densification of the material. Francioso et al. [124] incorporated *Ag* particles (5–6 μm) in screen printing inks produced with Sb_2Te_3 and Bi_2Te_3 (load of 56% and 60% for *p*- and *n*-type TE materials) with polystyrene nanoparticles and alpha-terpineol. Compounded inks were hot-pressed at 250 °C for 1 h under nitrogen gas flow with the aim to densify the final film. In this study, σ decreases with addition of *Ag* and this decrease is explained by possible formation of new interfaces and defects between grains of bulk matrix. On the contrary, Madan et al. [125] studied the effect of silver nanoparticles dispersed in Bi_2Te_3 composites and observed an increase in σ, concluding that metal micro-inclusions act as paths between grain boundaries.

4.2.4 Ionized Defect Engineering

Another approach to enhance *ZT* is to control organic contamination and physical and chemical defects in TE materials after printing [102]. This approach was studied by Kim et al. [102] using *post*-ionized defect engineering, which employs FGA process (4% H_2 + 96% *Ar*), and maximum *ZT* of 0.90 was reached at RT in screen printed *n*-type $Bi_2Te_{2.7}Se_{0.3}$ films. TE paste was composed of 84.5% of $Bi_2Te_{2.7}Se_{0.3}$, 2.7% of glass frites (SiO_2, BiO_2, Al_2O_3, and ZnO), and 12.8% of organic binder solvent. All the compounds were mixed using ball milling equipment and then printed and dried at 120 °C for 10 min to remove the solvent. There was performed a two-steps annealing, the first one being at 500 °C for 80 min with *Te*-rich environment, to promote recrystallization and densification, followed by FGA treatment at 450 °C for 2 h. The authors believe that annealing above 450 °C accelerates creation of ionized defects in TE matrix, resulting in low concentration of charge carriers.

4.2.5 Low Concentration of Binder Additives

In most additive manufacturing processes, high concentrations of binder additives were used that reducing percentage of TE material, which is usually toxic. However, it is responsible for negatively affecting electrical transport properties of printed films. Shin et al. [126] endorsed that to achieve high ZT in printed films, low amount of binder needs to be used, but high enough to allow high-quality printing and low decomposition temperature to avoid oxidation of TE component. Hence, they produced promising screen printed TE layer with high ZT of 0.65 and 0.81 at RT for p-type $Bi_{0.5}Sb_{1.5}Te_3$ and n-type $Bi_2Te_{2.7}Se_{0.3}$, respectively. The inks have only 0.45–0.60 wt. % of methyl cellulose, commonly known as Methocel. This decomposes in Ar at 260–300 °C (temperature below typical hot-pressing temperature for $Bi - Te$ alloys, 400–450 °C), and has a final viscosity of 3.3–3.5 Pa × s. Fig. 9 illustrates schematic of the process.

The authors studied also the influence of chitosan layer between substrate and TE ink with the aim to improve σ of produced inks (Fig. 10). The influence of substrate roughness and porosity was decreased by buffer layer, improving σ and leading to uniform printed layer in the end.

4.3 Inkjet-Printed Thermoelectric Films

Inkjet printing technique allows to pattern solution-based materials at high speed and resolution. The main issue in this printing technique remains in the persistent problems like nozzle plate flooding, nozzle clogging, and erratic droplet ejection [97]. To overcome this issue, physical and rheological properties of the inks like

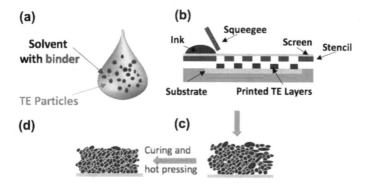

Fig. 9 (**a**) Schematic illustration of screen printing process [126]. TE particles produced by spark erosion process were mixed with binder solvent, i.e., with commercial Methocel HG 90 mixture with ethanol and water. (**b**) Then, the inks were printed in fiberglass fabric substrate. (**c**) Curing at a temperature of 250–300 °C for 30 min was used to burn polymeric binder and then (**d**) hot-pressed at 90 MPa at 450 °C for 5 min in Ar-filled glove box. (Reprinted with permission from [126], copyright 2017, Springer Nature) (Color figure online)

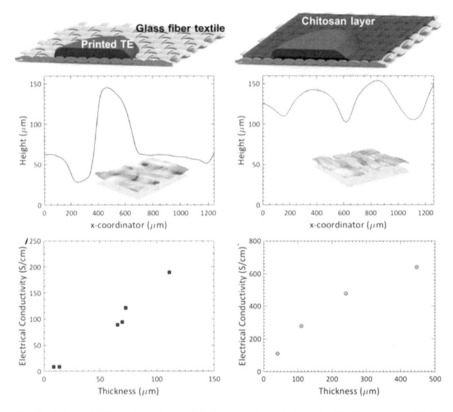

Fig. 10 Schematic illustration of printed TE layer without (left) and with chitosan layer (right). Also, optical surface profiles that show decrease in roughness and porosity with chitosan layer and plots of σ as a function of the thickness for both cases are shown. It is evident that increase in σ is caused by chitosan layer. (Reprinted with permission from Ref. [126], copyright 2017, Springer Nature) (Color figure online)

viscosity, surface tension, and evaporation should be consistent with the values described in Table 1. Thus, printability (*ZO*) of the ink is usually determined by inverse of *Ohnesorge* number, given by the following equation:

$$ZO = \frac{(\gamma d a)^{1/2}}{\eta_v},\tag{8}$$

where γ is surface tension (mN/m), d is density (g/cm^3), η_v is viscosity of the ink (mPa × s), and α is diameter of inkjet nozzle (µm) [127]. Typically, layer of inkjet printing achieves thickness around ~100–500 nm, and TE film is usually produced with multiple layers interleaved with drying times.

To the best of our knowledge, only in 2014, Lu et al. [128] reported first use of inkjet printing as a technique to fabricated TE-printed films. The authors produced

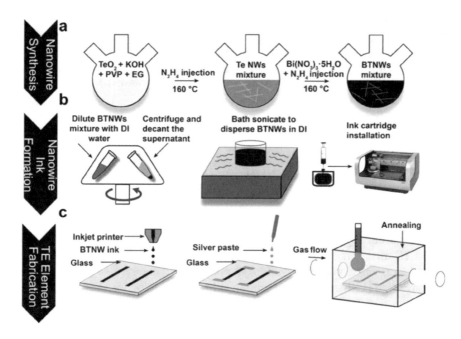

Fig. 11 Schematic representation of Bi_2Te_3 nanowire syntheses via solution phase processing, ink formulation, and printing process adopted by Chen et al. (Reprinted with permission from [129], copyright 2017, John Wiley and Sons, Inc.) (Color figure online)

aqueous dispersion, with γ and η_v of 32 mN/m and 6 cP (1 cP = 10^{-3} Pa × s), composed by p-type $Bi_{0.5}Sb_{1.5}Te_3$ and n-type $Bi_2Te_{2.7}Se_{0.3}$ nanoparticles with dipropylene glycol monomethylether. Maximum PF of 77 and 183 µW × m^{-1} × K^{-2} at 75 °C was achieved for p-type and n-type, respectively, for 150 layers after annealing in Ar/H_2 gas at 400 °C for 30 min to decompose polymeric stabilizers and sinter nanoparticles. In step forward, Chen et al. [129] produced first inkjet printed Bi_2Te_3 nanowire solution (1440 nm of length and 30 nm of average diameter) without any additive (conductive elements). TE ink must have η_v of 10–12 cP to avoid clogging piezoelectric printer nozzle. Figure 11 shows schematic illustration of fabrication process. FGA treatment at 400 °C for 2 h was performed to improve TE properties. The printed nanowires achieved PF of 163 µW × m^{-1} × K^{-2}, and estimated ZT of 0.26 at RT was determined through modified Eq. (1). Accounting for thermal diffusivity of the sample we get:

$$ZT = \frac{(PF)T}{xdC_p\alpha}, \qquad (9)$$

where x is fill factor of nanowires (25% in [129]), d is density of Bi_2Te_3 (7.7 g/cm^3), C_p is heat capacity (124.65 J/mol), and $\alpha = \left(\dfrac{\kappa}{dC_p}\right)$ is intrinsic thermal diffusivity of the sample.

Concerning printability, graphene ink produced by Juntunen et al. [127] has shown $ZO = 9.6$ (Eq. (8)) without formation of long filament or secondary/satellite droplets. In [127], graphene films were fabricated by ultrasonic-assisted liquid phase exfoliation dispersed in isopropyl alcohol as solvent, to promote fast drying and low toxicity, combined with PVP as stabilizer, obtaining PF of 18.7 ± 3.3 $\mu W \times m^{-1} \times K^{-2}$ at RT. Likewise, most of the reports in TE films produced by inkjet printing lack the report on κ measurements; thus, ZT is not presented.

4.4 Other Printing Techniques

4.4.1 Aerosol Jet Printing

In a more recent approach, aerosol jet printing has been gaining attention to produce printed TE films. This technique allows to achieve higher printing resolution (2–4 times higher) when compared with inkjet printing and, however, seems to be more expensive (see [130]). Aerosol jet printing allows developing inks with large range of η_v values (0.5–2000 cP) and particle sizes [131, 132]. To obtain inks for aerosol jet printing, Ou et al. [132] synthesized inorganic nanoparticles of Bi_2Te_3 and Sb_2Te_3 by solvothermal method, then compounded with commercial PEDOT:PSS (with 5 wt. % EG). However, lower PF was achieved when compared with the other approaches (PF of ~ 28.3 $\mu W \times m^{-1} \times K^{-2}$). Oven cure at 130 °C for 30 min was adopted during printing process to remove water and other organic solvents from printing ink, followed by a surface treatment by dipping within a de-doping agent EG submersion for 2 h at 80 °C to improve transport properties and quality of the ink. SEM images of 5 layers achieved by aerosol jet printing with 50 wt. % of Bi_2Te_3 and 85 wt. % of Sb_2Te_3 nanoflakes can be observed in Fig. 12a, b.

The same group reported on aerosol jet printing DMSO-treated Sb_2Te_3-MWCNTs-PVP-PEDOT:PSS [131]. As mentioned in Sect. 2.2.1, MWCNTs have

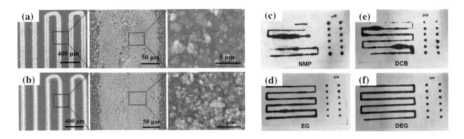

Fig. 12 (**a, b**) SEM images of five layers achieved by aerosol jet printing with PEDOT:PSS-based nanocomposites with 50 wt. % of Bi_2Te_3 and 85 wt. % of Sb_2Te_3 nanoflakes, respectively. (Reprinted with permission from [132], copyright 2018, ACS publication (**c–f**) Photographs showing the aspect of SWCNT inks produced by different solvents during the ball milling process. Solvents: NMP, DCB, EG, and DEG. Reprinted with permission from [133], copyright 2018, The Royal Society of Chemistry)

very poor PF (~ 0.1 μW $\times$ m^{-1} $\times$ K^{-2}) due to low σ (1 S/cm). However, PEDOT:PSS, in this work, operates as electrical contact substance between different components of the nanocomposite, displaying PF of ~41 μW $\times$ m^{-1} $\times$ K^{-2}. The preparation of PEDOT:PSS based inks requires typically surface treatments with polar solvents such as EG, DMSO, and glycerol. On the other hand, using CNTs requires surfactants such as sodium dodecyl sulfate (SDS) and PVP to promote stability of suspension. Usually, CNTs tend to aggregate into bundles or ropes due to high van der Waals attraction. Aerosol jet printing when compared with inkjet printing is taking its first steps in TE area, but it is ongoing with promising results.

4.4.2 Painting Printing and 3D Printing

Park et al. [134] fabricated paint ink with high ZT of 0.97 for p-type and 0.51 for n-type TE materials at RT, which enables to fabricate devices with direct brush paint. Inorganic TE paint ink included dispersion of Sb_2Te_3 chalcogenidometalate (ChaM) (20 wt. % of TE particles) in a mixed viscous solvent of glycerol and EG containing n-type $Bi_2Te_{2.7}Se_{0.3}$ (BTS) or p-type $Bi_{0.4}Sb_{1.6}Te_3$ (BST) TE microparticles (<45 μm). Viscosity η_v and evaporation temperature T_b of TE paint ink were controlled by adjusting the ratio of glycerol ($\eta_v \approx 934$ mP $\times$ s, $T_b \approx 290$ °C) and ethylene glycol ($\eta_v \approx 62$ mP $\times$ s, $T_b \approx 197$ °C). Also, we highlight that molecular Sb_2Te_3 ChaM ions act as sintering aid to fill up the void space among ball-milled particles and promote grain growth and densification. Since the whole process must be proceeded in N_2-filled glove box to prevent oxidation, this limits min cost of final TEG. Thus, through this approach, unlimited shape structures can be painted without losing TE performance. More recently, the same group reported on extrusion-based 3D printing method to produce TE materials with specific geometry to be applied in heat sources, namely, to exhaust pipes in automotive combustion systems for energy harvesting [135]. TE inks were compounded with Bi_2Te_3 based TE materials (BTS, n-type, or BST, p-type) composed with 20–25 wt. % of ChaM and 2 g of glycerol that will serve as humectant to help build a single structure. Inorganic ChaM ions act as stabilizer of the particles in solution via electrostatic interactions because those act like surface ligands in nano- and microparticles. However, inks produced for 3D printing compared with inks produced for brush printing show ZT lower at 25 °C, i.e., around ~0.38 for n-type and ~ 0.7 for p-type materials.

4.4.3 Roll-to-Plate, Dispenser Printing, and Spray Printing

A few examples of TE printed films production using roll-to-plate, dispenser printing, and spray printing can be found in the literature. For example, roll-to-plate printing was implemented by Kim et al. [121] to print $Bi_2Te_3 - In_2Te_3 - Ga_2Te_3$ TE ink with low melting point (93.2 °C). $Bi - In - Ga - Te$ powders (10–45 μm) were mixed with epoxy resin and polyethylene glycol using three-roll milling and then printed. The authors incorporated *In* and *Ga* with the aim to achieve low melting point, i.e., to promote the shift on eutectic temperature. Thus, possible damages can be avoided in conductive polymers and flexible substrates, as reported in a previous study [136]. These elements are also responsible for reduction in sintering temperature. However, lower *ZT* of 0.3 was reached at 298.15 K when compared with conventional Bi_2Te_3 TE films, due to large κ achieved.

Park et al. [133] have processed TE materials by dispenser printing where inks viscosity is crucial parameter. If viscosity is too high, it can block the nozzle; on the other hand, if viscosity it is too low, the ink will not maintain its dispersion, and it can drip due to the gravity. Thus, applied shear rate $\dot{\gamma}$ is a parameter that can be determined by the following equation:

$$\dot{\gamma} = \frac{4Q}{\pi r_{noz}^3}, \tag{10}$$

where Q is flow rate of the ink and r_{noz} is radius of the nozzle [122, 137].

Park et al. [133] produced CNT-based ink to be applied in dispenser printing. Several solvents were analyzed with the aim to achieve appropriate rheological properties to be used in 3D printing such as NMP, DCB, EG, and DEG. As it can be observed in Fig. 12c–f, EG and DEG showed stable dispersion due to high viscosity (1.32, 1.67, 16.1, and 30.2 cP for DCB, NMP, EG, and DEG, respectively).

Also, Jung et al. [122] produced TE inks for dispenser printing, as represented in Fig. 13a. Their TE inks shows $\dot{\gamma}$ of 117.2 s^{-1} and η_v of 1730 cP. TE inks were produced by mixing *p*-type $Bi_{0.4}Sb_{1.6}Te_3$ and *n*-type $Bi_2Te_{2.7}Se_{0.3}$ powders with Sb_2Te_3 based solder, and glycerol as a solvent. Sb_2Te_3 based solder was added to TE ink to improve σ. Excess of *Te* is used to help sintering, being responsible for the dense microstructure without voids by inducing liquid phase sintering to fill pores between the grains, i.e., to densely, as previously explained in Sect. 4.1. The *p*- and *n*-type legs printed by dispenser printing on polyimide substrates have similar $\sigma \sim 25{,}000$ S/m and *S* of 166.37 and -116.38 µV/K, respectively.

Non-contact dispenser printing combined with SLM was used by Wu et al. [113], besides the densification, to prepare *n*-type $Bi_2Te_{2.7}Se_{0.3}$ TE materials. $Bi - Te - Se$ powder was fabricated by self-propagating high-temperature synthesis followed by ball milling process. Tween 20 (CAS:9005-64-5) and antifoam AR (CAS: P11172) were used to compose ink and to adjust rheological properties as well as to remove bubbles out of final ink having viscosity of 350 mPa×s at 20 vol. % of solid content. The ink was then introduced in dispenser with nozzle with diameter of 120 µm. Figure 13b shows scheme of printing process. The printing conditions are also important to achieve a droplet with high quality (see Fig. 13c) that results from combination between ejection period and distance between nozzle and substrate [113]. Annealing temperature effect in *S* was studied using potential Seebeck microprobe technique (PSM) as shown in Fig. 13d. The sample without annealing showed a random distribution of *S* values caused by highly inhomogeneous distribution of *Te* in the matrix. *Te* tends to precipitate during SLM process, producing anti-site

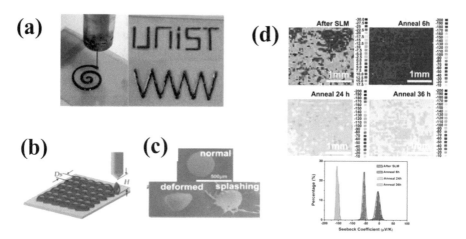

Fig. 13 (**a**) Dispenser printing patterns using TE ink on flexible polyimide [122]. (Reprinted with permission from [122], copyright 2017, Elsevier Science). (**b**) Schematic illustration of non-contact dispenser printing method with distance between the nozzle and the substrate H (3 mm optimal distance) [113]. The quality of dispenser printing is linked to ejection period (4.8 ms), height of the nozzle from the substrate, and spacing between two adjacent droplets (Ds, 350 μm). (**c**) SEM images of layers printed with three typical shapes of droplets: normal, deformed, and splashing depending on printing quality and (**d**) 2D mapping of S of the sample: (**d1**) prepared after SLM, not annealed; (**d2**) annealed at 673 K for 6 h; (**d3**) annealed at 673 K for 24 h; (**d4**) annealed at 673 K for 36 h; and (**d5**) percentages distribution of S achieved from PSM technique. (Reprinted with permission from [113], copyright 2017, John Wiley and Sons, Inc.) (Color figure online)

defects and increasing in hole concentration (p-type regions), and in n-type areas of the surface (Te-rich regions). By annealing the samples, surface distribution of S becomes more uniform and S increases after 36 h to -155 μV/K (after 6 h, S is equal to only -57 μV/K).

Another approach to increase σ of (Te nanorods)-PEDOT:PSS TE inks is reported by Bae et al. [138], aiming to be applied in spray printing. A small amount of nanocarbon-based materials, such as graphene nanoparticles (GNPs) or bundled SWCNTs, was incorporated into Te-PEDOT:PSS systems. Both GNPs and small bundled SWCNTs, increase in PF when compared with pristine Te-PEDOT:PSS (49 μW × m^{-1} × K^{-2}). Optimized powders reached 94 and 206 μW × m^{-1} × K^{-2} with 0.8 wt. % of GNPs and 0.3 wt. % of SWCNTs, respectively.

5 Practical Application of Flexible Micro-TEG

TEG represents promising opportunity to power wearable devices. Typical TEG contains p- and n-type legs electrically connected in series but thermally in parallel. When temperature difference is applied, then majority carriers migrate from hot to cold end, and TE voltage is generated. TEG performance depends on temperature difference between hot and cold sides, TE materials performance, configuration, and arrangements of TE legs [139]. Given poor TE properties of n-type materials, some authors build TEG with single p-type leg.

In this section, different configurations and arrangements of TEG legs and recent developments of organic-inorganic TEG produced with printing techniques will be presented.

5.1   Device Architecture

It is imperative, in the production of TEG, that the materials were keeping performance when included in the device since there is additional influence of thermal and electrical contact layers [139]. Consequently, the device configuration becomes essential to achieve optimized TEG with high output power. Concerning configuration and arrangements of TE legs, geometric parameters including n- or p-type leg area ratio, leg length, area of individual leg, and distance between adjacent legs can directly affect TE performance [140]. Figure 14 shows schematically possible structures of leg and TEG configurations.

TEGs can be designed in planar, vertical, radial, or origami configuration. In planar configuration, TE legs are patterned onto the substrate surface. However, in the worst case scenario, typical high internal resistance can reach up to MOhm. Vertical configuration (Fig. 14b) is created from massive alternating p- and n-type semiconductor bulk TE materials, commonly with traditional cuboid structure. The semiconductors are electrically connected in series using a stack metallic contact layer. TE legs geometry can adopt different shapes such as cuboid [141], pyramid [139, 142], exponential [143, 144], and quadratic [145] as depicted in Fig. 14b. In most of the studies on influence of legs geometry, authors presented only theoretical analysis due to difficulties involved in fabrication of TE legs with complex shapes. Recently, Mijangos et al. [139] produced a prototype consisting of pyramid-shaped leg based on Bi_2Te_3 material and proved the importance of geometric configuration of TE legs in device performance. Harman method [146] was used to determine ZT. Thus, correction of parasitic electrical resistance was performed using ZT determined as a function of adiabatic resistance R_{ad}, isothermal resistance R_{iso}, and parasitic resistance R_p, resulting in the following equation:

$$ZT = \frac{R_{ad} - R_p}{R_{iso} - R_p} - 1. \tag{11}$$

For example, ZT of 0.79 and 1.02 was reached for nine TE modules with cuboid and pyramid legs geometry, respectively, using a correction of R_p.

Another pattern to produce TEGs is cylindrical (Fig. 14c) where heat flows in radial direction [123, 147, 148]. Usually, this geometry is applied in oil pipelines, cooling channels for power station transformers, vehicle exhaust pipes, etc.

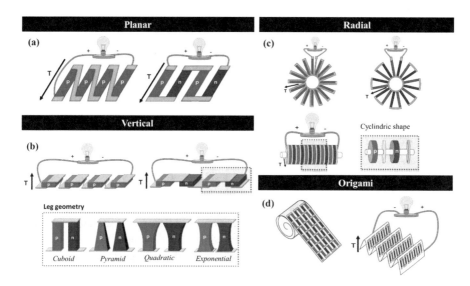

Fig. 14 Schematic illustration of possible structures of leg and TEG configurations such as (**a**) planar, (**b**) vertical, (**c**) radial, and (**d**) origami configuration. Legend: *p*-type material (blue modules) and *n*-type material (red modules) (Color figure online)

Additionally, a more innovative configuration, i.e., origami configuration (see Fig. 14d), was reported by Horike et al. [149]. Origami configuration TE modules were fabricated by inkjet printing TE material onto polyethylene naphthalate foil, which is then folded in origami configuration.

In TEG evaluation, two crucial metrics need to be taken into account, such as open circuit voltage V_{OC} and maximum output power P_{max} [150]. The following equation establishes V_{OC}:

$$V_{OC} = \frac{N}{2}\left(\left|S_p\right| + \left|S_n\right|\right)\Delta T, \tag{12}$$

where N is number of $p - n$ couples (pairs of legs) and ΔT is temperature difference. P_{max} is measured with external load and is defined by the following equation:

$$P_{max} = \frac{N\left(\left|S_p\right| + \left|S_n\right|\right)^2 \Delta T^2}{16R_{int}}, \tag{13}$$

where R_{int} is effective resistance of legs, metal electrodes, and contacts on one unit of the module.

Concerning contact resistance, P_{out} and conversion efficiency η could be written as [151]:

$$P_{out} = \frac{\left(|S_p|+|S_n|\right)^2}{2\rho} \times \frac{AN\left(T_h - T_c\right)^2}{\left(n_R + l\right)\left(1 + \dfrac{2rl_c}{l}\right)^2}, \tag{14}$$

$$\eta = \left(\frac{T_h - T_c}{T_h}\right) \Bigg/ \left\{\left(1 + \frac{2rl_c}{l}\right)^2 \left[2 - \frac{1}{2}\left(\frac{T_h - T_c}{T_h}\right) + \left(\frac{4}{ZT_h}\right)\left(\frac{l + n_R}{l + 2rl_c}\right)\right]\right\}, \tag{15}$$

where $n_R = 2\rho_c/\rho$ and $r = \kappa/\kappa_c$.

In Eqs. (14) and (15), T_h and T_c are hot and cold side temperatures; N is number of $p - n$ couples (pairs of legs) in TE module; A and l are cross-sectional area and length of TE leg; l_c is thickness of contact layer; and ρ_c and κ_c are electrical contact resistivity and thermal conductivity. Low contact resistance is required for good TEG performance. He et al. [151] suggested the requirements to establish good contact layer: (1) high σ_c and κ_c to minimize internal electrical and heat power losses, (2) high ductility, (3) low electrical and thermal contact resistance with TE material, and (4) high mechanical and thermal stabilities upon thermal cycling.

5.2 Printed Thermoelectric Devices

Recent developments of TEGs using TE materials with high ZT will be addressed in this section. Figure 15 summarizes the most relevant thermoelectric information regarding n- and p-type of Bi - Te based printed materials found in the literature. Moreover, when applied, corresponding output power result from the printed devices is inscribed.

According to Fig. 15, the highest ZT was achieved by Choi et al. [116, 119]. The authors optimized screen printed $Bi_{0.5}Sb_{1.5}Te_3$ (p-type) and $Bi_2Te_{2.7}Se_{0.3}$ (n-type) thick films through *post*-annealing process with mechanical pressure. In 2017, they reported on use of TE-printed materials with ZT of 0.89 and 0.57 for p- and n-type materials. TEG composed by arrays of 72 and 200 couples were produced with above mentioned TE materials, achieving P_{out} of 0.1 W and 0.31 W, respectively, for ΔT of 28 K (as shown in Fig. 16) [116]. The authors observed that output voltage of TEG with 200 TE couples is nine times larger than with 72 couples and attributed this achievement to combination of printing technique with annealing step with mechanical pressure. Thus, P_{out} density of 7.34 mW/cm² and conversion efficiency of 0.014 were achieved at the same ΔT. Concerning internal resistances, these two devices show 0.51 Ohm and 1.49 Ohm for 72 and 200 TE couples, respectively.

Later, the same authors [119] used identical TE inks but with simplified crystallization process. With this approach, lower P_{out} density of 5.23 ± 0.2 mW/cm² was achieved for TEG with 200 couples at $\Delta T = 25$ K. However, P_{out} density of these devices remains lower than reported for bulk TE material devices which is 8.4 mW/cm² [119].

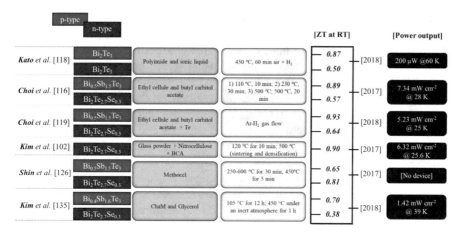

Fig. 15 List of the highest ZT values found in the literature, at room temperature, and corresponding power output of TEGs. Legend: p-type material (blue modules) and n-type material (red modules) (Color figure online)

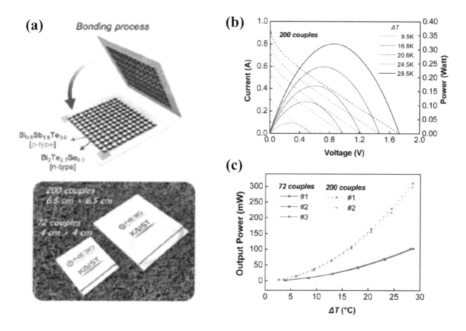

Fig. 16 (**a**) Schematic illustration of TEG and photo of TEGs composed of 72 and 200 TE couples. (**b**) Output power and current-voltage curves of TEG and (**c**) maximum output power of TEG vs ΔT. (Reprinted with permission from [116], copyright 2010, Royal Society of Chemistry) (Color figure online)

In a new strategy, Kim et al. [102] used screen printed films with ZT of 0.9 (for n-type material) to produce TEG using ionized defect engineering process. Flexible TEG (40 mm × 40 mm × 0.8 mm) with 72 pairs of legs of p-type $Bi_{0.5}Sb_{1.5}Te_3$ and reduced n-type $Bi_2Te_{2.7}Se_{0.3}$ was fabricated using screen printing and laser multi-scanning (LMS) techniques. This TEG reached P_{out} density of 6.32 mW/cm^2 at ΔT = 25.6 °C, being 30% higher than the control sample produced without FGA. Kato et al. [118] have produced planar TEG by screen printing using low-cost process and affirmed that 5 cm × 5 cm module could generate 200 µW (maximum P_{out}). A higher value of 308 µW was achieved at ΔT = 127 K by Iezzi et al. [152] with TEG composed of 420 TE couples by screen printing.

Kim et al. [135] used 3D printing technique to produce half rings mounted on alumina pipe forming cylindrical TEG design. TEG resistance was 105 mOhm, five times higher than calculated resistance of 20 mOhm from electrical contact resistance between legs and Ag interlayers. This influence is also reported in other works [124, 134]. Cylindrical TEG showed linear and quadratic increase in output voltage and P_{out}, respectively, achieving P_{out} density of 1.42 mW/cm^2 for ΔT = 39 °C (27.0 mV and 1.62 mW) [135].

Park et al. [133] produced bracelet-type TEG by dispenser-type printing using CNTs. The bracelet was composed by 60 $p - n$ couples produced with CNT ink showing maximum P_{out} of 0.2 and 1.95 µW for ΔT = 10 and 30 K, respectively. This achievement is one of higher reported P_{out} for TEG produced with CNT inks, and the authors also performed mechanical durability tests, and circuit resistance of the bracelet is rarely changed after 3500 bending cycles.

Generally speaking, output voltage depends not only on ZT of TE material but also on geometric dimensions of TE couples and operating conditions such as ΔT and electrical load [153, 154]. An outstanding work studying this influence was performed by Park et al. [134] where it was studied TEG in different configurations. The authors reported different output performances with different TEG configurations, such as (1) in-plane TE devices painted on flat substrates achieved P_{out} density of 2.43 mW/cm^2, (2) painted on concave or convex configuration achieved P_{out} density of 0.7 mW/cm^2, and (3) a hemisphere configuration reached P_{out} density of 0.073 mW/cm^2. To optimize TEG geometry, analytical and numerical models are used, aiming to minimize expenses and time-consumption of experimental part. Several models can be found in the literature, namely, one-dimensional [155, 156] and complex three-dimensional models [157, 158].

Finally, Table 2 compiles the most relevant examples of TEGs produced by printing techniques in different configurations and corresponding P_{out}.

Table 2 Compilation of recent examples of TEGs produced by printed techniques and the corresponding output powers (Color table online)

Printing technique	Type	Device	TE materials	Substrate	N legs	ΔT	Output power		Pub. year	REF
Screen printing	**Planar**		Bi_2Te_3 and polyimide with ionic liquid	Polyimide	-	60 °C	1.4 mW		2018	[118]
	Radial		n-type $Bi_2Te_{2.7}Se_{0.3}$ and p-type Sb_2Te_3	Polyimide	5 pairs	~48.6 °C [1.5 W]	237.7 μW/cm²	6.31 μW	2018	[103]
			p-type $Bi_{0.5}Sb_{1.5}Te_3$ and n-type $Bi_2Te_{2.7}Se_{0.3}$	Al_2O_3	200 individual	25 K	5.02 – 5.45 mW/cm²		2018	[119]
	Vertical		p-type $Bi_{0.5}Sb_{1.5}Te_3$ and reduced n-type $Bi_2Te_{2.7}Se_{0.3}$	Alumina	72 pairs	25.6 °C	6.32 mW/cm²		2017	[102]
			p-type Sb_2Te_3 and n-type Bi_2Te_3+Ag	PDMS	45 pairs	10 K	-	27 nW	2017	[124]
	Planar		p-type PEDOT:PSS and n-type nickel	Polyimide	288 legs	65 K	2.35 μW/cm²	46 μW	2017	[150]

(continued)

Table 2 (continued)

Printing technique	Type	Device	TE materials	Substrate	N legs	AT	Output power		Pub. year	REF
Painting printing	**In plane**		p-type $Bi_{0.5}Sb_{1.5}Te_3$, n-type $Bi_2Te_{2.7}Se_{0.3}$ + Sb_2Te_3	Polyimide		50 °C	2.43 mW/cm²	60.8 µW		
	Concave				5 couples					
	Convex			Flat glass		30 °C	0.70 mW/cm²	17-18 µW	2016	[134]
	Hemisphere			Alumina substrate diameter of ~70 mm	5.5 couples of triangular TE layers	20 °C	0.073 mW/cm²	3.0 µW		
3D-printed	**Cylinder**		p-type $Bi_{0.5}Sb_{1.5}Te_3$ and n-type $Bi_2Sb_{2.7}Se_{0.3}$ + Sb_2Te_4 ChaM.	Alumina pipe (inner and outer d ~ of 5 mm and 8 mm)	3 pairs of n-type and p-type TE half-rings materials	30 °C	1.42 mW/cm²	1.62 mW	2018	[135]
Aerosol-jet printing	**Planar**		DMSO-treated Sb_2Te_3. MWCNTs-PVP-PEDOT.PSS	Polyimide		–	–	–	2018	[132]
Dispenser printing	**Cylinder**		n- and p-type doped CNT	PU cable	60 pairs	30 K	–	1.95 µW	2018	[133]
	Planar		p-type $Bi_{0.4}Sb_{1.6}Te_3$ and n-type $Bi_2Se_{0.3}Te_{2.7}$ + Te	Polyimide	10 pairs	20.9 °C	–	44 µW	2017	[122]
Spray printing	**Planar**		SWCNT-mixed Te-PEDOT:PSS	Polyarylate	28 legs (2 rows)	20 °C	–	126 nW	2017	[138]

Reprinted with permission from [102, 103, 118, 119, 122, 124, 132–135, 138, 150]

6 New Challenges and Future Prospects on Thermoelectric Research

With the evolution of printed electronics in our living style, present in several new gadgets emerging in the market, in response to the societal challenges related with energy consumption, in the last years, printed TEGs have gained relevance due to singular ability of directly converting heat into electrical energy. However, up to now, low conversion efficiency and unavailability of cost-effective materials remain as major drawbacks in TE area. A possible solution consists of mixing organic with inorganic materials to develop hybrid TE composites which have been thoroughly discussed in this chapter. Nevertheless, several points should be addressed toward efficient materials and, consequently, to achieve efficient TE devices.

The precise doping control in composites seems to be a key point to bridge the gap between distinct performance of organic and inorganic materials. The inclusion of nanomaterials in the matrix of TE composites is of primordial importance to increase in ZT due to nanoconfinement effects. The strategy relies on precise control of morphology, structure, porosity, composition, colloidal stability, and electrical transport properties of nanomaterials. Particular attention should be devoted to raw materials (e.g., Te remains as the rarest element) used in hybrid TE fabrication, which also represents a significant drawback for this technology to revolutionize the energy market. To overcome this issue, materials must be improved but also manufacturing costs should be reduced.

In comparison terms, gold has an abundance of 0.004 ppm on earth, higher than tellurium element (only 0.001 ppm). Thus, other materials constituted by more abundant elements could be used due to cost-effectiveness, even if presenting lower performance. This requirement drives over new quantification on TE materials, namely, by taking into account cost/performance balance. In this sense, in chapter, new parameter concerning ZT was considered, consisting of ratio between ZT and price (\$/kg). This relation is explored in Fig. 17 where it can be verified that Si nanowires, $MnSi$ nanobulk, thin film superlattices, and organic polymers are the materials that present greater potential to low-temperature applications concerned this parameter. We highlight the case of carbon nanotubes that, despite low ZT value, present appealing price when compared with $Bi - Te$ based materials. This factor is increasingly relevant and could reveal strong substitutes for inorganic materials. The balance between ZT and price needs to be considered, since the mix, for instance, of PEDOT:PSS with SWCNTs and nanowires can increase in ZT and reduce in material cost. The idea is to increase in ZT of cost-effective materials, making those attractive for industrial applications.

Moreover, on ink formulation and thin film production, appropriate flexibility of printed films is promoted by using small spherical particles (<50 μm) and/or plates. The adhesion of TE materials to the substrates should be addressed, and new substrates and/or adhesion-promoting additives on the ink could be further investigated.

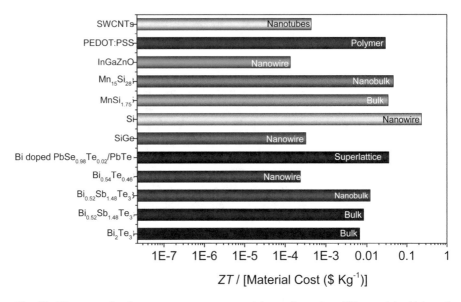

Fig. 17 *ZT* measured at low temperature, per material cost for various TE materials. (Adapted from [159]) (Color figure online)

Need for films post-deposition treatment is evident and is related to densification or calcination of the sample to increase significantly in TE performance. To reduce in production cost and make the system simpler, new strategies to substantially reduce in annealing temperature and even eliminate that process will contribute significantly to reduction in costs and process simplification and should be addressed in future studies.

Concerning the production techniques, screen, roll-to-plate, inkjet, aerosol, and spray printing are preferred techniques given reduced cost, massive production capacity, and scalability. To the best of our knowledge, there is only one company that is dedicated to study/manufacture of TE devices using printing techniques (*Otego company*). Thus, large scale production of TE materials and devices becomes vital to increase in offering that goods in the market. More start-ups and spin-offs should be encouraged to explore this field to promote intensive search for TE applications, through the process of design thinking, which will automatically result in new optimized architectures of TE devices.

Finally, we consider that new characterization techniques should be introduced. In particular, since κ depends strongly on σ, precise experimental methods and line-up of correct measurement procedures are crucial in TE materials research. Mainly in in-plane κ measurement of thin films, there is urgent necessity to standardize such procedure. Traditional 3ω method has been raising uncertainties that are subsequently associated with calculated *ZT*, and this challenge should be overcome.

Acknowledgments This work was financially supported by Fundação para a Ciência e a Tecnologia (FCT)/MEC and FEDER under Program PT2020 through the projects PTDC/CTM-NAN/5414/2014, PTDC/CTMTEX/31271/2017, UIDB/04968/2020, and UIDB/50006/2020. ALP, JAS, and S.S thanks FCT for the grant under the project PTDC/CTM-NAN/5414/2014. This work was also financially supported by European Union's Horizon 2020 Research and Innovation Programme under Grant Agreement No. 863307 (Ref. H2020-FETOPEN-2018-2019-2020-01). ALP thanks the funding from the WiPTherm project (Grant Agreement No. 863307). MMM is thankful to FCT for grant SFRH/BD/144229/2019. C.P. thanks FCT for the FCT Investigator contract IF/01080/2015.

Bibliography

1. IDTechX, Company provides research, consulting, advisory and events services, (2017). Available at https://www.idtechex.com/research/reports/flexible-printed-and-organic-electronics-2019-2029-forecasts-players-and-opportunities-000639.asp
2. M. Prudenziati, J. Hormadaly, Technologies for printed films, in *Printed Films: Materials Science and Applications in Sensors, Electronics and Photonics*, ed. by M. Prudenziati, J. Hormadaly, (Woodhead Publishing Books, 2012), Cambridge, UK, pp. 3–29
3. M. Mäntysalo, L. Xie, F. Jonsson, Y. Feng, A.L. Cabezas, L.-R. Zheng, System integration of smart packages using printed electronics, 2012 IEEE 62nd Electron. Components Technol. Conf., 997–1002 (2012)
4. M. Finn, C.J. Martens, A.V. Zaretski, B. Roth, R.R. Søndergaard, F.C. Krebs, D.J. Lipomi, Mechanical stability of roll-to-roll printed solar cells under cyclic bending and torsion. Sol. Energy Mater. Sol. Cells **174**, 7–15 (2018)
5. T. Ahmadraji, L. Gonzalez-Macia, T. Ritvonen, A. Willert, S. Ylimaula, D. Donaghy, S. Tuurala, M. Suhonen, D. Smart, A. Morrin, V. Efremov, R.R. Baumann, M. Raja, A. Kemppainen, A.J. Killard, Biomedical diagnostics enabled by integrated organic and printed electronics. Anal. Chem. **89**, 7447–7454 (2017)
6. K.H. Choi, D.B. Ahn, S.Y. Lee, Current status and challenges in printed batteries: Toward form factor-free, monolithic integrated power sources. ACS Energy Lett. **3**, 220–236 (2018)
7. A.E. Ostfeld, A.M. Gaikwad, Y. Khan, A.C. Arias, High-performance flexible energy storage and harvesting system for wearable electronics. Sci. Rep. **6**, 1–10 (2016)
8. R. Zhang, B. Wang, W. Zhu, C. Li, H. Wang, Preparation and luminescent performances of transparent screen-printed Ce3+: Y3Al5O12 phosphors-in-glass thick films for remote white LEDs. J. Alloys Compd. **720**, 340–344 (2017)
9. F.K. Shaikh, S. Zeadally, Energy harvesting in wireless sensor networks: A comprehensive review. Renew. Sust. Energ. Rev. **55**, 1041–1054 (2016)
10. L. Tocci, T. Pal, I. Pesmazoglou, B. Franchetti, Small scale Organic Rankine Cycle (ORC): A techno-economic review. Energies **10**, 413 (2017)
11. O.H. Ando Junior, A.L.O. Maran, N.C. Henao, A review of the development and applications of thermoelectric microgenerators for energy harvesting. Renew. Sust. Energ. Rev. **91**, 376–393 (2018)
12. M. Karunanithy, G. Prabhavathi, A.H. Beevi, B.H.A. Ibraheem, K. Kaviyarasu, S. Nivetha, N. Punithavelan, A. Ayeshamariam, M. Jayachandran, Nanostructured metal tellurides and their heterostructures for thermoelectric applications – A review. J. Nanosci. Nanotechnol. **18**, 6680–6707 (2018)
13. D.S. Patil, R.R. Arakerimath, P.V. Walke, Thermoelectric materials and heat exchangers for power generation – A review. Renew. Sust. Energ. Rev. **95**, 1–22 (2018)
14. Y. Du, J. Xu, B. Paul, P. Eklund, Flexible thermoelectric materials and devices. Appl. Mater. Today **12**, 366–388 (2018)

15. Z.G. Chen, X. Shi, L.D. Zhao, J. Zou, High-performance SnSe thermoelectric materials: Progress and future challenge. Prog. Mater. Sci. **97**, 283–346 (2018)
16. R.O. Fitriani, B.D. Long, M.C. Barma, M. Riaz, M.F.M. Sabri, S.M. Said, R. Saidur, A review on nanostructures of high-temperature thermoelectric materials for waste heat recovery. Renew. Sust. Energ. Rev. **64**, 635–659 (2016)
17. A. Bulusu, D.G. Walker, Review of electronic transport models for thermoelectric materials. Superlattice. Microst. **44**, 1–36 (2008)
18. A.F. Ioffe, L.S. Stil'bans, E.K. Iordanishvili, T.S. Stavitskaya, A. Gelbtuch, G. Vineyard, Semiconductor thermoelements and thermoelectric cooling. Phys. Today **12**, 42–42 (1959)
19. G.J. Snyder, E.S. Toberer, Complex thermoelectric materials. Nat. Mater. **7**, 105–114 (2008)
20. K. Biswas, J. He, I.D. Blum, C. Wu, T.P. Hogan, D.N. Seidman, V.P. Dravid, M.G. Kanatzidis, High-performance bulk thermoelectrics with all-scale hierarchical architectures. Nature **489**, 414–418 (2012)
21. S.I. Kim, K.H. Lee, H.A. Mun, H.S. Kim, S.W. Hwang, J.W. Roh, D.J. Yang, W.H. Shin, X.S. Li, Y.H. Lee, G.J. Snyder, S.W. Kim, Dense dislocation arrays embedded in grain boundaries for high-performance bulk thermoelectrics. Science **348**, 109–114 (2015)
22. J.P. Heremans, V. Jovovic, E.S. Toberer, A. Saramat, K. Kurosaki, A. Charoenphakdee, S. Yamanaka, G.J. Snyder, Enhancement of thermoelectric efficiency in PbTe by distortion of the electronic density of states. Science **321**, 554–557 (2008)
23. Y. Pei, X. Shi, A. LaLonde, H. Wang, L. Chen, G.J. Snyder, Convergence of electronic bands for high performance bulk thermoelectrics. Nature **473**, 66–69 (2011)
24. S.A. Yamini, D.R.G. Mitchell, Z.M. Gibbs, R. Santos, V. Patterson, S. Li, Y.Z. Pei, S.X. Dou, G. Jeffrey Snyder, Heterogeneous distribution of sodium for high thermoelectric performance of p-type multiphase lead-chalcogenides. Adv. Energy Mater. **5**, 1501047 (2015)
25. J.R. Szczech, J.M. Higgins, S. Jin, Enhancement of the thermoelectric properties in nanoscale and nanostructured materials. J. Mater. Chem. **21**, 4037–4055 (2011)
26. L. Hu, T. Zhu, X. Liu, X. Zhao, Point defect engineering of high-performance bismuth-telluride-based thermoelectric materials. Adv. Funct. Mater. **24**, 5211–5218 (2014)
27. Z. Fan, H. Wang, Y. Wu, X.J. Liu, Z.P. Lu, Thermoelectric high-entropy alloys with low lattice thermal conductivity. RSC Adv. **6**, 52164–52170 (2016)
28. Y. Zheng, Q. Zhang, X. Su, H. Xie, S. Shu, T. Chen, G. Tan, Y. Yan, X. Tang, C. Uher, G.J. Snyder, Mechanically robust BiSbTe alloys with superior thermoelectric performance: A case study of stable hierarchical nanostructured thermoelectric materials. Adv. Energy Mater. **5**, 1401391 (2015)
29. G.S. Nolas, J. Sharp, H.J. Goldsmid, *Thermoelectrics: Basic Principles and New Materials Developments* (Springer, Berlin Heidelberg, 2001)
30. R. Boston, W.L. Schmidt, G.D. Lewin, A.C. Iyasara, Z. Lu, H. Zhang, D.C. Sinclair, I.M. Reaney, Protocols for the fabrication, characterization, and optimization of n-type thermoelectric ceramic oxides. Chem. Mater. **29**, 265–280 (2017)
31. P.J. Taylor, J.R. Maddux, W.A. Jesser, F.D. Rosi, Room-temperature anisotropic, thermoelectric, and electrical properties of n-type (Bi2Te3)90 (Sb2Te3)5 (Sb2Se3)5 and compensated p-type (Sb2Te3)72 (Bi2Te3)25 (Sb2Se3)3 semiconductor alloys. J. Appl. Phys. **85**, 7807–7813 (1999)
32. H.J. Goldsmid, Introduction to Thermoelectricity Springer Series in Materials Science, Review of Thermoelectric Materials, Springer Berlin Heidelberg, 153–195 (2016)
33. K. Wada, K. Tomita, M. Takashiri, Fabrication of bismuth telluride nanoplates via solvothermal synthesis using different alkalis and nanoplate thin films by printing method. J. Cryst. Growth **468**, 194–198 (2017)
34. W. Guo, J. Ma, W. Zheng, Bi2Te3 nanoflowers assembled of defective nanosheets with enhanced thermoelectric performance. J. Alloys Compd. **659**, 170–177 (2016)
35. L. Yang, Z.G. Chen, M. Hong, G. Han, J. Zou, Enhanced thermoelectric performance of nanostructured Bi2Te3 through significant phonon scattering. ACS Appl. Mater. Interfaces **7**, 23694–23699 (2015)

36. X. Li, C. Chen, W. Xue, S. Li, F. Cao, Y. Chen, J. He, J. Sui, X. Liu, Y. Wang, Q. Zhang, N-type bi-doped SnSe thermoelectric nanomaterials synthesized by a facile solution method. Inorg. Chem. **57**, 13800 (2018)

37. M. Gharsallah, F. Serrano-Sánchez, J. Bermúdez, N.M. Nemes, J.L. Martínez, F. Elhalouani, J.A. Alonso, Nanostructured Bi2Te3 prepared by a straightforward arc-melting method. Nanoscale Res. Lett. **11**, 142 (2016)

38. M.M. Nassary, H.T. Shaban, M.S. El-Sadek, Semiconductor parameters of Bi2Te3 single crystal. Mater. Chem. Phys. **113**, 385–388 (2009)

39. O. Yamashita, S. Tomiyoshi, High performance n-type bismuth telluride with highly stable thermoelectric figure of merit. J. Appl. Phys. **95**, 6277–6283 (2004)

40. M. Zakeri, M. Allahkarami, G. Kavei, A. Khanmohammadian, M.R. Rahimipour, Synthesis of nanocrystalline Bi2Te3 via mechanical alloying. J. Mater. Process. Technol. **209**, 96–101 (2009)

41. R. Venkatasubramanian, E. Siivola, T. Colpitts, B. O'Quinn, Thin-film thermoelectric devices with high room-temperature figures of merit. Nature **413**, 597–602 (2001)

42. M. Scheele, N. Oeschler, K. Meier, A. Kornowski, C. Klinke, H. Weller, Synthesis and thermoelectric characterization of Bi2Te3 nanoparticles. Adv. Funct. Mater. **19**, 3476–3483 (2009)

43. Y. Choi, Y. Kim, S.G. Park, Y.G. Kim, B.J. Sung, S.Y. Jang, W. Kim, Effect of the carbon nanotube type on the thermoelectric properties of CNT/Nafion nanocomposites. Org. Electron. **12**, 2120–2125 (2011)

44. Y. Ouyang, J. Guo, A theoretical study on thermoelectric properties of graphene nanoribbons. Appl. Phys. Lett. **94**, 263107 (2009)

45. Y. Li, Z. Zhou, P. Shen, Z. Chen, Structural and electronic properties of Graphane nanoribbons. J. Phys. Chem. C **113**, 15043–15045 (2009)

46. J.L. Blackburn, A.J. Ferguson, C. Cho, J.C. Grunlan, Carbon-nanotube-based thermoelectric materials and devices. Adv. Mater. **30**, 1704386 (2018)

47. H. Sevinçli, G. Cuniberti, Enhanced thermoelectric figure of merit in edge-disordered zigzag graphene nanoribbons. Phys. Rev. B **81**, 113401 (2010)

48. N. Xiao, X. Dong, L. Song, D. Liu, Y. Tay, S. Wu, L.-J. Li, Y. Zhao, T. Yu, H. Zhang, W. Huang, H.H. Hng, P.M. Ajayan, Q. Yan, Enhanced thermopower of graphene films with oxygen plasma treatment. ACS Nano **5**, 2749–2755 (2011)

49. W. Zhou, Q. Fan, Q. Zhang, K. Li, L. Cai, X. Gu, F. Yang, N. Zhang, Z. Xiao, H. Chen, S. Xiao, Y. Wang, H. Liu, W. Zhou, S. Xie, Ultrahigh-power-factor carbon nanotubes and an ingenious strategy for thermoelectric performance evaluation. Small **12**, 3407–3414 (2016)

50. T.M. Tritt, Thermoelectric phenomena, materials, and applications. Annu. Rev. Mater. Res. **41**, 433–448 (2011)

51. M. Culebras, C. Gómez, A. Cantarero, Review on polymers for thermoelectric applications. Materials (Basel). **7**, 6701–6732 (2014)

52. M.P. Gordon, E.W. Zaia, P. Zhou, B. Russ, N.E. Coates, A. Sahu, J.J. Urban, Soft PEDOT:PSS aerogel architectures for thermoelectric applications. J. Appl. Polym. Sci. **134**, 44070 (2017)

53. L. Wang, D. Wang, G. Zhu, J. Li, F. Pan, Thermoelectric properties of conducting polyaniline/graphite composites. Mater. Lett. **65**, 1086–1088 (2011)

54. Y. Du, K.F. Cai, S.Z. Shen, P.S. Casey, Preparation and characterization of graphene nanosheets/poly(3-hexylthiophene) thermoelectric composite materials. Synth. Met. **162**, 2102–2106 (2012)

55. H. Kaneko, T. Ishiguro, A. Takahashi, J. Tsukamoto, Magnetoresistance and thermoelectric power studies of metal-nonmetal transition in iodine-doped polyacetylene. Synth. Met. **57**, 4900–4905 (1993)

56. L. Wang, Y. Liu, Z. Zhang, B. Wang, J. Qiu, D. Hui, S. Wang, Polymer composites-based thermoelectric materials and devices. Compos. Part B Eng. **122**, 145–155 (2017)

57. C. Yu, Y.S. Kim, D. Kim, J.C. Grunlan, Thermoelectric behavior of segregated-network polymer nanocomposites. Nano Lett. **8**, 4428–4432 (2008)

58. L.M. Cowen, J. Atoyo, M.J. Carnie, D. Baran, B.C. Schroeder, Review—Organic materials for thermoelectric energy generation. ECS J. Solid State Sci. Technol. **6**, N3080–N3088 (2017)

59. F. Ely, C.O. Avellaneda, P. Paredez, V.C. Nogueira, T.E.A. Santos, V.P. Mammana, C. Molina, J. Brug, G. Gibson, L. Zhao, Patterning quality control of inkjet printed PEDOT:PSS films by wetting properties. Synth. Met. **161**, 2129–2134 (2011)

60. G.H. Kim, L. Shao, K. Zhang, K.P. Pipe, Engineered doping of organic semiconductors for enhanced thermoelectric efficiency. Nat. Mater. **12**, 719–723 (2013)

61. Y. Sun, P. Sheng, C. Di, F. Jiao, W. Xu, D. Qiu, D. Zhu, Organic thermoelectric materials and devices based on p- and n-type poly(metal 1,1,2,2-ethenetetrathiolate)s. Adv. Mater. **24**, 932–937 (2012)

62. Y. Du, S.Z. Shen, K. Cai, P.S. Casey, Research progress on polymer–inorganic thermoelectric nanocomposite materials. Prog. Polym. Sci. **37**, 820–841 (2012)

63. C. Guo, F. Chu, P. Chen, J. Zhu, H. Wang, L. Wang, Y. Fan, W. Jiang, Effectively enhanced thermopower in polyaniline/Bi0.5Sb1.5Te3 nanoplate composites via carrier energy scattering. J. Mater. Sci. **53**, 6752–6762 (2018)

64. K. Zhang, J. Qiu, S. Wang, Thermoelectric properties of PEDOT nanowire/PEDOT hybrids. Nanoscale **8**, 8033–8041 (2016)

65. Y. Chen, M. He, B. Liu, G.C. Bazan, J. Zhou, Z. Liang, Bendable n-type metallic nanocomposites with large thermoelectric power factor. Adv. Mater. **29**, 1604752 (2017)

66. H. Wang, S. Yi, X. Pu, C. Yu, Simultaneously improving electrical conductivity and thermopower of polyaniline composites by utilizing carbon nanotubes as high mobility conduits. ACS Appl. Mater. Interfaces **7**, 9589–9597 (2015)

67. Q. Yao, Q. Wang, L. Wang, L. Chen, Abnormally enhanced thermoelectric transport properties of SWNT/PANI hybrid films by the strengthened PANI molecular ordering. Energy Environ. Sci. **7**, 3801–3807 (2014)

68. Q. Yao, L.D. Chen, W.Q. Zhang, S.C. Liufu, X.H. Chen, Enhanced thermoelectric performance of single-walled carbon nanotubes/polyaniline hybrid nanocomposites. ACS Nano **4**, 2445–2451 (2010)

69. K. Chatterjee, M. Mitra, K. Kargupta, S. Ganguly, D. Banerjee, Synthesis, characterization and enhanced thermoelectric performance of structurally ordered cable-like novel polyaniline–bismuth telluride nanocomposite. Nanotechnology **24**, 215703 (2013)

70. Y. Wang, S.M. Zhang, Y. Deng, Flexible low-grade energy utilization devices based on high-performance thermoelectric polyaniline/tellurium nanorod hybrid films. J. Mater. Chem. A **4**, 3554–3559 (2016)

71. C.J. An, Y.H. Kang, A.Y. Lee, K.S. Jang, Y. Jeong, S.Y. Cho, Foldable thermoelectric materials: Improvement of the thermoelectric performance of directly spun CNT webs by individual control of electrical and thermal conductivity. ACS Appl. Mater. Interfaces **8**, 22142–22150 (2016)

72. J. Choi, J.Y. Lee, S.S. Lee, C.R. Park, H. Kim, High-performance thermoelectric paper based on double carrier-filtering processes at nanowire heterojunctions. Adv. Energy Mater. **6**, 1502181 (2016)

73. D. Yoo, J. Kim, S.H. Lee, W. Cho, H.H. Choi, F.S. Kim, J.H. Kim, Effects of one- and two-dimensional carbon hybridization of PEDOT:PSS on the power factor of polymer thermoelectric energy conversion devices. J. Mater. Chem. A **3**, 6526–6533 (2015)

74. J. Xiong, F. Jiang, H. Shi, J. Xu, C. Liu, W. Zhou, Q. Jiang, Z. Zhu, Y. Hu, Liquid exfoliated graphene as dopant for improving the thermoelectric power factor of conductive PEDOT:PSS nanofilm with hydrazine treatment. ACS Appl. Mater. Interfaces **7**, 14917–14925 (2015)

75. F. Li, K. Cai, S. Shen, S. Chen, Preparation and thermoelectric properties of reduced graphene oxide/PEDOT:PSS composite films. Synth. Met. **197**, 58–61 (2014)

76. G.P. Moriarty, S. De, P.J. King, U. Khan, M. Via, J.A. King, J.N. Coleman, J.C. Grunlan, Thermoelectric behavior of organic thin film nanocomposites. J. Polym. Sci. Part B Polym. Phys. **51**, 119–123 (2013)

77. H. Ju, J. Kim, Chemically exfoliated SnSe nanosheets and their SnSe/poly(3,4-ethylenedioxy thiophene):Poly(styrenesulfonate) composite films for polymer based thermoelectric applications. ACS Nano **10**, 5730–5739 (2016)

78. K. Wei, T. Stedman, Z.H. Ge, L.M. Woods, G.S. Nolas, A synthetic approach for enhanced thermoelectric properties of PEDOT:PSS bulk composites. Appl. Phys. Lett. **107**, 153301 (2015)

79. Y. Du, K.F. Cai, S. Chen, P. Cizek, T. Lin, Facile preparation and thermoelectric properties of Bi2Te3 based alloy nanosheet/PEDOT:PSS composite films. ACS Appl. Mater. Interfaces **6**, 5735–5743 (2014)

80. B. Zhang, J. Sun, H.E. Katz, F. Fang, R.L. Opila, Promising thermoelectric properties of commercial PEDOT:PSS materials and their Bi2Te3 powder composites. ACS Appl. Mater. Interfaces **2**, 3170–3178 (2010)

81. E. Jin Bae, Y. Hun Kang, K.-S. Jang, S. Yun Cho, Enhancement of thermoelectric properties of PEDOT:PSS and tellurium-PEDOT:PSS hybrid composites by simple chemical treatment. Sci. Rep. **6**, 18805 (2016)

82. N.E. Coates, S.K. Yee, B. McCulloch, K.C. See, A. Majumdar, R.A. Segalman, J.J. Urban, Effect of interfacial properties on polymer-nanocrystal thermoelectric transport. Adv. Mater. **25**, 1629–1633 (2013)

83. K.C. See, J.P. Feser, C.E. Chen, A. Majumdar, J.J. Urban, R.A. Segalman, Water-Processable polymer–nanocrystal hybrids for thermoelectrics. Nano Lett. **10**, 4664–4667 (2010)

84. H. Yao, Z. Fan, H. Cheng, X. Guan, C. Wang, K. Sun, J. Ouyang, Recent development of thermoelectric polymers and composites. Macromol. Rapid Commun. **39**, 1700727 (2018)

85. T. Zhang, K. Li, C. Li, S. Ma, H.H. Hng, L. Wei, Mechanically durable and flexible thermoelectric films from PEDOT:PSS/PVA/Bi0.5Sb1.5Te3 nanocomposites. Adv. Electron. Mater. **3**, 1600554 (2017)

86. Y.Y. Wang, K.F. Cai, J.L. Yin, B.J. An, Y. Du, X. Yao, In situ fabrication and thermoelectric properties of PbTe–polyaniline composite nanostructures. J. Nanopart. Res. **13**, 533–539 (2011)

87. M. Mitra, C. Kulsi, K. Kargupta, S. Ganguly, D. Banerjee, Composite of polyaniline-bismuth selenide with enhanced thermoelectric performance. J. Appl. Polym. Sci. **135**, 46887 (2018)

88. Y. Nonoguchi, M. Nakano, T. Murayama, H. Hagino, S. Hama, K. Miyazaki, R. Matsubara, M. Nakamura, T. Kawai, Simple salt-coordinated n-type nanocarbon materials stable in air. Adv. Funct. Mater. **26**, 3021–3028 (2016)

89. Y. Nonoguchi, K. Ohashi, R. Kanazawa, K. Ashiba, K. Hata, T. Nakagawa, C. Adachi, T. Tanase, T. Kawai, Systematic conversion of single walled carbon nanotubes into n-type thermoelectric materials by molecular dopants. Sci. Rep. **3**, 3344 (2013)

90. X. Zhou, C. Pan, A. Liang, L. Wang, T. Wan, G. Yang, C. Gao, W.Y. Wong, Enhanced figure of merit of poly(9,9-di-n-octylfluorene-alt-benzothiadiazole) and SWCNT thermoelectric composites by doping with FeCl 3. J. Appl. Polym. Sci. **136**, 47011 (2018)

91. B. Dörling, S. Sandoval, P. Kankla, A. Fuertes, G. Tobias, M. Campoy-Quiles, Exploring different doping mechanisms in thermoelectric polymer/carbon nanotube composites. Synth. Met. **225**, 70–75 (2017)

92. C.K. Mai, J. Liu, C.M. Evans, R.A. Segalman, M.L. Chabinyc, D.G. Cahill, G.C. Bazan, Anisotropic thermal transport in thermoelectric composites of conjugated polyelectrolytes/single-walled carbon nanotubes. Macromolecules **49**, 4957–4963 (2016)

93. J. Choi, Y. Jung, S.J. Yang, J.Y. Oh, J. Oh, K. Jo, J.G. Son, S.E. Moon, C.R. Park, H. Kim, Flexible and robust thermoelectric generators based on all-carbon nanotube yarn without metal electrodes. ACS Nano **11**, 7608–7614 (2017)

94. M. Piao, M.R. Alam, G. Kim, U. Dettlaff-Weglikowska, S. Roth, Effect of chemical treatment on the thermoelectric properties of single walled carbon nanotube networks. Phys. Status Solidi **249**, 2353–2356 (2012)

95. D.D. Freeman, K. Choi, C. Yu, N-type thermoelectric performance of functionalized carbon nanotube-filled polymer composites. PLoS One **7**, e47822 (2012)

96. J. Chang, T. Ge, E. Sanchez-Sinencio, Challenges of printed electronics on flexible substrates. Midwest Symp. Circuits Syst., 582–585 (2012)
97. M. Orrill, S. LeBlanc, Printed thermoelectric materials and devices: Fabrication techniques, advantages, and challenges. J. Appl. Polym. Sci. **134**, 44256 (2017)
98. E. Cantatore, *Applications of Organic and Printed Electronics*, 1st edn. (Springer, Eindhoven, 2013)
99. S. Khan, L. Lorenzelli, R.S. Dahiya, Technologies for printing sensors and electronics over large flexible substrates: A review. IEEE Sensors J. **15**, 3164–3185 (2015)
100. R. Abbel, Y. Galagan, P. Groen, Roll-to-roll fabrication of solution processed electronics. Adv. Eng. Mater. **20**, 1–30 (2018)
101. A. Hobby, Screen Printing for the Industrial User, Available at http://www.gwent.org/gem_thick_film.html
102. S.J. Kim, H. Choi, Y. Kim, J.H. We, J.S. Shin, H.E. Lee, M.W. Oh, K.J. Lee, B.J. Cho, Post ionized defect engineering of the screen-printed Bi2Te2.7Se0.3 thick film for high performance flexible thermoelectric generator. Nano Energy **31**, 258–263 (2017)
103. Z. Yuan, X. Tang, Z. Xu, J. Li, W. Chen, K. Liu, Y. Liu, Z. Zhang, Screen-printed radial structure micro radioisotope thermoelectric generator. Appl. Energy **225**, 746–754 (2018)
104. Z. Cao, E. Koukharenko, M.J. Tudor, R.N. Torah, S.P. Beeby, Flexible screen printed thermoelectric generator with enhanced processes and materials. Sensors Actuators A Phys. **238**, 196–206 (2016)
105. A. Al-Halhouli, H. Qitouqa, A. Alashqar, J. Abu-Khalaf, Inkjet printing for the fabrication of flexible/stretchable wearable electronic devices and sensors. Sens. Rev. **38**, 438–452 (2018)
106. D. Jang, D. Kim, J. Moon, Influence of fluid physical properties on ink-jet printability. Langmuir **25**, 2629–2635 (2009)
107. J. Izdebska, S. Thomas, *Printing on Polymers: Fundamentals and Applications*, 1st edn. (William Andrew, Amsterdam, 2015)
108. J. Perelaer, P.J. Smith, D. Mager, D. Soltman, S.K. Volkman, V. Subramanian, J.G. Korvink, U.S. Schubert, Printed electronics: The challenges involved in printing devices, interconnects, and contacts based on inorganic materials. J. Mater. Chem. **20**, 8446 (2010)
109. R.K. Bordia, S.J.L. Kang, E.A. Olevsky, Current understanding and future research directions at the onset of the next century of sintering science and technology. J. Am. Ceram. Soc. **100**, 2314–2352 (2017)
110. S.J.L. Kang, *Sintering: Densification, Grain Growth and Microstructure*, (Elsevier, Amsterdam, 2004)
111. H. Ham, N.H. Park, S.S. Kim, H.W. Kim, Evidence of Ostwald ripening during evolution of micro-scale solid carbon spheres. Sci. Rep. **4**, 3579 (2015)
112. W.D. Kingery, Densification during sintering in the presence of a liquid phase. I. Theory. J. Appl. Phys. **30**, 301–306 (1959)
113. K. Wu, Y. Yan, J. Zhang, Y. Mao, H. Xie, J. Yang, Q. Zhang, C. Uher, X. Tang, Preparation of n-type Bi2Te3 thermoelectric materials by non-contact dispenser printing combined with selective laser melting. Phys. Status Solidi Rapid Res. Lett. **11**, 1700067 (2017)
114. D. Jafari, W.W. Wits, The utilization of selective laser melting technology on heat transfer devices for thermal energy conversion applications: A review. Renew. Sust. Energ. Rev. **91**, 420–442 (2018)
115. M.W. Oh, J.H. Son, B.S. Kim, S.D. Park, B.K. Min, H.W. Lee, Antisite defects in n-type Bi2(Te,Se)3: Experimental and theoretical studies. J. Appl. Phys. **115**, 133706 (2014)
116. H. Choi, S.J. Kim, Y. Kim, J.H. We, M.W. Oh, B.J. Cho, Enhanced thermoelectric properties of screen-printed Bi 0.5Sb1.5Te3 and Bi2Te2.7Se0.3 thick films using a post annealing process with mechanical pressure. J. Mater. Chem. C **5**, 8559–8565 (2017)
117. Z. Yuan, X. Tang, Y. Liu, Z. Xu, K. Liu, Z. Zhang, W. Chen, J. Li, A stacked and miniaturized radioisotope thermoelectric generator by screen printing. Sensors Actuators A Phys. **267**, 496–504 (2017)
118. K. Kato, K. Kuriyama, T. Yabuki, K. Miyazaki, Organic-inorganic thermoelectric material for a printed generator. J. Phys. Conf. Ser. **1052**, 012008 (2018)

119. H. Choi, Y.J. Kim, C.S. Kim, H.M. Yang, M.-W. Oh, B.J. Cho, Enhancement of reproducibility and reliability in a high-performance flexible thermoelectric generator using screen-printed materials. Nano Energy **46**, 39–44 (2018)
120. C. Han, G. Tan, T. Varghese, M.G. Kanatzidis, Y. Zhang, High-performance PbTe thermoelectric films by scalable and low-cost printing. ACS Energy Lett. **3**, 818–822 (2018)
121. S.H. Kim, T. Min, J.W. Choi, S.H. Baek, J.-P. Choi, C. Aranas, Ternary Bi2Te3In2Te3Ga2Te3 (n-type) thermoelectric film on a flexible PET substrate for use in wearables. Energy **144**, 607–618 (2018)
122. Y.S. Jung, D.H. Jeong, S.B. Kang, F. Kim, M.H. Jeong, K.-S. Lee, J.S. Son, J.M. Baik, J.-S. Kim, K.J. Choi, Wearable solar thermoelectric generator driven by unprecedentedly high temperature difference. Nano Energy **40**, 663–672 (2017)
123. A.K. Menon, O. Meek, A.J. Eng, S.K. Yee, Radial thermoelectric generator fabricated from n- and p-type conducting polymers. J. Appl. Polym. Sci. **134**, 44060 (2017)
124. L. Francioso, C. De Pascali, V. Sglavo, A. Grazioli, M. Masieri, P. Siciliano, Modelling, fabrication and experimental testing of an heat sink free wearable thermoelectric generator. Energy Convers. Manag. **145**, 204–213 (2017)
125. Q. Zhang, X. Ai, L. Wang, Y. Chang, W. Luo, W. Jiang, L. Chen, Improved thermoelectric performance of silver nanoparticles-dispersed Bi2Te3 composites deriving from hierarchical two-phased heterostructure. Adv. Funct. Mater. **25**, 966–976 (2015)
126. S. Shin, R. Kumar, J.W. Roh, D.S. Ko, H.S. Kim, S. Il Kim, L. Yin, S.M. Schlossberg, S. Cui, J.-M. You, S. Kwon, J. Zheng, J. Wang, R. Chen, High-performance screen-printed thermoelectric films on fabrics. Sci. Rep. **7**, 7317 (2017)
127. T. Juntunen, H. Jussila, M. Ruoho, S. Liu, G. Hu, T. Albrow-Owen, L.W.T. Ng, R.C.T. Howe, T. Hasan, Z. Sun, I. Tittonen, Inkjet printed large-area flexible few-layer graphene thermoelectrics. Adv. Funct. Mater. **28**, 1800480 (2018)
128. Z. Lu, M. Layani, X. Zhao, L.P. Tan, T. Sun, S. Fan, Q. Yan, S. Magdassi, H.H. Hng, Fabrication of flexible thermoelectric thin film devices by inkjet printing. Small **10**, 3551–3554 (2014)
129. B. Chen, S.R. Das, W. Zheng, B. Zhu, B. Xu, S. Hong, C. Sun, X. Wang, Y. Wu, J.C. Claussen, Inkjet printing of single-crystalline Bi2Te3 thermoelectric nanowire networks. Adv. Electron. Mater. **3**, 1600524 (2017)
130. T. Seifert, E. Sowade, F. Roscher, M. Wiemer, T. Gessner, R.R. Baumann, Additive manufacturing technologies compared: Morphology of deposits of silver ink using inkjet and aerosol jet printing. Ind. Eng. Chem. Res. **54**, 769–779 (2015)
131. C. Ou, A.L. Sangle, T. Chalklen, Q. Jing, V. Narayan, S. Kar-Narayan, Enhanced thermoelectric properties of flexible aerosol-jet printed carbon nanotube-based nanocomposites. APL Mater. **6**, 096101 (2018)
132. C. Ou, A.L. Sangle, A. Datta, Q. Jing, T. Busolo, T. Chalklen, V. Narayan, S. Kar-Narayan, Fully printed organic–inorganic nanocomposites for flexible thermoelectric applications. ACS Appl. Mater. Interfaces **10**, 19580–19587 (2018)
133. K.T. Park, J. Choi, B. Lee, Y. Ko, K. Jo, Y.M. Lee, J.A. Lim, C.R. Park, H. Kim, High-performance thermoelectric bracelet based on carbon nanotube ink printed directly onto flexible cable. J. Mater. Chem. A **6**, 19727–19734 (2018)
134. S.H. Park, S. Jo, B. Kwon, F. Kim, H.W. Ban, J.E. Lee, D.H. Gu, S.H. Lee, Y. Hwang, J.S. Kim, D.-B. Hyun, S. Lee, K.J. Choi, W. Jo, J.S. Son, High-performance shape-engineerable thermoelectric painting. Nat. Commun. **7**, 13403 (2016)
135. F. Kim, B. Kwon, Y. Eom, J.E. Lee, S. Park, S. Jo, S.H. Park, B.S. Kim, H.J. Im, M.H. Lee, T.S. Min, K.T. Kim, H.G. Chae, W.P. King, J.S. Son, 3D printing of shape-conformable thermoelectric materials using all-inorganic Bi2Te3-based inks. Nat. Energy **3**, 301–309 (2018)
136. S. Kim, M. Son, V. Nguyen, T.-S. Lim, D.Y. Yang, M.H. Kim, K. Kim, Y. Kim, J. Lee, Y. Kim, I. Kim, T.M. Lee, Y.J. Kim, S. Yang, Preparation of property-controlled bi-based solder powders by a ball-milling process. Metals (Basel). **6**, 74 (2016)
137. Y. Chisti, M. Moo-Young, On the calculation of shear rate and apparent viscosity in airlift and bubble column bioreactors. Biotechnol. Bioeng. **34**, 1391–1392 (1989)

138. E.J. Bae, Y.H. Kang, C. Lee, S.Y. Cho, Engineered nanocarbon mixing for enhancing the thermoelectric properties of a telluride-PEDOT:PSS nanocomposite. J. Mater. Chem. A **5**, 17867–17873 (2017)
139. A. Fabián-Mijangos, G. Min, J. Alvarez-Quintana, Enhanced performance thermoelectric module having asymmetrical legs. Energy Convers. Manag. **148**, 1372–1381 (2017)
140. H. Fateh, C.A. Baker, M.J. Hall, L. Shi, High fidelity finite difference model for exploring multi-parameter thermoelectric generator design space. Appl. Energy **129**, 373–383 (2014)
141. S. Ferreira-Teixeira, A.M. Pereira, Geometrical optimization of a thermoelectric device: Numerical simulations. Energy Convers. Manag. **169**, 217–227 (2018)
142. S. Oki, R.O. Suzuki, Performance simulation of a flat-plate thermoelectric module consisting of square truncated pyramid elements. J. Electron. Mater. **46**, 2691–2696 (2017)
143. H. Ali, A.Z. Sahin, B.S. Yilbas, Thermodynamic analysis of a thermoelectric power generator in relation to geometric configuration device pins. Energy Convers. Manag. **78**, 634–640 (2014)
144. B.S. Yilbas, H. Ali, Thermoelectric generator performance analysis: Influence of pin tapering on the first and second law efficiencies. Energy Convers. Manag. **100**, 138–146 (2015)
145. Y. Shi, D. Mei, Z. Yao, Y. Wang, H. Liu, Z. Chen, Nominal power density analysis of thermoelectric pins with non-constant cross sections. Energy Convers. Manag. **97**, 1–6 (2015)
146. H. Iwasaki, M. Koyano, H. Hori, Evaluation of the figure of merit on thermoelectric materials by Harman method. Jpn. J. Appl. Phys. **41**, 6606–6609 (2002)
147. A. Schmitz, C. Stiewe, E. Müller, Preparation of ring-shaped thermoelectric legs from PbTe powders for tubular thermoelectric modules. J. Electron. Mater. **42**, 1702–1706 (2013)
148. G. Min, D.M. Rowe, Ring-structured thermoelectric module. Semicond. Sci. Technol. **22**, 880–883 (2007)
149. S. Horike, T. Fukushima, T. Saito, T. Kuchimura, Y. Koshiba, M. Morimoto, K. Ishida, Highly stable n-type thermoelectric materials fabricated via electron doping into inkjet-printed carbon nanotubes using oxygen-abundant simple polymers. Mol. Syst. Des. Eng. **2**, 616–623 (2017)
150. H. Fang, B.C. Popere, E.M. Thomas, C.K. Mai, W.B. Chang, G.C. Bazan, M.L. Chabinyc, R.A. Segalman, Large-scale integration of flexible materials into rolled and corrugated thermoelectric modules. J. Appl. Polym. Sci. **134**, 44208 (2017)
151. R. He, G. Schierning, K. Nielsch, Thermoelectric devices: A review of devices, architectures, and contact optimization. Adv. Mater. Technol. **3**, 1700256 (2018)
152. B. Iezzi, K. Ankireddy, J. Twiddy, M.D. Losego, J.S. Jur, Printed, metallic thermoelectric generators integrated with pipe insulation for powering wireless sensors. Appl. Energy **208**, 758–765 (2017)
153. R. Kishore, R. Mahajan, S. Priya, Combinatory finite element and artificial neural network model for predicting performance of thermoelectric generator. Energies **11**, 2216 (2018)
154. J. Chen, K. Li, C. Liu, M. Li, Y. Lv, L. Jia, S. Jiang, Enhanced efficiency of thermoelectric generator by optimizing mechanical and electrical structures. Energies **10**, 1329 (2017)
155. G. Min, D.M. Rowe, Design theory of thermoelectric modules for electrical power generation. IEE Proc. Sci. Meas. Technol. **143**, 351–356 (1996)
156. D. Mitrani, J. Salazar, A. Turó, M.J. García, J.A. Chávez, One-dimensional modeling of TE devices considering temperature-dependent parameters using SPICE. Microelectron. J. **40**, 1398–1405 (2009)
157. K. Teffah, Y. Zhang, X. Mou, Modeling and experimentation of new thermoelectric cooler–thermoelectric generator module. Energies **11**, 576 (2018)
158. X.D. Wang, Y.X. Huang, C.H. Cheng, D. Ta-Wei Lin, C.H. Kang, A three-dimensional numerical modeling of thermoelectric device with consideration of coupling of temperature field and electric potential field. Energy **47**, 488–497 (2012)
159. S. LeBlanc, S.K. Yee, M.L. Scullin, C. Dames, K.E. Goodson, Material and manufacturing cost considerations for thermoelectrics. Renew. Sust. Energ. Rev. **32**, 313–327 (2014)

Novel Organic Polymer Composite-Based Thermoelectrics

Zimeng Zhang and Shiren Wang

1   Introduction

Polymer TE materials are in the spotlight due to exceptional properties such as low or nontoxicity, ease of process, lightweight, flexibility, and intrinsic low thermal conductivity compared to traditional inorganic materials [1–5]. However, similar to traditional inorganic materials, electrical conductivity σ, Seebeck coefficient S, and thermal conductivity κ are closely correlated as well [6–8]. Doping is a popular method to improve σ of polymer-based thermoelectrics, but heavily doped polymers show a sharply decreased S due to increased concentrations of charge carriers [9–13] (detailed explanation is presented in Sect. 3.4). Therefore, it is challenging to tune one parameter without compromising another; hence, TE performance of the majority of polymer materials becomes unsatisfactory.

The emergence of fillers in organic matrix provides effective way to tune the dimensionless figure-of-merit value ZT, which is as follows: $ZT = \sigma S^2 T/\kappa$, where T is absolute temperature; $PF = \sigma S^2$ is defined as power factor for detaching interacting effect among TE parameters [14]. The composite thus obtained consists of polymer matrix and incorporated fillers and is not just a simple combination of two components but also a system that contributes to the interfacial thermal/electrical transport phenomena [1, 15–19]. As nanofillers incorporation is promising, organic based composite thermoelectrics have been widely explored in recent decades, including organic–organic, organic–carbon, and organic–inorganic TE composites [20–22]. A previous review summarized the state-of-the-art research of TE properties of selected polymer-based composites at room temperature [23]. A list of previous work and the literature search of recent research have shown higher ZT values for organic–inorganic composites. For example, ZT value of organic–inorganic

Z. Zhang · S. Wang (✉)
Texas A&M University, College Station, TX, USA
e-mail: s.wang@tamu.edu

© Springer Nature Switzerland AG 2021
S. Skipidarov, M. Nikitin (eds.), *Thin Film and Flexible Thermoelectric Generators, Devices and Sensors*, https://doi.org/10.1007/978-3-030-45862-1_5

Table 1 Detailed thermoelectric properties of the state-of-the-art organic materials and organic–inorganic composites at room temperature [44]

Materials	σ (S/cm)	S (μV/K)	κ (W×m^{-1}×K^{-1})	ZT	Ref.
PEDOT-Tos	80	200	0.37	0.25	[45]
Poly[K$_x$(Ni-ett)]	46	−122	0.21	0.1	[46]
PEDOT:PSS	880	73	0.336	0.42	[28]
TiS_2/organic	790	−78	0.69	0.2	[25]
Poly(Ni-ett)	230	−125	0.5	0.21	[44]
PEDOT:PSS/$SnSe$	320	110	0.36	0.32	[47]
Polymer/SWCNTs	170	64	1.7	0.012	[48]
$C_6H_4NH_2CuBr_2I$	3600	−70	3.63	0.14	[49]
PVDF/Ni	4701	−21	0.55	0.11	[16]
PEI-doped SWCNTs	1780	−82	18	0.02	[50]
PEDOT/Bi_2Te_3	483	168	0.71	0.58	[24]

composites can reach about ~0.58 for p-type materials [24] and ~ 0.2 for n-type materials [25], as shown in Table 1. Additionally, organic–inorganic composites also exhibit excellent mechanical flexibility, which can be utilized in a wide range of applications such as production of wearable electronic power generators [26] and biomedical sensors [23].

It has been proved that surface-to-volume ratio due to creation of interfaces between surfaces of introduced inorganic particles and basic organic material has a great impact on TE properties of organic–inorganic composites due to energy filtering and phonon scattering effects [27]. However, attempts to improve TE performance have focused on increasing interfacial surface-to-volume ratio. For example, Kim et al. improved PF of composites by substituting nanoparticle fillers with nanowire fillers in poly(3-hexylthiophene-2,5-diyl) (P3HT) matrix, considering that nanowires have a higher specific surface area [28]. However, incorporating inorganic nanofillers is challenging due to severe aggregation, which causes a large reduction in the interface area and further affects TE performance [29–32]. Most of previous organic–inorganic composites, such as Bismuth Telluride dispersed polymers, encountered aforementioned problems, so a more effective fabrication method for incorporating nanofillers is highly in need. Specifically, nanofillers are needed in a periodic way with tunable nanostructure and controlled organic–inorganic material interface.

In order to obtain high ZT value, efforts were also made to reduce in κ of organic polymers by incorporating nanofillers enhance phonon scattering [33, 34]. In this case, engineering organic–inorganic composites with carefully tuned nanostructure will provide an effective way of achieving ultralow κ [4, 35–37]. Yu et al. and Sun et al. reported on reduced κ by incorporating organized nanostructures into silica- and metal–organic framework, respectively [38, 39].

It is quite promising to reach high ZT value by fabricating organic–inorganic composites with monodispersed and periodic nanostructures because proposed nanostructure has a potential to improve PF and ZT simultaneously [40]. However, periodic nanostructure has not been reported yet because of the difficulty in fabrication process [41–43].

Here, we introduce innovative method of fabricating composites with monodispersed and periodic nanostructures characterized by appropriate tunable size, nanoparticle distance, and nanopattern. The main idea is based on the nanosphere lithography of pre-prepared periodic nanocavities as templates for nanofillers. After preparation of periodic nanofillers, organic matrix is applied. For demonstration purpose were chosen Bi_2Te_3 as inorganic filler material and PEDOT as organic matrix. The as-fabricated PEDOT/Bi_2Te_3 organic-inorganic composites thus obtained demonstrated high PF due to combination of high electrical conductive tosylate-doped PEDOT, high S of Bi_2Te_3, and enhanced phonon scattering effect at interfaces between PEDOT and Bi_2Te_3 nanoparticles. Fabrication method discussed in this chapter is a general method and it can be adapted to many other organic–inorganic material systems.

2 Fabrication

The key challenge lies in fabrication of monodispersed and periodic TE organic–inorganic composites. Most of the efforts focused on engineering polymer TE composites with random dispersion of nanofillers. Attempts have been made to fabricate organic–inorganic composites by using different methods such as solution mixing and nanolithography, including either top-down or bottom-up nanostructure fabrication process. In this section, we consider different fabrication methods of nanostructured organic–inorganic composites and discuss results of characterization of obtained composite samples.

2.1 Solution-Based Fabrication Process

Solution-based fabrication is a simple process and can be utilized in printing process; for example, it can be utilized in novel additive manufacturing and scale-up process like roll-to-roll manufacturing, which is deemed to be time-efficient and energy-efficient way of production in industrial sectors [51–56]. To date, many efforts have focused on fabricating organic–inorganic composites by using solution-based process of a number of material systems [13, 57]. For example, Yu et al. [57] dispersed carbon nanotubes (CNTs) with PEDOT:PSS and injected the solution into a mold to form a hybrid composite film. CNTs act as highly conductive filler in organic matrix, CNTs connected and formed electron pathway for electrons and synergistically enhanced phonon scattering at interfaces between CNTs surfaces and surrounding PEDOT:PSS organic substance. As-fabricated CNTs-PEDOT:PSS composite film exhibited PF of ~ 100 $\mu W \times m^{-1} \times K^{-2}$. More recently, organic–inorganic composites formulated with intercalation of 0-dimension (0D) fullerene and two-dimension (2D)-like polycrystalline TiS_2 nanosheets displayed exceptional TE performance as n-type material. The resultant ZT value of organic–inorganic composites reached ~0.3 at room temperature, which is comparable to single crystal TiS_2 nanosheet [14].

However, one of the drawbacks of solution-based fabrication process is that nanofillers are randomly introduced into polymer matrix without or with little control of aggregation. Aggregation is not desired scenario for dispersion of nanofiller since it greatly reduces in interfacial area between nanofiller and organic matrix, and interfacial surface-to-volume ratio is closely related to σ and κ because of specific transport of charge carriers through the interface.

2.2 Fabrication of Composites with Periodic Nanostructure

Realizing concept to control periodicity of nanofillers pattern can achieve independent control of material's κ and thus separate κ from σ, and, for that, attempts have been made using different materials and fabrication methods as well. In principle, κ is more sensitive to size of nanostructures for phonons scattering at both the surface and the interface, while σ is less sensitive to decrease in size of nanostructures. Based on previous attempts, fabrication process can be categorized into two classes: top-down and bottom-up methods. In top-down method, templates with desired nanostructures are generated for assisting the nanostructure formation. In bottom-up method, nanostructures are formed by progressively introducing individual elements into the whole system.

2.2.1 Introduction to Top-Down and Bottom-up Methods

Yu et al. [57] fabricated silicon nanomesh on silicon wafer via transferring platinum nanowire pattern (grey) to silicon epilayer (yellow) through the technique of superlattice nanowire pattern transfer (SNAP) using top-down approach as shown in Fig. 1. Very low κ was observed when nanostructures were tuned with a spacing that was smaller than or equivalent to phonon mean free path.

Later, bottom-up fabrication of TE composites involving self-assembly strategy to fabricate microporous materials such as metal-organic frameworks (MOF) was reported. For example, Sun et al. [39] prepared n-type TE materials based on Ni3(2,3,6,7,10,11-hexaiminotriphenylene)$_2$ MOF structure and fabricated pellets based on TE material which displayed high σ and ultralow κ. The working principle of electron and phonon transport of designed microporous materials with MOF structure is shown in Fig. 2. This is an illustration of bottom-up fabrication method.

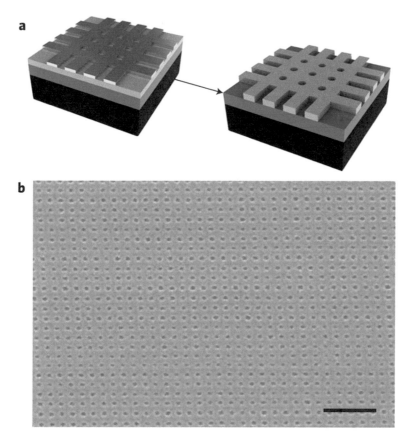

Fig. 1 Silicon nanomesh fabrication [38]. (**a**) Schematic of silicon nanomesh fabrication. (**b**) SEM (scanning electron microscopy) image of fabricated silicon nanomesh

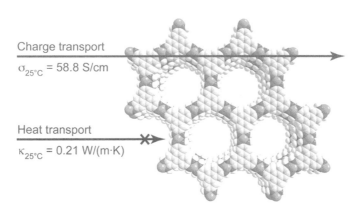

Fig. 2 Schematic illustration of electron and phonon transport of designed microporous materials with MOF structure [39]

2.2.2 Nanolithography-Based Top-Down Method with Polystyrene (PS) Nanosphere

TE performance, such as *ZT* value of two materials mentioned above, remains unknown. A more versatile method of engineering flexible composites with controlled and organized nanophase pattern via lithography of nanosphere mask has been reported recently. Nanosphere lithography is a well-studied method suitable for fabrication large area periodic nanoarrays on various kinds of substrates including flexible and rigid ones, and conductive and non-conductive ones in cost-effective manner. It has been successfully used in a number of research works for synthesizing well-ordered nanostructures that are reproducible and controllable. Briefly, the nanosphere mask is formed in the following sequence: deposition of monolayer of nanospheres on the top of the substrate with a sacrifice layer, coating of nanoparticles, tuning of nanoparticles size and removal of nanoparticles, etching of sacrifice layer, and formation of periodic pattern.

Periodic nanoarrays were prepared by following the reported process with some adaptions. Fabrication procedure is explained in more detail as follows, see Fig. 3. *Si* wafer covered with thin *SiO₂* layer is chosen as supporting substrate. 1st step, monolayer of polystyrene (PS) nanospheres deposits on *SiO₂* surface. Fig. 4 shows SEM images of closely packed monolayer of PS nanospheres with diameter of 100 nm, 300 nm, and 600 nm. 2nd step, size of PS nanospheres tunes by reactive ion etching (RIE). 3rd step, thin layer of *Cr* deposits onto PS nanoparticles and *SiO₂* surface by thermal evaporation deposition, 4th step, PS nanospheres lift-off. 5th step, nanomesh is created, thin layer of *Cr* serves as protection of nanomesh. Cylindrical

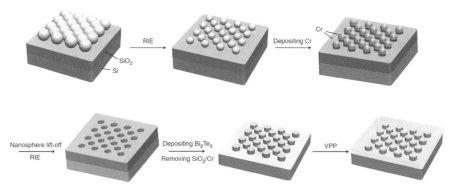

Fig. 3 Scheme of controllable fabrication of organic–inorganic composite film. Fabrication procedure for poly(3,4-ethylenedioxythiophene) PEDOT/*Bi₂Te₃* composite films, including nanospheres deposition, nanospheres lift-off and reactive ion etching (RIE) for nanohole arrays template, filling *Bi₂Te₃* into template by thermal evaporation deposition, removing template, and compositing *Bi₂Te₃* nanoparticle arrays with PEDOT using chemical vapor phase polymerization (VPP) deposition. Geometrical parameters of resultant nanostructure can be tuned by diameter of polystyrene (PS) nanospheres and RIE etching time of PS nanospheres [24]

Fig. 4 Morphology of
polystyrene (PS)
nanospheres [24]. SEM
images of closely packed
monolayer PS nanospheres
with diameter of (**a**)
100 nm, (**b**) 300 nm, and
(**c**) 600 nm

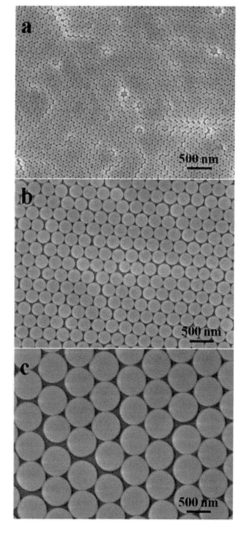

Fig. 4 Morphology of polystyrene (PS) nanospheres [24]. SEM images of closely packed monolayer PS nanospheres with diameter of (**a**) 100 nm, (**b**) 300 nm, and (**c**) 600 nm

nanoholes are formed by removing exposed SiO_2 with CH_3/O_2 plasma etching. 6th step, cylindrical nanoarrays of p-type Bi_2Te_3 form by thermal evaporation deposition and followed by removing SiO_2 sacrifice layer in strong acid. At last, 7th step, pre-prepared Bi_2Te_3 nanoparticle arrays cover by PEDOT layer using method of modified chemical vapor phase polymerization (VPP) – VPP PEDOT. The corresponding structure in each fabrication step was characterized by SEM. Fig. 5a–b show digital and SEM images of original closely packed monolayer of 100 nm diameter PS nanospheres; Fig. 5c–d show digital and SEM images of same PS nanospheres after RIE etching for 22 seconds. Reduction in size of nanospheres after RIE etching is

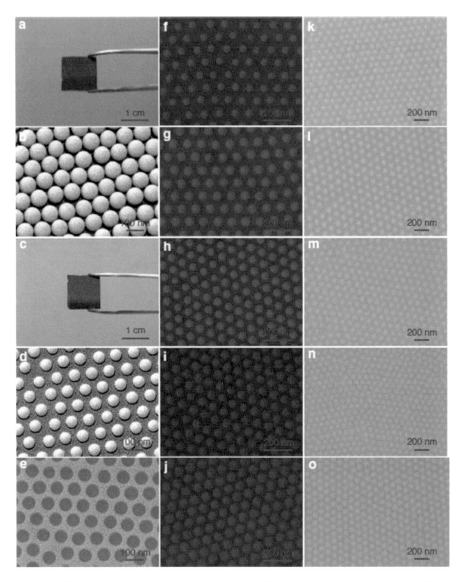

Fig. 5 (**a**) Digital photo of closely packed monolayer of PS nanospheres with diameter of 100 nm on *SiO₂/Si* substrate. (**b**) SEM image of closely packed monolayer of PS nanospheres with diameter of 100 nm. (**c**) Digital photo of PS nanospheres on *SiO₂/Si* substrate after etching for 22 s. (**d**) SEM image of PS nanospheres after etching for 22 s. (**e**) SEM image of ready nanoholes array template. (**f-j**) SEM images of *Bi₂Te₃* nanoparticle arrays prepared through nanoholes template with 100 nm PS nanospheres as pattern. (**k-o**) SEM images of PEDOT/*Bi₂Te₃* (100) organic–inorganic composite films [24]

evident. Fig. 5e shows SEM image of ready nanoholes array template after lift-off of PS nanospheres, which clearly indicates on successful removal of PS nanospheres. Fig. 5(f–j) show SEM images of Bi_2Te_3 nanoparticle arrays prepared through nanoholes template created with 100 nm PS nanospheres as pattern. Fig. 5k–o show SEM images of PEDOT/Bi_2Te_3 (100) organic–inorganic composite films. There is also a very clear difference in the SiO_2/Si substrate before and after PS nanosphere monolayer coating. Surface of SiO_2/Si substrate displays colorful just after monolayer of PS nanospheres is deposited on it (Fig. 5a, c). Fig. 6a–b show schematic dispersion of PS nanospheres on substrate surface. Monodispersed and periodic nanoparticles in VPP PEDOT continuous organic matrix were clearly demonstrated in aforementioned images, which indicates the successful fabrication of PEDOT/Bi_2Te_3 composites films.

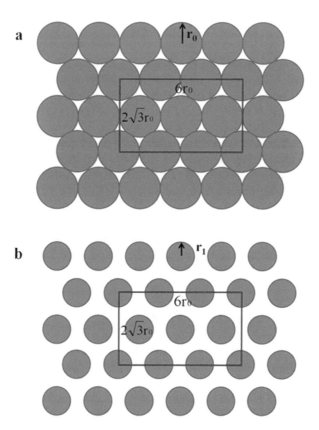

Fig. 6 Schematic diagram [24]. (**a**) Closely packed PS nanospheres and (**b**) obtained Bi_2Te_3 nanoparticle array

2.2.3 **Structure Tuning and Nanofiller Volume Calculation**

Using described fabrication method of PEDOT/Bi_2Te_3 organic–inorganic composites, the structures of Bi_2Te_3 nanoparticle arrays, including size and spacing, can be easily tuned by modulating different fabrication parameters. For example, original diameter of PS nanospheres can be changed from 100 nm to several hundred nanometers. Here, we select 100, 300, and 600 nm PS nanospheres to tune the patterns. As-fabricated PEDOT/Bi_2Te_3 composite films prepared with templates created on the base of PS nanospheres of original diameter 100, 300, 600 nm are denoted as PEDOT/Bi_2Te_3 (100), (300), (600), respectively, Size of PS nanospheres and, hence, spacing in Bi_2Te_3 nanoparticle array can be easily tuned by varying etching time. Volume fraction V_d of Bi_2Te_3 nanoparticles in composite film has been calculated based on nanoparticle packing model (Fig. 6), and used equation is shown below:

$$V_d = \frac{\pi r_1^2 d_1}{2\sqrt{3} r_0^2 d_0},$$
(1)

where r_0 is original radius of PS nanospheres in closely packed monolayer, d_0 is thickness of composite film, r_1 radius of Bi_2Te_3 nanoparticles fabricated with template obtained after RIE etching of PS nanospheres with radius r_0, and d_1 is height of Bi_2Te_3 nanoarrays. Determination of parameters in Eq. (1) is illustrated in Fig. 6. Here, the centers of nanoparticles do not change after RIE.

Since Bi_2Te_3 nanoparticle arrays were fabricated using thermal evaporation deposition of Bi_2Te_3 thin film on pre-prepared nanohole templates, height d_1 can be precisely controlled by carefully monitoring thermal evaporation process. Further, exact height of Bi_2Te_3 nanoparticle was measured. The measurement results confirmed that the thickness of Bi_2Te_3 nanoparticle equals to ~70 nm. A similar method was used to determine thickness of spin coated VPP PEDOT thin films. Thickness of VPP PEDOT films was determined by spin coating speed and process time. At least five different areas were measured, and average value was calculated. Here, measured average value of PEDOT thin film thickness is equal to ~75 nm.

2.2.4 **X-Ray Diffraction Characterization**

X-ray diffraction (XRD) patterns were used to characterize Bi_2Te_3 crystallinity of as-prepared composites as shown in Fig. 7. Here, three types of thin film were compared including PEDOT/Bi_2Te_3 (100) composites with ~31 vol. % of Bi_2Te_3 nanoparticles, VPP PEDOT thin film and thermal evaporation deposited Bi_2Te_3 thin film. VPP PEDOT thin film displays a broad diffraction peak because of amorphous nature of fabricated polymer compared to other two thin films. For thermal evaporation deposited Bi_2Te_3 film and composite thin film, the peaks are well indexed (Fig. 7).

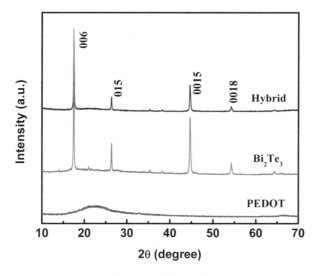

Fig. 7 XRD patterns of composites, Bi_2Te_3, and PEDOT thin films [24]

3 Thermoelectric Performance Characterization

In this section, we introduce state-of-the-art techniques to measure TE properties of composites. Generally, TE performance is characterized by dimensionless figure-of-merit value, $ZT = \sigma S^2 T/\kappa$. These are four key parameters in determining TE performance.

For inorganic–organic composites, σ, S, and κ are calculated with modification, considering incorporation of interface between organic and inorganic materials. Parallel and series-connected models are conventional methods to calculate σ and S of composites. Specifically, models have been broadly used in studying and calculating the filler loading effect on σ and S. Two phases of binary composites are interpreted as filler and matrix, which herein stand for parallel and series components, respectively. The equations used to calculate parameters of organic–inorganic composites using parallel and series-connected models [58] are given below:

$$\sigma\left(\text{parallel}\right) = \sigma_{\text{filler}}x + \sigma_{\text{matrix}}\left(1-x\right), \tag{2}$$

$$\sigma\left(\text{series}\right) = \frac{\sigma_{\text{filler}}\sigma_{\text{matrix}}}{\sigma_{\text{filler}}\left(1-x\right)+\sigma_{\text{matrix}}x}, \tag{3}$$

$$S\left(\text{parallel}\right) = \frac{S_{\text{filler}}\sigma_{\text{filler}}x + S_{\text{matrix}}\sigma_{\text{matrix}}\left(1-x\right)}{\sigma_{\text{filler}}x + \sigma_{\text{matrix}}\left(1-x\right)}, \tag{4}$$

$$S\left(\text{series}\right) = \frac{S_{\text{filler}}\kappa_{\text{matrix}}x + S_{\text{matrix}}\kappa_{\text{filler}}\left(1-x\right)}{\kappa_{\text{matrix}}x + \kappa_{\text{filler}}\left(1-x\right)}, \tag{5}$$

where σ(parallel) and S(parallel) are calculated σ and S of composites using parallel model, σ(series) and S(series) are calculated σ and S of composites using series model, σ_{filler}, σ_{matrix}, S_{filler}, S_{matrix}, κ_{filler}, κ_{matrix} and x are σ, S, κ of filler material and polymer matrix and volume ratio of nanofillers to organic polymer matrix. From Eqs. (2)–(5) one can conclude as follows. Parallel model represents a situation where charge carriers are transported through either filler or matrix, with minimal transport between two phases. On the other hand, series model represents a situation where charge carriers are constantly transferring through two phases. Between these two extremes lie all intermediate situations with varying degrees of transport between phases. So, σ and S of organic–inorganic composites are also dependent on contribution of parallel and series charge carriers transport components. In experiments, average value of at least three specimens is preferred for calculation using the above equations.

In PEDOT/Bi_2Te_3-based composites, Bi_2Te_3 is filler material and PEDOT is matrix material. So, in Eqs. (2)–(5), symbols σ_{filler}, σ_{matrix}, S_{filler}, S_{matrix}, κ_{filler} and κ_{matrix} can be changed to $\sigma_{Bi_2Te_3}$, σ_{PEDOT}, $S_{Bi_2Te_3}$, S_{PEDOT}, $\kappa_{Bi_2Te_3}$ and κ_{PEDOT}, respectively.

The following section will discuss the details of TE performance measurement including σ, S, and κ using as example of fabricated monodispersed and periodic inorganic–organic PEDOT/Bi_2Te_3 composites.

3.1   Electrical Conductivity Measurement

Van de Pauw method is typical method for electrical conductivity characterization and measurement setup is shown in Fig. 8 [59, 60]. Thin film specimens were prepared for σ measurement as shown in Fig. 9. To eliminate uncertainties due to contact

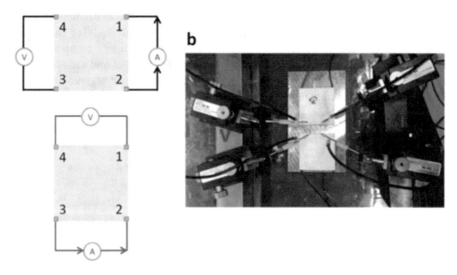

Fig. 8 Electrical conductivity measurement (Van de Pauw method). (Manufacturing Lett) Schematic illustration of Van de Pauw method; (**b**) photo of measurement setup [61]

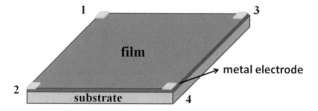

Fig. 9 Illustration of the sample configuration for electrical conductivity measurement [24]

resistance between probe and specimen surface, four gold electrodes were deposited by thermal evaporation on each corners of the specimen. The idea is to measure in-plane resistance of as-fabricated thin film in each direction by sequentially applying voltage to adjacent two deposited metal electrodes, noted as $V_{1,2}$, while in the meantime, measuring current between remaining two electrodes at opposite side, noted as $I_{3,4}$ as shown in Fig. 8a, b. And σ is calculated by the following formula [61]:

$$\sigma = \left[\frac{\pi d}{\ln 2} \times \frac{R_A + R_B}{2} \times f\left(\frac{R_A}{R_B}\right) \right]^{-1}, \tag{6}$$

where d is thickness of composite thin films and $f(x) = \dfrac{1}{\cosh\left(\ln(x)/2.304\right)}$ is correction factor [61].

In Eq. (6), resistance R_A was measured by $R_A = \dfrac{1}{2}\left(\dfrac{V_{3,4}}{I_{1,2}} + \dfrac{V_{1,2}}{I_{3,4}}\right)$ and R_B by $R_B = \dfrac{1}{2}\left(\dfrac{V_{3,4}}{I_{1,2}} + \dfrac{V_{1,2}}{I_{3,4}}\right)\dfrac{1}{2}\left(V_{1,2}/I_{3,4} + V_{3,4}/I_{1,2}\right)$.

Figure 10 shows the measurement results of as-fabricated PEDOT/Bi_2Te_3 composite films with different structures. The green, blue, and red solid lines respond to composite films of PEDOT/Bi_2Te_3 (100), (300), (600), respectively. The dashed lines are theoretical calculations based on parallel and series models. Most of the experiment data are fitted within the theoretical calculation. The experiment data indicate an obvious size-dependent effect on σ varying from ~280 S/cm to ~1350 S/cm [24, 62, 63]. This phenomenon is observed because VPP PEDOT exhibits higher σ of ~1350 S/cm compared to σ of thermal evaporation deposited Bi_2Te_3 of ~143 S/cm as shown in Table 2. So intuitively, higher volume ratio of high σ component will result in increased overall σ of composite film and vice versa. In our case, PEDOT owns higher σ. Also, it is observed a slight difference among composite films with different structures where PEDOT/Bi_2Te_3 (600) is the highest. This phenomenon is due to the effect of interfacial transport on final TE properties of composite film rather than the single mixed effect.

The stability of as-fabricated PEDOT/Bi_2Te_3 composites (~ 31 vol. % Bi_2Te_3 nanoparticles) with (100), (300), (600) structure was analyzed and experimental results are shown in Fig. 11. Bi_2Te_3 content of ~31 vol. % is chosen due to

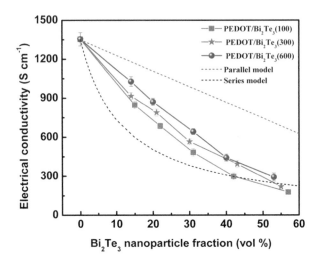

Fig. 10 Electrical conductivity measurement results of PEDOT/Bi_2Te_3 composite films [24]

Table 2 TE properties of Bi_2Te_3 film on SiO_2/Si substrate prepared under the same condition for composite films at room temperature

σ (S/cm)	S (μV/K)	κ (W$\times$m$^{-1}\times$K^{-1})	ZT
143	273	0.82	0.39

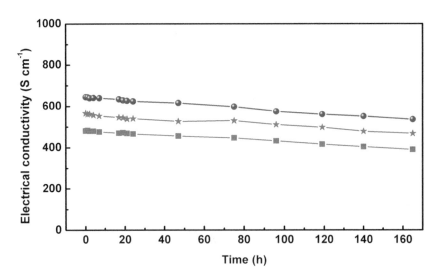

Fig. 11 Air stability test [24]

synergistic effect of σ, S, and κ, which yields the highest ZT value. This is explained in detail in Sect. 3.4. No obvious decrease in σ with increase in time exposed to air at room temperature was observed, which indicates that σ of prepared composite films is enough stable on air. The samples displayed slight changes in σ after 1 week.

The mechanical flexibility of as-prepared PEDOT/Bi_2Te_3 composite films with ~31 vol. % Bi_2Te_3 nanoparticles was further analyzed. In order to test mechanical flexibility, composite films were fabricated on soft substrates as illustrated in Fig. 12. Film resistance was measured by wrapping flexible PEDOT/Bi_2Te_3 (100) composite films with ~31 vol. % Bi_2Te_3 around glass tube under a range of curve radii (r) from 2 mm to 14 mm. Film resistance R was compared to the original electrical resistance R_0 with no bending, and ratio R/R_0 was defined. Experimental results indicate only a slight decrease in R with increase in bending angle. Even at a very small radius (3.5 mm), decrease in R is smaller than 5% compared to original thin film resistance as shown in Fig. 13. Control sample of all-inorganic Bi_2Te_3 thin film deposited by thermal evaporation was analyzed which exhibited significant increse in R of ~18% at bending radius of 3.5 mm. Comparison between as-fabricated composite thin films and all-inorganic Bi_2Te_3 thin film deposited by thermal evaporation is illustrated in Fig. 14. Except for electrical stability, we also measured mechanical stability of each type of thin films. Mechanical stability tests were performed via measuring electrical resistance after different numbers of bending up to 100 as shown in Fig. 14. With increased number of bending, changes in electrical resistance of composite films are merely negligible while a large increase in resistance was observed for Bi_2Te_3 inorganic films. This indicates a better mechanical stability of composite film compared to Bi_2Te_3 one. This can be explained by micro-cracks appearing in Bi_2Te_3 film after several bending cycles. But for composite films, inorganic phase, which is intrinsically rigid and easily breaks under

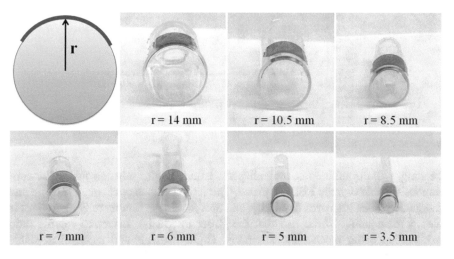

Fig. 12 Images of prepared composite films on flexible substrates for mechanical flexibility test [24]

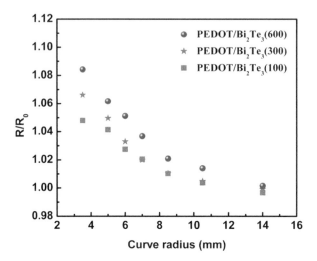

Fig. 13 Stability of electrical resistance. Changes of resistance for PEDOT/Bi_2Te_3 composite films as a function of bending radius r [24]

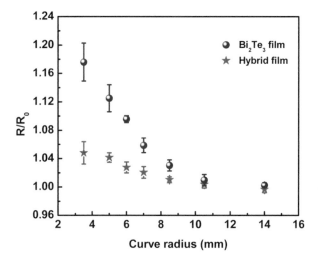

Fig. 14 Difference in flexibility between thermal evaporation deposited Bi_2Te_3 film and as-fabricated PEDOT/Bi_2Te_3 composite film [24]

bending, was discontinuous, monodispersed, and is uniformly surrounded by continuous flexible organic PEDOT phase. The intimate contact between organic and inorganic phases improves durability, and thus makes the film more flexible. This is also corresponding to the fact that the larger interfacial surface-to-volume ratio exhibited a better mechanical flexibility. Highly flexible and mechanical stable nature of composite film provides a great promise to use it as a material for flexible/wearable thermoelectric power generators.

3.2   Seebeck Coefficient Measurement

S of as-prepared composite films was measured with in-house self-built measurement setup. S measurement setup is shown in Fig. 15. Briefly, two Peltier devices were positioned on the two edges of thin film square shape sample. First Peltier device serves as heater on one edge and second serves as heatsink to cool opposite edge of thin film square shape sample. Before measurement, a pair of square-shaped gold electrodes with 5 mm spacing was deposited by thermal evaporation onto each thin film. These electrodes were used to define electrical measurement spacing. Two micro-thermocouples (TC) were placed on thin film next to prepared electrodes with the same spacing. The temperature difference ΔT is determined by displayed temperature difference between two thermocouples. Previous experiments and simulation have proved linear temperature gradient along direction between two thermocouples; for details, refer to Kun [40, 63]. Two gold micro-sized wires were soldered to gold electrodes using indium and voltage was determined as ΔV. Thus, S is calculated using the equation [24]:

$$S = -\frac{\Delta V}{\Delta T}. \tag{7}$$

In order to prove measurement accuracy used in-house self-built S measurement setup, two standard samples of n-type Bi_2Te_3 and constantan were analyzed and S of standard samples is illustrated in Fig. 16. Measured ΔV on ΔT graphs displayed linear dependences. This proved the accuracy of measurement. The slope of linear approximation represents S of measured sample. Figure 17 shows typical S plot of as-prepared PEDOT/Bi_2Te_3 (100) with 31 vol. % Bi_2Te_3 nanoparticles with linear ΔV on ΔT dependence. The slope of plot linear approximation was calculated to be $-168\mu V/K$, representing S of prepared sample.

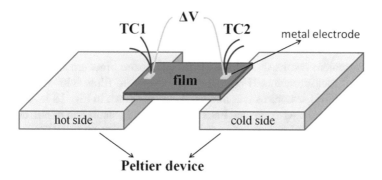

Fig. 15 Illustration of setup to measure Seebeck coefficient [24]

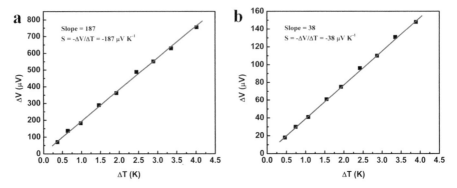

Fig. 16 Measured S of standard samples. (**a**) n-type Bi_2Te_3 and (**b**) constantan [24]

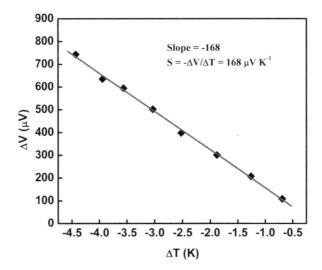

Fig. 17 Typical S measurement plot. The sample is PEDOT/Bi_2Te_3 (100) composite film with ~31 vol. % Bi_2Te_3 [24]

S measurement results of PEDOT/Bi_2Te_3 composite films are shown in Fig. 18. Similar to σ measurements (Fig. 10), S of composite films with PEDOT/Bi_2Te_3 (100), (300), (600) structures was analyzed, and is shown in Fig. 18 in green, blue, and red, respectively. Also, dashed lines are theoretical calculation of S in composite films based on parallel and series models. Unlike σ, S of some PEDOT/Bi_2Te_3 composite thin films exceeded the upper extremum of theoretical calculation via series-connected model. Obviously from experimental data that S depends on filler content and varying in absolute value from ~40 μV/K to ~180 μV/K with increase in volume content of incorporated Bi_2Te_3 nanoparticles. This dependence is due to excellent S of Bi_2Te_3 thin film deposited by thermal evaporation Table 2. So, intuitively, higher

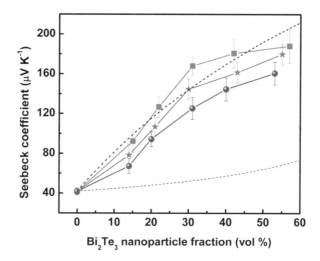

Fig. 18 Seebeck coefficient measurement results of PEDOT/*Bi₂Te₃* composites [24]

volume content of Bi_2Te_3 component will result in increased overall S of composite films and vice versa. Also, it is observed a slight difference among composite films with different structures where PEDOT/Bi_2Te_3 (100) exhibits a higher S. This is due to the effect of interfacial transport on the final TE properties of the composite rather than the single mixed effect. Additionally, previous literature reported on positive effect of energy filtering on TE performance of organic–inorganic composites. Specifically, S was proved to be improved because organic–inorganic interfaces form energy barrier that is prone to scatter low-energy charge carriers [64, 65]. This can further influence on relaxation time and make it more dependent on energy and simultaneously result in a more asymmetric charge carriers transport near Fermi level. That explains S enhancement. Size of inorganic nanoparticles in array affects S as well. Composite films with smaller inorganic nanoparticles as filler will possess larger interfacial surface-to-volume ratio, so, increased number of interfaces will act as energy barrier scattering selectively low-energy charge carriers, thus, resulting in larger S. In our experiment, PEDOT/Bi_2Te_3 (100) composite films are with the smallest 100 nm nanoparticles. Moreover, clear saturated trend of S exhibited for PEDOT/Bi_2Te_3 (100) composite films. Maximum S value reaching −180 μV/K when adding 55 vol. % Bi_2Te_3 100 nm nanoparticles.

Mechanical flexibility test of prepared composite films was performed in concern S stability. Similar to σ test, bending effect on S was analyzed by attaching flexible composite films onto glass tube with various radii, and the results are demonstrated as ratio of S measured after bending to original one, as shown in Fig. 19. Experimental result shows only 10% decrease in S under bending with small radius of 3.5 mm, which exhibits the high mechanical flexibility of fabricated composite films. Compared to all-inorganic thin films which are vulnerable to break and generate micro cracks even under small bending degree, fabricated organic–inorganic composite films show a great promise as flexible TE materials for powering wearable electronics.

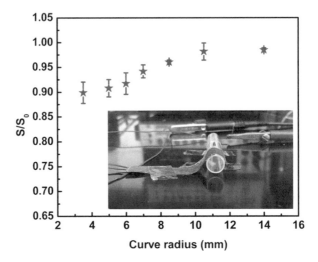

Fig. 19 Mechanical stability test over Seebeck coefficient [24]

3.3 Power Factor Calculation

As mentioned in the very beginning of this section, $PF = S^2\sigma$ is indicator of σ and S, in $ZT = \sigma S^2 T/\kappa$. So, larger PF will result in higher ZT value while maintaining at the same κ and has the same environmental temperature. From experimental data of σ and S shown in Figs. 10 and 18, PF of each specimen can be calculated and the results are shown in Fig. 20. When comparing σ and S, one can observe that specimens with higher σ often demonstrated a relatively lower S and vice versa. This is due to interrelation between σ and S in TE material. The original equation of S is shown below:

$$S = \frac{8\pi^2 k_B^2}{3eh^2} \times m_e T \left(\frac{\pi}{3n} \right)^{2/3}, \tag{8}$$

where n is concentration of charge carriers, e is electron charge, m_e is effective mass of charge carrier, h is Planck constant, and k_B is Boltzmann constant. S is inversely proportional to concentration of charge carriers n, and σ is proportional to n:

$$\sigma = e\mu n, \tag{9}$$

where μ is mobility of charge carriers.

PF of PEDOT/Bi_2Te_3 (100) composite thin film reaches the peak value of ~1350 $\mu W \times m^{-1} \times K^{-2}$ with ~ 31 vol. % Bi_2Te_3 nanoparticles.

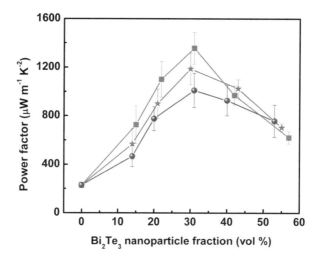

Fig. 20 Power factor measurement results of PEDOT/Bi_2Te_3 composite films [24]

3.4 Thermal Conductivity Measurement

To further study phonon transport through nanophase, we measured in-plane thermal conductivity $\kappa_\parallel$ of fabricated thin films with configuration shown in Fig. 21.

Thickness of as-fabricated samples for σ and S measurements equals to around 75 nm. The differential 3ω method was applied to measure κ for ZT value estimation [S 5–7]. In this method, κ is measured in a similar manner like σ and S are measured. Prior to κ measurement, specimens with thicker films are prepared. First, a thick layer of silicon dioxide was deposited onto the substrate (Silicon wafer) via Plasma-enhanced chemical vapor deposition (PECVD). Then silicon dioxide layer is etched from half of substrate by RIE. In the meantime, the thickness of thermal evaporated Bi_2Te_3 film was increased to 0.7 μm. At last, in VPP step, polymer is coated onto prepared nanoarrays with increased coating time of up to 25 s at a lower speed of 1500 rpm. To get a uniform thick polymer layer, VPP process was repeated three times. Ultimately, thick composites films with uniform thickness around 0.75 μm were prepared. The total κ includes both electronic κ_e and lattice κ_L thermal conductivity. For prepared composites films, κ_L is dominated by the morphology of the film. Thus, it is important to guarantee that thick composites films prepared for κ analysis exhibit a very similar morphology with the corresponding composites thin films. Figure 22 shows SEM of fabricated PEDOT/Bi_2Te_3 (100) with 31 vol. % Bi_2Te_3 nanoparticles evidence a similar morphology of thick composites films.

Another factor affecting the accuracy of κ measurement is that σ will change with different sample thicknesses. Specifically, κ_e is changed with the increase in the sample thickness. In general, σ decreases with the increase in the sample thickness. In our case, σ of thick films of PEDOT/Bi_2Te_3 (100) with 31 vol. % Bi_2Te_3

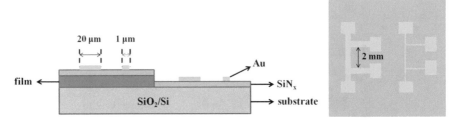

Fig. 21 Schematic illustration of differential 3ω method for thermal conductivity [24]

Fig. 22 SEM image of
prepared PEDOT/Bi_2Te_3
(100) with 31 vol. % Bi_2Te_3
nanoparticles film for
thermal conductivity
measurement [24]

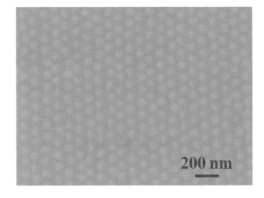

nanoparticles decreased from ~483 S/cm to ~391 S/cm compared with the corresponding thin film. Since there is no dramatic change in σ, we then approximated κ measured on thick composites films to estimate κ of corresponding thin composites film considering limited erroneous this may cause. For control sample of PEDOT thick films composed by VPP, the speed of spin coating was kept the same as the previous ones and repeated three times during VPP process for κ analysis. As-fabricated film has a thickness of ~0.62 μm.

To measure $\kappa_\parallel$, modifications were done to prepare thicker composites films. Figure 23 shows specimen structure and electrodes used for thermal conductivity measurement.

Initially, a layer of SiN_x (100 nm) was coated onto the silicon substrate via PECVD. Gold metal strips with thickness of 100 nm were patterned onto SiN_x surface with 2 μm of length, 1 μm/20 μm of width, using photolithography. Finally, 5 nm Cr layer was coated to improve adhesion of the film and gold metal strip heater.

The width of gold metal strip heater matters to κ measurement. For example, if width of metal strip heater is much larger than thickness of the film, in our case, 20 μm, heat conduction is approximately 1D in vertical direction. Thus, cross-plane $\kappa_\perp$ instead of $\kappa_\parallel$ was obtained. Vertical temperature drop, defined as ΔT_f, is calculated by the following equation:

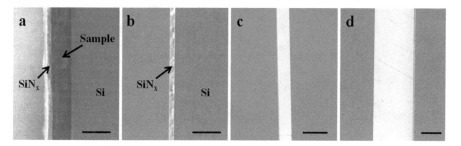

Fig. 23 Specimen structure for thermal conductivity measurement [24]

$$\frac{\Delta T_f}{P_s} = \frac{\Delta T_s}{P_s} - \frac{\Delta T_r}{P_r}, \tag{10}$$

where ΔT_s and ΔT_r are temperature oscillations of as-fabricated specimens and gold metal strip heaters, P_s and P_r are corresponding power dissipations. ΔT_s and ΔT_r are determined by:

$$\Delta T_{s,r} = 2R\frac{dT}{dR} \times \frac{V_{3\omega}}{V_{1\omega}}, \tag{11}$$

where R and dT/dR are resistance and temperature coefficient of resistance of gold metal strip heater, $V_{3\omega}$ and $V_{1\omega}$ are third and first harmonic voltages. Value $\kappa_\perp$ of the film is calculated by:

$$\kappa_\perp = \frac{Pd_f}{bL\Delta T_f}, \tag{12}$$

where d_f is film thickness, b is width of heater strip, and L is length of heater strip.

On the other hand, if width of the metal strip is similar to film thickness, e.g., 1 μm in our case, then both $\kappa_\perp$ and $\kappa_\parallel$ will be influenced by heat transport. The ratio of reduced temperatures related to narrow (ΔT_{2D}) and wide (ΔT_{1D}) gold metal heater strips can be shown as:

$$\frac{\Delta T_{2D}}{\Delta T_{1D}} = \left(\frac{\kappa_\perp}{\kappa_\parallel}\right)^{1/2} \times \frac{b_n}{d_f} \times \frac{K(\lambda)}{2K'(\lambda)}, \tag{13}$$

where b_n is width of narrow metal heater strip line, $K(\lambda)$ is complete elliptical integral of first kind, and $K'(\lambda)$ is its complementary integral. The argument λ is given by:

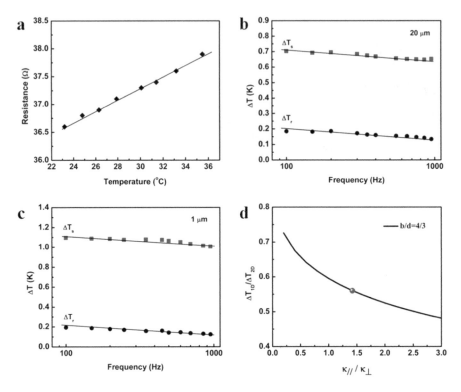

Fig. 24 Typical results of thermal conductivity measurement [24]. (**a**) Temperature dependence of resistance of metal strip heater. (**b**) Temperature oscillation amplitudes of 20 μm strip heater for measurements of ΔT_{1D}. (**c**) Temperature oscillation amplitudes of 1 μm strip heater for measurements of ΔT_{2D} (**d**) Relationship between $\Delta T_{1D}/\Delta T_{2D}$ and $\kappa_\parallel / \kappa_\perp$ under b/d = 4/3

$$\frac{1}{\lambda} = \cosh\left[\frac{\pi}{4}\frac{b_n}{d_f}\left(\frac{\kappa_\perp}{\kappa_\parallel}\right)^{1/2}\right] \tag{14}$$

Figure 24 gives an example for testing $\kappa_\parallel$ of PEDOT/Bi_2Te_3 composite film with ~31 vol. % Bi_2Te_3 nanoparticles. For wide metal strip heater (20 μm), ΔT_f was calculated as ~0.51 K. $\kappa_\perp$ was equal to 0.54 W × m^{-1} × K^{-1} based on Eq. S4. For narrow metal strip heater (1 μm), ΔT_f was calculated as ~0.91 K. The ratio of $\Delta T_{1D}/\Delta T_{2D}$ was 0.56 and fitted to aforementioned equation, determining value of $\kappa_\parallel/\kappa_\perp$ to be 1.42. Thus, $\kappa_\parallel$ was equal to 0.77 W × m^{-1} × K^{-1}.

Value $\kappa_\parallel$ measured by 3ω method equals to 1.52 W × m^{-1} × K^{-1} in all-inorganic Bi_2Te_3 film at room temperature. In comparison, as-fabricated PEDOT/Bi_2Te_3 composite films exhibited much lower κ because of introduction of Bi_2Te_3 fillers into PEDOT matrix, κ decreases with increase in Bi_2Te_3 nanoparticles volume fraction. The lowest $\kappa \sim 0.5$ W × m^{-1} × K^{-1} was observed in composite films with Bi_2Te_3 volume fraction of 52% for (100) structure that is around 300% decrease compared to original VPP PEDOT films and ~ 64% reduction compared to deposited by

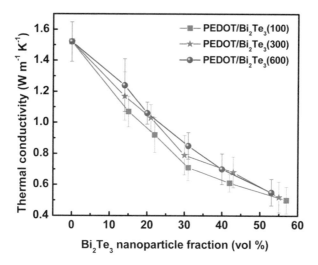

Fig. 25 Thermal conductivity measurement results of PEDOT/Bi_2Te_3 composite films [24]

thermal evaporation Bi_2Te_3 thin film (κ of Bi_2Te_3 thin film is shown in Table 2). This is interesting phenomenon, since κ of composites is lower than either κ of any of two components. This phenomenon indicates that reduction in κ of composites is not only affected by introduction of lower κ of Bi_2Te_3 fillers, but also due to introduction of organic–inorganic interface when incorporated Bi_2Te_3 fillers. Measurements of κ displayed a difference for composite films with different sizes of Bi_2Te_3 nanoarrays which further evidences of interfacial effect on κ of as-fabricated composite films. As shown in Fig. 25, at the same level of Bi_2Te_3 nanoparticles volume fraction, larger Bi_2Te_3 nanoparticles exhibit a higher κ while smaller sized Bi_2Te_3 nanoparticles own a relatively lower κ. Smaller sized Bi_2Te_3 nanoparticles yield a higher interfacial surface-to-volume ratio, thus exhibiting a relatively lower κ.

To theoretically estimate κ_{eff} of composite films, the following equation was used:

$$\kappa_{\text{eff}} = \kappa_m \frac{\left[\left(\dfrac{\kappa_d}{\kappa_m} - \dfrac{\kappa_d}{r_d h_c} - 1\right)V_d + \left(\dfrac{\kappa_d}{\kappa_m} + \dfrac{\kappa_d}{r_d h_c} + 1\right)\right]}{\left[\left(\dfrac{\kappa_d}{r_d h_c} - \dfrac{\kappa_d}{\kappa_m} + 1\right)V_d + \left(\dfrac{\kappa_d}{\kappa_m} + \dfrac{\kappa_d}{r_d h_c} + 1\right)\right]} \tag{15}$$

where κ_{eff} is overall thermal conductivity of composite film, κ_m is matrix's thermal conductivity (VPP PEDOT), κ_d is filler's thermal conductivity of Bi_2Te_3 dispersions, V_d is volume fraction of filler, r_d is radius of circular cylindrical nanoparticles, and h_c is interfacial thermal conductance. If $h_c = \infty$, Eq. (15) coincides with Maxwell equation for measuring composites' κ neglecting interfacial thermal resistance. Effects of nanoparticle size, volume fraction of nanoparticles, and interfacial effect of two components are all considered in the equation presented above. Based on Eq. (15), for a given Bi_2Te_3 fraction, theoretical values are calculated and plotted as dash lines in Fig. 26.

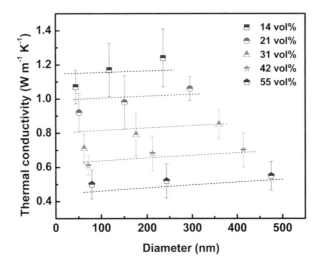

Fig. 26 Thermal conductivity of PEDOT/*Bi₂Te₃* composite films with various diameters of *Bi₂Te₃* nanoarrays and *Bi₂Te₃* volume ration [24]

3.5 Dimensionless Figure-of-Merit (ZT)

Based on previous measurements of σ, S, and κ, *ZT* value of as-fabricated PEDOT/ *Bi₂Te₃* composites is calculated at room temperature. Figure 27 shows *ZT* plots of composites as a function of *Bi₂Te₃* nanoparticles volume fraction with different PS patterns: green square, blue star, and red dot represent PEDOT/*Bi₂Te₃* (100), (300) and (600) patterns, respectively. Peak of *ZT* value reaches ~0.58 with 31 vol. % *Bi₂Te₃* nanoparticles with (100) pattern, the highest value of inorganic–organic composites and organic material reported to date [66–70]. High *ZT* value results due to synergistic contribution of high σ provided by VPP PEDOT component, high S provided by *Bi₂Te₃* and relatively lower κ for reduced thermal transport, as well as the enhanced phonon scattering at organic–inorganic interface.

Further, we make a comparison of *PF* and *ZT* values of VPP PEDOT thin films and deposited by thermal evaporation *Bi₂Te₃* thin films with fabricated PEDOT/*Bi₂Te₃* composite films with ~31 vol. % *Bi₂Te₃* nanoparticles. The measurement results are shown in Fig. 28. Both *PF* and *ZT* values are significantly increased for as-fabricated composite films. *PF* has increased ~5 times and ~ 1.2 times compared to PEDOT and *Bi₂Te₃* thin films, respectively. Both VPP PEDOT and *Bi₂Te₃* are well known as high-performance TE materials, but still not as good as composite films with monodispersed and periodic nanopatterns.

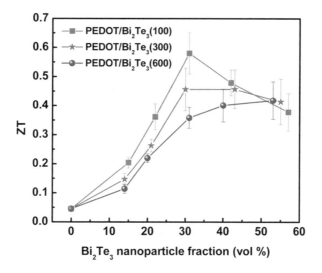

Fig. 27 Figure-of-merit *ZT* measurement results of PEDOT/*Bi₂Te₃* composite films [24]

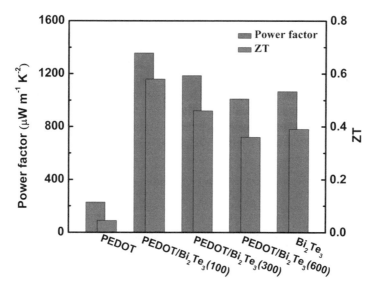

Fig. 28 Comparisons of power factor *PF* and *ZT* of as-prepared PEDOT film, PEDOT/*Bi₂Te₃* composite films with ~31 vol. % *Bi₂Te₃*, and *Bi₂Te₃* film [24]

4 Discussion

Originally, VPP PEDOT has superior electrical conductivity over p-type organic materials with intrinsic low κ. When combining PEDOT with p-type Bi_2Te_3, which is known as the most efficient inorganic material with superior S at room temperature, the resultant composites are expected to exhibit exceptional TE performance. In our design, we achieve the nanoscale control of inorganic nanoparticle arrays to precise control on phonon transport and electron transport. This contributes to high overall TE performance of as-fabricated composite films. With introduction of organic–inorganic interface in the composites, ultimate performance of composite films is not simple combination of two single materials.

The data listed in Table 1 are used to make a direct comparison of TE performance between PEDOT/Bi_2Te_3 composites and other peers' work around room temperature. Results obtained on fabricated PEDOT/Bi_2Te_3 composite films presented in this chapter yields the highest ZT value among all the listed organic materials and organic–inorganic composites.

In conclusion, fabricated PEDOT/Bi_2Te_3 composite films using proposed nanofabrication method achieved controlled nanoscale patterns and displayed exceptional TE performance. ZT value reaches as high as ~0.58 at room temperature. The detailed fabrication process of composite films is illustrated in Sect. 2. Briefly, PS nanoparticles were introduced to generate the nanocavities for later Bi_2Te_3 nanocylinders formation before coating VPP PEDOT thin film onto Bi_2Te_3 nanocylinders. Detailed explanation of ultrahigh ZT value achieved in fabricated composite thin films is given. Additionally, composite films with periodically patterned nanostructure exhibited well-performed mechanical stability. The proposed fabrication method is not restricted to PEDOT/Bi_2Te_3 composite films, but a number of other materials can be adapted in such composite films with monodispersed and periodic nanostructures. The method we propose is a platform for novel organic–inorganic composites fabrication for different applications.

References

1. J. Chen et al., Strong anisotropy in thermoelectric properties of CNT/PANI composites. Carbon **114**, 1–7 (2017)
2. J. Yang, H.L. Yip, A.K.Y. Jen, Rational design of advanced thermoelectric materials. Adv. Energy Mater. **3**(5), 549–565 (2013)
3. K. Zhang, Y. Zhang, S. Wang, Enhancing thermoelectric properties of organic composites through hierarchical nanostructures. Sci. Rep. **3**, 3448 (2013)
4. Q. Zhang, Y. Sun, W. Xu, D. Zhu, Organic thermoelectric materials: Emerging green energy materials converting heat to electricity directly and efficiently. Adv. Mater. **26**(40), 6829–6851 (2014)
5. Z. Zhang, J. Qiu, S. Wang, Roll-to-roll printing of flexible thin-film organic thermoelectric devices. Manuf. Lett. **8**, 6–10 (2016)

6. J. Choi, J.Y. Lee, S.S. Lee, C.R. Park, H. Kim, High-performance thermoelectric paper based on double carrier-filtering processes at nanowire heterojunctions. Adv. Energy Mater. **6**(9), 1502181 (2016)
7. K.C. See, J.P. Feser, C.E. Chen, A. Majumdar, J.J. Urban, R.A. Segalman, Water-processable polymer– nanocrystal hybrids for thermoelectrics. Nano Lett. **10**(11), 4664–4667 (2010)
8. E.W. Zaia et al., Carrier scattering at alloy nanointerfaces enhances power factor in PEDOT: PSS hybrid thermoelectrics. Nano Lett. **16**(5), 3352–3359 (2016)
9. L. Brownlie, J. Shapter, Advances in carbon nanotube n-type doping: Methods, analysis and applications. Carbon **126**, 257–270 (2018)
10. S. Horike et al., Highly stable n-type thermoelectric materials fabricated via electron doping into inkjet-printed carbon nanotubes using oxygen-abundant simple polymers. Mol. Syst. Design. Eng. **2**(5), 616–623 (2017)
11. Y. Nonoguchi et al., Systematic conversion of single walled carbon nanotubes into n-type thermoelectric materials by molecular dopants. Sci. Rep. **3**, 3344 (2013)
12. K. Oshima, Y. Yanagawa, H. Asano, Y. Shiraishi, N. Toshima, Improvement of stability of n-type super growth CNTs by hybridization with polymer for organic hybrid thermoelectrics. Synth. Met. **225**, 81–85 (2017)
13. C. Yu, A. Murali, K. Choi, Y. Ryu, Air-stable fabric thermoelectric modules made of N- and P-type carbon nanotubes. Energy Environ. Sci. **5**(11), 9481 (2012)
14. L. Wang et al., Solution-printable fullerene/TiS 2 organic/inorganic hybrids for high-performance flexible n-type thermoelectrics. Energy Environ. Sci. **11**(5), 1307–1317 (2018)
15. J.L. Blackburn, A.J. Ferguson, C. Cho, J.C. Grunlan, Carbon-nanotube-based thermoelectric materials and devices. Adv. Mater. **30**(11),1704386 (2018)
16. Y. Chen, M. He, B. Liu, G.C. Bazan, J. Zhou, Z. Liang, Bendable n-type metallic nanocomposites with large thermoelectric power factor. Adv. Mater. **29**(4), 1604752 (2017)
17. Z.-G. Chen, G. Han, L. Yang, L. Cheng, J. Zou, Nanostructured thermoelectric materials: Current research and future challenge. Prog. Nat. Sci. Mater. Int. **22**(6), 535–549 (2012)
18. M. He, F. Qiu, Z. Lin, Towards high-performance polymer-based thermoelectric materials. Energy Environ. Sci. **6**(5), 1352 (2013)
19. D. Kim, Y. Kim, K. Choi, J.C. Grunlan, C. Yu, Improved Thermoelectric Behavior of Nanotube-Filled Polymer Composites with Poly(3,4-ethylenedioxythiophene) Poly(styrenesulfonate). ACS Nano **4**(1), 513–523 (2010)
20. J.B. Neaton, Single-molecule junctions: thermoelectricity at the gate. Nat. Nanotechnol. **9**(11), 876 (2014)
21. C. Wan et al., Flexible thermoelectric foil for wearable energy harvesting. Nano Energy **30**, 840–845 (2016)
22. J.J. Urban, Prospects for thermoelectricity in quantum dot hybrid arrays. Nat. Nanotechnol. **10**(12), 997 (2015)
23. L. Wang et al., Polymer composites-based thermoelectric materials and devices. Compos. Part B **122**, 145–155 (2017)
24. L. Wang et al., Exceptional thermoelectric properties of flexible organic-inorganic hybrids with monodispersed and periodic nanophase. Nat. Commun. **9**(1), 3817 (2018)
25. C. Wan et al., Flexible n-type thermoelectric materials by organic intercalation of layered transition metal dichalcogenide TiS 2. Nat. Mater. **14**(6), 622 (2015)
26. S.J. Kim, J.H. We, B.J. Cho, A wearable thermoelectric generator fabricated on a glass fabric. Energy Environ. Sci. **7**(6), 1959 (2014)
27. M. He, F. Qiu, Z. Lin, Towards high-performance polymer-based thermoelectric materials. Energy Environ. Sci. **6**(5), 1352–1361 (2013)
28. G.-H. Kim, L. Shao, K. Zhang, K.P. Pipe, Engineered doping of organic semiconductors for enhanced thermoelectric efficiency. Nat. Mater. **12**(8), 719 (2013)
29. M. He et al., Thermopower enhancement in conducting polymer nanocomposites via carrier energy scattering at the organic–inorganic semiconductor interface. Energy Environ. Sci. **5**(8), 8351–8358 (2012)

30. J. Liu, X. Wang, D. Li, N.E. Coates, R.A. Segalman, D.G. Cahill, Thermal conductivity and elastic constants of PEDOT: PSS with high electrical conductivity. Macromolecules **48**(3), 585–591 (2015)
31. A. Weathers et al., Significant electronic thermal transport in the conducting polymer poly (3, 4-ethylenedioxythiophene). Adv. Mater. **27**(12), 2101–2106 (2015)
32. B. Zhang, J. Sun, H.E. Katz, F. Fang, R.L. Opila, Promising thermoelectric properties of commercial PEDOT: PSS materials and their Bi2Te3 powder composites. ACS Appl. Mater. Interfaces **2**(11), 3170–3178 (2010)
33. D.J. Singh, I. Terasaki, Thermoelectrics: Nanostructuring and more. Nat. Mater. **7**(8), 616 (2008)
34. P.-a. Zong et al., Skutterudite with graphene-modified grain-boundary complexion enhances zT enabling high-efficiency thermoelectric device. Energy Environ. Sci. **10**(1), 183–191 (2017)
35. H. Lee, D. Vashaee, D.Z. Wang, M.S. Dresselhaus, Z.F. Ren, G. Chen, Effects of nanoscale porosity on thermoelectric properties of SiGe. J. Appl. Phys. **107**(9), 094308 (2010)
36. A. Minnich, M.S. Dresselhaus, Z.F. Ren, G. Chen, Bulk nanostructured thermoelectric materials: Current research and future prospects. Energy Environ. Sci. **2**(5), 466–479 (2009)
37. L. Sun, M.G. Campbell, M. Dincă, Electrically conductive porous metal–organic frameworks. Angew. Chem. Int. Ed. **55**(11), 3566–3579 (2016)
38. J.-K. Yu, S. Mitrovic, D. Tham, J. Varghese, J.R. Heath, Reduction of thermal conductivity in phononic nanomesh structures. Nat. Nanotechnol **5**(10), 718 (2010)
39. L. Sun et al., A microporous and naturally nanostructured thermoelectric metal-organic framework with ultralow thermal conductivity. Joule **1**(1), 168–177 (2017)
40. K. Zhang et al., Effect of host-mobility dependent carrier scattering on thermoelectric power factors of polymer composites. Nano Energy **19**, 128–137 (2016)
41. Y. Zhang et al., A mesoporous anisotropic n-type Bi2Te3 monolith with low thermal conductivity as an efficient thermoelectric material. Adv. Mater. **24**(37), 5065–5070 (2012)
42. B. Xu et al., Highly porous thermoelectric nanocomposites with low thermal conductivity and high figure of merit from large-scale solution-synthesized Bi2Te2. 5Se0. 5 hollow nanostructures. Angew. Chem. Int. Ed. **56**(13), 3546–3551 (2017)
43. H.-C. Zhou, J. R. Long, and O. M. Yaghi, Introduction to Metal–Organic Frameworks. Chem. Rev. **112**(2), 673–674(2012)
44. L. Yan, M. Shao, H. Wang, D. Dudis, A. Urbas, B. Hu, High Seebeck effects from hybrid metal/polymer/metal thin-film devices. Adv. Mater. **23**(35), 4120–4124 (2011)
45. O. Bubnova et al., Optimization of the thermoelectric figure of merit in the conducting polymer poly (3, 4-ethylenedioxythiophene). Nat. Mater. **10**(6), 429 (2011)
46. T. Zhan, L. Fang, Y. Xu, Prediction of thermal boundary resistance by the machine learning method. Sci. Rep. **7**(1), 7109 (2017)
47. H. Ju, J. Kim, Chemically exfoliated SnSe nanosheets and their SnSe/poly (3, 4-ethylenedioxythiophene): Poly (styrenesulfonate) composite films for polymer based thermoelectric applications. ACS Nano **10**(6), 5730–5739 (2016)
48. D. Huang et al., Conjugated-backbone effect of organic small molecules for n-type thermoelectric materials with ZT over 0.2. J. Am. Chem. Soc. **139**(37), 13013–13023 (2017)
49. W. Shi, Z. Shuai, D. Wang, Tuning thermal transport in chain-oriented conducting polymers for enhanced thermoelectric efficiency: a computational study. Adv. Funct. Mater. **27**(40), 1702847 (2017)
50. D.P.H. Hasselman, L.F. Johnson, Effective thermal conductivity of composites with interfacial thermal barrier resistance. J. Compos. Mater. **21**(6), 508–515 (1987)
51. I. Chowdhury et al., On-chip cooling by superlattice-based thin-film thermoelectrics. Nat. Nanotechnol. **4**(4), 235 (2009)
52. D.X. Crispin, Retracted article: Towards polymer-based organic thermoelectric generators. Energy Environ. Sci. (2012)

53. X. Liu et al., Low electron scattering potentials in high performance Mg2Si0. 45Sn0. 55 based thermoelectric solid solutions with band convergence. Adv. Energy Mater. **3**(9), 1238–1244 (2013)

54. Y. Pei, X. Shi, A. LaLonde, H. Wang, L. Chen, G.J. Snyder, Convergence of electronic bands for high performance bulk thermoelectrics. *Nature* **473**(7345), 66 (2011)

55. A. Sotelo, S. Rasekh, M.A. Madre, E. Guilmeau, S. Marinel, J.C. Diez, Solution-based synthesis routes to thermoelectric Bi2Ca2Co1. 7Ox. J. Eur. Ceram. Soc. **31**(9), 1763–1769 (2011)

56. K.H. Yim et al., Controlling electrical properties of conjugated polymers via a solution-based p-type doping. Adv. Mater. **20**(17), 3319–3324 (2008)

57. C. Yu, K. Choi, L. Yin, J.C. Grunlan, Light-weight flexible carbon nanotube based organic composites with large thermoelectric power factors. ACS Nano **5**(10), 7885–7892 (2011)

58. Z. Liang, M.J. Boland, K. Butrouna, D.R. Strachan, K.R. Graham, Increased power factors of organic–inorganic nanocomposite thermoelectric materials and the role of energy filtering. J. Mater. Chem. A **5**(30), 15891–15900 (2017)

59. O. Bierwagen, T. Ive, C.G. Van de Walle, J.S. Speck, Causes of incorrect carrier-type identification in van der Pauw–Hall measurements. Appl. Phys. Lett. **93**(24), 242108 (2008)

60. P.M. Hemenger, Measurement of high resistivity semiconductors using the van der Pauw method. Rev. Sci. Instrum. **44**(6), 698–700 (1973)

61. K. Zhang, High-Efficiency Organic Thermoelectric Materials for Energy Harvesting. PhD, Texas Tech University, Texas Tech University, 2014

62. O. Bubnova et al., Semi-metallic polymers. Nat. Mater. **13**(2), 190 (2014)

63. K. Zhang, J. Qiu, S. Wang, Thermoelectric properties of PEDOT nanowire/PEDOT hybrids. Nanoscale **8**(15), 8033–8041 (2016)

64. R. Nunna et al., Ultrahigh thermoelectric performance in Cu 2 Se-based hybrid materials with highly dispersed molecular CNTs. Energy Environ. Sci. **10**(9), 1928–1935 (2017)

65. T. Zhang, K. Li, C. Li, S. Ma, H.H. Hng, L. Wei, Mechanically durable and flexible thermoelectric films from PEDOT: PSS/PVA/Bi0. 5Sb1. 5Te3 nanocomposites. Adv. Electr. Mater. **3**(4), 1600554 (2017)

66. A.D. Avery et al., Tailored semiconducting carbon nanotube networks with enhanced thermoelectric properties. Nat. Energy **1**(4), 16033 (2016)

67. Y. Liu, X. Li, J. Wang, L. Xu, B. Hu, An extremely high power factor in Seebeck effects based on a new n-type copper-based organic/inorganic hybrid C 6 H 4 NH 2 CuBr 2 I film with metal-like conductivity. J. Mater. Chem. A **5**(26), 13834–13841 (2017)

68. Y. Sun et al., Flexible n-type high-performance thermoelectric thin films of poly (nickelethylenetetrathiolate) prepared by an electrochemical method. Adv. Mater. **28**(17), 3351–3358 (2016)

69. Q. Yao, Q. Wang, L. Wang, L. Chen, Abnormally enhanced thermoelectric transport properties of SWNT/PANI hybrid films by the strengthened PANI molecular ordering. Energy Environ. Sci. **7**(11), 3801–3807 (2014)

70. W. Zhou et al., High-performance and compact-designed flexible thermoelectric modules enabled by a reticulate carbon nanotube architecture. Nat. Commun. **8**, 14886 (2017)

Synergistic Strategies to Boost Lead Telluride as Prospective Thermoelectrics

Yong Yu, Haijun Wu, and Jiaqing He

1   A Brief Introduction of PbTe Material

PbTe is a mid-temperature thermoelectric (TE) material (performance peaks at 600–800 K)[1]. The figure of merit ($ZT=S^2\sigma T/\kappa$) is the most important parameter to evaluate performance of thermoelectric materials, where S, σ, κ and T are Seebeck coefficient, electrical conductivity, total thermal conductivity (which equals to the sum of electrons thermal conductivity κ_e and lattice thermal conductivity κ_L: $\kappa=\kappa_e+\kappa_L$) and absolute temperature, respectively. Crystal lattice structure of *PbTe* is *NaCl* type with *Pb* atoms occupying cation sites and *Te* on anion sites [2]. Figure 1a shows the crystal structure of *PbTe*. Figure 1b–d show projections of lattice cell planes (100), (110), and (111), respectively. The band gap E_g = 0.32 eV. *PbTe* can be doped into either *n*- or *p*-type by proper dopants [3–8]. TE properties of *PbTe* have been intensively studied [9–24]. The maximum ZT value of *PbTe* has been reported to be 1.8 for *n*-type [25] and 2.5 for *p*-type [26]. The trend in study on *PbTe* TE materials is shown in Fig. 2. Figure 2a summarizes the number of papers about *PbTe* TE materials from 1959 to 2019. The numbers of papers are lower than 10 per year from 1959 to 1998. After that, the number increases rapidly from 10 to 80 per year, and still increases more dramatically in recent years. This trend indicates that *PbTe* TE materials are still a research hotspot. Figure 2b illustrates the variations of ZT from 1960

Y. Yu · H. Wu
Department of Physics, South University of Science and Technology of China, Shenzhen, China

Department of Materials Science and Engineering, National University of Singapore, Singapore, Singapore

J. He (✉)
Department of Physics, South University of Science and Technology of China, Shenzhen, China
e-mail: hejq@sustc.edu.cn

© Springer Nature Switzerland AG 2021
S. Skipidarov, M. Nikitin (eds.), *Thin Film and Flexible Thermoelectric Generators, Devices and Sensors*, https://doi.org/10.1007/978-3-030-45862-1_6

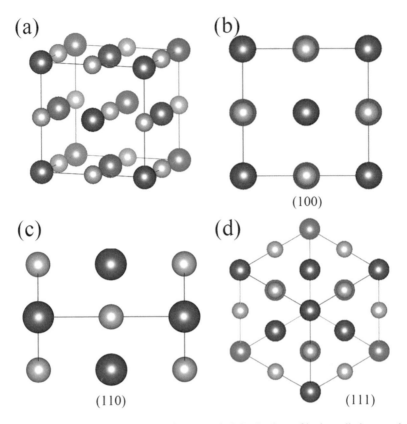

Fig. 1 (**a**) The crystal structure of *PbTe*. (**b**), (**c**), and (**d**) Projections of lattice cell planes – planes (100), (110) and (111)

to 2018 [27]. For *p*-type *PbTe*, ZT value experiences a slight growth from 1960 to 2010 and then boosts rapidly. However, for *n*-type *PbTe*, ZT value has not grown much in a long time until now.

2 Engineering Concentration of Charge Carriers to Peak ZT in PbTe

2.1 Peaking ZT by Adjusting Concentration of Charge Carriers

The strong interdependencies of S, σ, and κ to n (concentration of charge carriers) are a prime challenge to design high ZT thermoelectric materials [28]. So, n is a key parameter which can be controlled by rational doping. For degenerated

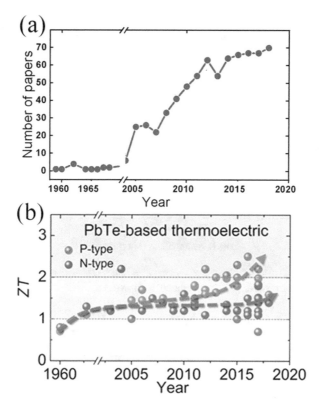

Fig. 2 The trend in study on *PbTe* TE material. (**a**) Publications versus year from 1959 to 2019. (**b**) Timeline of ZT values achieved in representative *PbTe* based TE materials, including both *p*-type (orange ball) and *n*-type *PbTe* (green ball) [27]. Reproduced with permission [27]. Copyright 2018, Springer Nature Publishing

semiconductor with parabolic band dispersion, by neglecting the influence of dopant on scattering relaxation time or band structure, S is given by Eq. (1) [29]:

$$S = \frac{8\pi^2 k_B^2}{3eh^2} m^* T \left(\frac{\pi}{3n} \right)^{2/3},$$
(1)

where k_B is Boltzmann constant, e is electron charge, h is Plank constant, and m^* is density of states (DOS) effective mass of charge carriers.

Obviously, high S is usually accompanied by low n, and with increasing in n S drops rapidly (Fig. 3). On the contrary, to obtain large σ, n should be as high as possible, see Eq. (2) and Fig. 3 [28]:

$$\sigma = en\mu,$$
(2)

where μ is electron mobility. As a consequence, power factor ($S^2\sigma$) maximum can be reached by balancing S and σ within proper range of charge carriers concentration (e.g., 10^{19}–10^{20} cm^{-3} for most semiconductors [31]) [28].

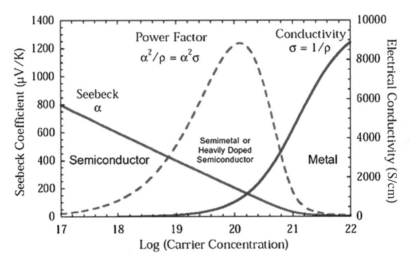

Fig. 3 Dependencies of electrical properties of TE materials on charge carriers concentration [30]. (Reproduced with permission [30]. Copyright 2016, Elsevier)

As Eq. (3) shows, thermal conductivity κ comprises two major components: electronic contribution κ_e and lattice contribution κ_L (among all related to ZT physical parameters, κ_L is the only one with weak dependence on n (Fig. 3a) and potentially can be minimized to amorphous limit through independent crystal structure and/or microstructure design [32]):

$$\kappa = \kappa_e + \kappa_L, \tag{3}$$

and κ_e is influenced by n through Wiedemann-Franz law (Eq. (4)) [33]:

$$\kappa_e = L_0 \sigma T = L_0 e n \mu T, \tag{4}$$

where L_0 is Lorenz number. Eq. (4) links electrical to thermal transport and makes optimization of ZT even tougher. However, one can always achieve maximum ZT at a certain temperature by fine-tuning n to optimal level by appropriate doping [28].

Figure 4a shows that most of advanced p-type $PbTe$ TE materials are doped with Na and K, optimal hole concentration is in the range of $(3–40) \times 10^{19}$ cm^{-3}. In n-type $PbTe$, dopants I and Br (doping on Te site), Sb, Bi, Al, Ga, and In (doping on Pb sites) are usually chosen as donors, and electron concentration in superior n-type $PbTe$ is in the range of $(4–40) \times 10^{18}$ cm^{-3} (Fig. 4b). Notably, optimal hole concentration in p-type $PbTe$ is an order of magnitude higher than that in n-type $PbTe$ [27].

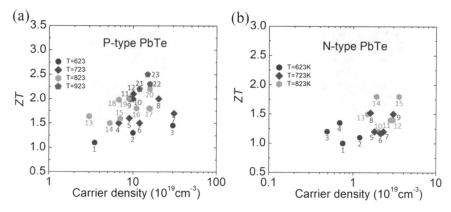

Fig. 4 (**a**) Peak ZT values as a function of optimized hole concentration in *p*-type *PbTe* systems [27]: (1) *PbTe − Ge : Na* [34], (2) *PbTe : (Na, Bi)* [35], (3) *PbTe − Sn : (Ag, Sb)* [36], (4) *PbTe − Mn : Na* [37], (5) *PbTe − Se : K* [38], (6) *PbTe − Ca : Na* [39], (7) *PbTe − Cd : Na* [40], (8) *PbTe − S : K* [41], (9) *PbTe : Na* [42], (10) *PbTe : Na* (nanoprecipitates) [43], (11) *PbTe − Ge* [44], (12) *PbTe − Se − S : Na* [7], (13) *PbTe − Hg : Na* [45], (14) *PbTe : Tl* [23], (15) *PbTe − Sn − Se − S : Na* [46], (16) *PbTe − Yb : Na* [47], (17) *PbTe − Se : Na* [48], (18) *PbTe − Mn − Sr : Na* [49], (19) *PbTe − Mg : Na* [50], (20) *PbTe − Eu : Na* [51], (21) *PbTe − Sr : Na* [52], (22) *PbTe − S : Na* [53], (23) *PbTe − Sr : Na* (non-equilibrium) [26]; (**b**) peak ZT values as a function of optimized electron concentration in *n*-type *PbTe* systems: (1) *PbTe − Se : Cr* [54], (2) *PbTe − Se − S : I* [55], (3) *PbTe − NaCl* [56], (4) *PbTe − Zn : I* [57], (5) *PbTe − Cd : I* [58], (6) *PbTe − Sn − Se : I* [59], (7) *PbTe − Mg : I* [60], (8) *PbTe − Se − S : Cl* [61], (9) *PbTe : Cd* [62], (10) *PbTe : La* [63], (11) *PbTe : (I, In)* [64], (12) *PbTe − S : Sb* [65], (13) *PbTe − Cu : I* [66], (14) *PbTe : (Sb, I)* [25], (15) *PbTe : (Sb, In)* [67]. Data show the large difference in optimized concentrations of charge carriers between *p*-type $(3–40) \times 10^{19}$ cm^{-3} and *n*-type $(0.4 \sim 4) \times 10^{19}$ cm^{-3} [27]. (Reproduced with permission [27]. Copyright 2018, Springer Nature Publishing)

2.2 Reaching Maximum ZT Within Temperature Range

In the case of conventional doping when impurity atoms are homogeneously distributed in solid solutions, concentrations of charge carriers are generally constant in relation to temperature. However, optimal concentration of charge carriers n^* of TE semiconductor material usually increases rapidly with rising temperature, roughly obeying power law of $T^{3/2}$. The consequence of this is that maximum ZT cannot be reached at every working temperature. Figure 5 illustrates approaches to overcome problem. Figure 5a shows schematic diagram of conventional doping and its possible realizations: functionally graded doping [68, 69] and temperature dependent solubility doping (*T*-dependent doping) [28, 70, 71]. Functionally graded doping is to integrate two or multiple segments with dissimilar *n*, see green lines in Fig. 5a, b. One common approach developed in the late 2000s of preparing functionally graded material is so-called spark plasma sintering or hot pressing of compacted stack of powder layers; each of those has different concentration of charge carriers [69]. Alternatively, by use of temperature dependent solubility of some specific dopants (Fig. 5a, b), one can create gradient of *n* within single material controlled only by

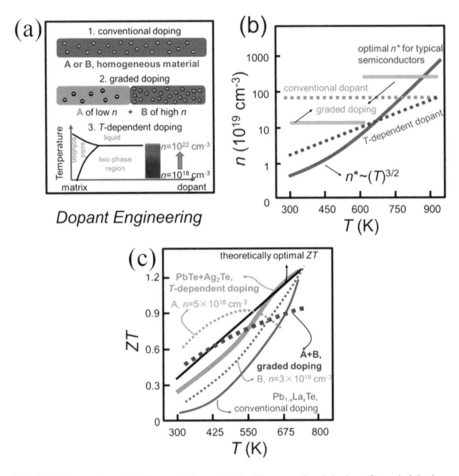

Fig. 5 (**a**) Comparison of different doping methods: (1) conventional doping, (2) graded doping, and (3) *T*-dependent doping. (**b**) Strategies for stabilizing optimal concentration of charge carriers (n^*, denoted by red line, usually shows $T^{3/2}$ dependence). For most conventional dopants, the resultant concentration of charge carriers (denoted by orange dotted line) is almost temperature independent, functionally graded doping (green lines) by use of samples with dissimilar n, and use of temperature dependent solubility dopant, namely, n has a strong temperature dependence (purple line). (**c**) Enhancement of ZT values over a broad temperature range through stabilizing n^* in comparison to conventional doping approach [28]. (Reproduced with permission [28]. Copyright 2016, ACS Publications)

temperature and its gradient (see purple line in Fig. 5a, b) [28]. Well-known dopants are *Ag*, *Cu*, and excess *Pb*, which have limited solubility within *PbTe* around room temperature but have much higher solubility at elevated temperatures [72, 73]. This temperature dependent doping is reversible in heating–cooling process and gets rid of diffusion problem with graded doping, making it better candidate for actual application [6, 70].

Figure 5c compares temperature dependent ZT values of n-type $PbTe$ using different doping methods [28]. Green and orange dotted lines represent A and B materials of conventional n-type $PbTe$ with $n = 5 \times 10^{18}$ and 3×10^{18} cm^{-3}, respectively, and dotted purple line is for graded doping by integrating two components [74]. Material A has low n and thus high ZT at low temperatures, while material B with larger n has significantly higher ZT at high temperatures. Graded doping holds ZT between A and B, but ZT can achieve much larger average value in the entire working temperature range. The solid black line shows theoretically optimal ZT for n-type $PbTe$ [6]. Conventional doping approaches, such as by heavy La doping in $PbTe$ (solid navy line), can allow to reach with this approach high ZT value optimal around 750 K, but cannot achieve high performance at all temperatures [6]. Temperature dependent solubility limit of Ag in $PbTe + Ag_2Te$ composite (T-dependent doping, solid green line) enables increase in n with increasing T and pushes ZT values close to optimal levels over the entire temperature range [6, 53, 71, 75]. T-dependent doping approach has also been successfully applied in many other systems including p-type Na doped $PbTe$. It was demonstrated that Na-rich precipitates located at the grain boundaries at low temperature are re-dissolved into $PbTe$ matrix at $T > 600$ K [6, 34]. This gives rise to the enhancement of hole concentration, contributing to the increase in electrical conductivity and power factor and suppression of bipolar conduction at elevated temperature achieving superior performance. Modifications of conventional doping method via including T-dependent doping are helpful to acquire larger average ZT values, which are especially important in technological applications [28].

However, above strategies have limits. For graded doping, after extended duration in service the initial gradient in concentration of charge carriers in graded material may fade or vanish due to the diffusion induced homogenization effect, thereby deteriorating conversion efficiency [76]. This method is easy to practice, however, not good enough because when certain segments reach peak ZT, other segments must be not peak due to different n. For T-dependent doping, doping level is decided by solubility of dopants, which may not obey power law of $T^{3/2}$ and cannot reach optimized ZT within all temperature ranges. In addition, dopants will form precipitates from over-saturated solid solution from high temperature to low temperature as the second half of a tested cycle. Some of precipitates will be coarser from time to time, resulting in a potential detrimental influence on electrical conductivity.

3   Band Engineering to Optimize Electrical Conductivity

Charge transports in crystalline materials are closely connected to electronic band structure, so manipulation on electronic band structure is effective strategy to optimize electrical transport properties. Manipulating approaches on electronic band structure comprise band convergence, DOS distortion and resonant state, band alignment between matrix and second phase, band flattening, and introducing impurity level, etc. [27].

3.1 Manipulations on Valence Band Structure in p-Type PbTe

3.1.1 Band Convergence (BC)

In p-type $PbTe$, small energy offset between light hole valence band L and heavy hole valence band Σ facilitates charge redistribution from single band to multiple bands. Elements alloying could enclose light and heavy hole valence bands, called band convergence. Converging valence bands could enhance effective mass through increasing in valley degeneracy numbers [77, 78]. Recently, the approach of band convergence was revisited along with nano-structuring technology, which is one powerful synergistic strategy to enhance TE performance of p-type $PbTe$ [48, 79]. Band convergence can achieve high valley degeneracy number N_v to enlarge effective mass m^*, defined as $m^* = N_v^{2/3} m_b^*$, where m_b^* denotes local band effective mass. Because mobility of charge carriers μ is proportional to m_b^*, μ is nominally unaffected by N_v. Even though μ may be deteriorated by inter-valley scattering, the enhancement in m^* overcomes the loss in μ, resulting in enhanced net ZT.

It should be noted that multiple electronic bands can be activated only when band energy offsets are small enough and comparable to several $k_B T$. In p-type $PbTe$, offset between L and Σ bands equals ~0.15–0.20 eV, and small energy offset makes multiple valence bands easily converged. Typically, several superior p-type $PbTe$ based systems are achieved through manipulating electronic band structures, and typical examples are $PbTe - PbSe$ [48], $PbTe - PbS$ [78, 80], $PbTe - MgTe$ [50], $PbTe - SrTe$ [26, 52], etc. As for valence bands of $PbTe$, N_v at L point is equal to 4, while N_v at Σ point equals to 12 [48]. Therefore, when two valence bands jointly carry charge carriers at a close energy range, m^* can get a dramatic enhancement.

A representative work of band convergence was done by Yanzhong Pei et al. [48]. In this work, a convergence of at least 12 valleys in doped $PbTe_{1-x}Se_x$ alloys was achieved, reaching a remarkable ZT value of 1.8 at about 850 K. As Fig. 6a shows, N_v is equal to 4 and 12 for L and Σ band. By properly doping, one can produce the convergence of many valleys at desired temperatures, and TE performance can be greatly enhanced then. They demonstrate this effect in $PbTe_{1-x}Se_x$, where L and Σ valence bands (Fig. 6b) can be converged, giving increased $N_v = 16$.

3.1.2 DOS Distortion and Resonant State (RL)

Another mechanism to increase in m^* and S is to put distortion energy level into conduction or valence band. The effect of locally increased DOS to improve S can be explained by Mott expression as follows [23]:

$$S = \frac{\pi^2 k_B^2 T}{3q} \left\{ \frac{d\left[\ln\left(\sigma\left(E\right)\right)\right]}{dE} \right\}_{E=E_f} = \frac{\pi^2 k_B^2 T}{3q} \left\{ \frac{1}{n} \frac{dn\left(E\right)}{dE} + \frac{1}{n} \frac{d\mu\left(E\right)}{dE} \right\}_{E=E_f}. \quad (5)$$

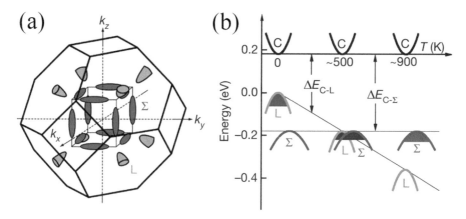

Fig. 6 Valence band structure of $PbTe_{1-x}Se_x$ [48]. (**a**) Brillouin zone showing low degeneracy hole pockets (orange) centered at L point, and high degeneracy hole pockets (blue) along Σ line. Schematic drawing shows 8 half-pockets at L point so that $N_v = 4$, while for Σ band $N_v = 12$. (**b**) Relative energy of valence bands in $PbTe_{0.85}Se_{0.15}$. At ~500 K, two valence bands converge, resulting in transport contributions from both L and Σ bands. C is conduction band; L is low degeneracy hole band; Σ is high degeneracy hole band [48]. (Reproduced with permission [48]. Copyright 2011, Springer Nature Publishing)

From Eq. (5), we can understand that the enhancement in S induced by resonant level arises from two mechanisms: (1) excess DOS $n(E)$ near Fermi level and (2) increased energy dependent $\mu(E)$, known as "resonant scattering" [81]. The first mechanism is intrinsic feature that is slightly temperature dependent, and it will predominate with rising temperature. The second mechanism works well only at cryogenic temperature when phonon–electron scattering is weak. When phonon–electron scattering dominates at elevated temperatures, then "resonant scattering" plays ignorable role in enhancing S [82]. Therefore, the resonant state is very sensitive to temperature and it can only play a significant role in enhancing S near and below room temperature range [81]. Figure 7 schematically presents effects of resonant level on electron energy distribution in $PbTe$ matrix, which could cause sharp increase in DOS near Fermi level. This phenomenon has been observed in p-type Tl-doped $PbTe$ system [83]. When coming to mobility of charge carriers, it should be noted that DOS distortion will cause larger deterioration in mobility of charge carriers than that caused by band convergence. DOS distortion enhances the total m^* through producing larger local band m^*; however, band convergence enhances the total m^* through increase in N_v.

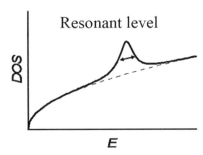

Fig. 7 Schematic presentation how resonant level effects on electron energy distribution in *PbTe* matrix, which could cause a sharp increase in DOS near Fermi level [83]. (Reproduced with permission [83]. Copyright 2011, Springer Nature Publishing)

3.1.3 Band Alignment (BA)

It is well known that introducing defects in a host material not only intensifies phonon scattering, but also deteriorates mobility of charge carriers. When nanostructured second phase exists in the matrix, it is crucial to maintain high mobility of charge carriers. One of the most effective approaches is to introduce exotic phase with small energy band offset compared with the matrix, namely band alignment. Actually, band alignment works very well in solar cell community to maintain mobility of charge carriers [84, 85], which was firstly well elucidated in *PbTe* system [86]. Figure 8a presents schematically band diagram of valence bands alignment between matrix and second phase precipitate [86]. Because energy of valence band edge in *SrTe* is comparable to that of *p*-type *PbTe*, when second phase *SrTe* precipitates out in *p*-type *PbTe* matrix, mobility of charge carriers will be slightly deteriorated, but phonon scattering is intensified by precipitates, which is considered as a typical example of well-known "phonon glass-electron crystal" proposed by Slack [87]. To quantitatively evaluate contribution of band alignment to TE performance, one can calculate the ratio μ/κ_L, as shown in Fig. 8b. The large increase in μ/κ_L indicates that *SrTe* phases in *p*-type *PbTe* matrix scatter phonon stronger than charge carrier.

3.1.4 Summary of BC, RL, and BA in *p*-Type PbTe

It is readily seen that m^* in $PbTe - PbSe$ and $PbTe - MgTe$ systems is larger than Pisarenko plot that is derived from the single parabolic band (SPB) model with $m^* = 0.3\ m_e$, evidencing larger m^* contributed by multiple valence bands (Fig. 9a).

The final enhancements in ZT values for *p*-type *PbTe* by synergistic approaches are well elucidated in Fig. 9b, which reveals that the methods of band convergence, resonant level, and band alignment are pretty effective to enhance electrical transport properties in *p*-type *PbTe*. More importantly, these approaches are also promising to be extensively applied in TE community as a general route.

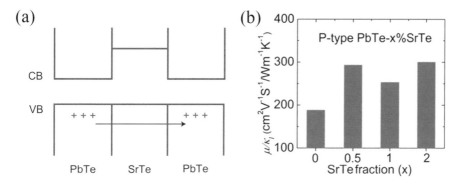

Fig. 8 (**a**) Graphic presentation of valence bands alignment in system *PbTe* matrix and *SrTe* precipitate [27]. (**b**) Ratio mobility of charge carriers to lattice thermal conductivity as function of *SrTe* content in *PbTe – SrTe* [27, 86]. (Reproduced with permission [27]. Copyright 2009, American Chemical Society)

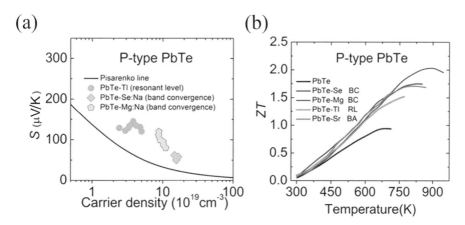

Fig. 9 (**a**) Theoretical Pisarenko plot for single parabolic band model with $m^* = 0.3\ m_e$ at 300 K, and experimental data include *PbTe – PbSe* : *Na* (band convergence) [48], *PbTe – MgTe* : *Na* (band convergence) [50], and *PbTe* : *Tl* (resonant level) [23, 27]. (**b**) ZT of *PbTe – PbSe* : *Na*, *PbTe – MgTe* : *Na*, *PbTe* : *Tl* and *PbTe – SrTe* : *Na* [27] as function of temperature. (Reproduced with permission [27]. Copyright 2009, American Chemical Society)

3.2 Manipulations on Conduction Band Structure in n-Type PbTe

3.2.1 Flattening Band Structure

In *n*-type *PbTe*, conduction band flattening can distinctly enlarge m^*, thereby optimizing S [78]. For a single parabolic band (SPB), local band effective mass m_b^* near band edge is sensitive to band shape. From energy dispersion relationship, m_b^* is defined by the following equation:

$$m_{\mathrm{b}}^{*} = \hbar^{2} \left(\frac{\partial^{2} E(k)}{\partial k^{2}} \right)^{-1}, \tag{6}$$

where $\hbar$ is reduced Planck constant, $E(k)$ and k denote energy dispersion function and wave vector, respectively, in reciprocal space. Eq. (6) demonstrates that flat band shape will lead to high m^*, as shown in Fig. 10a. Experimentally, alloying of *PbTe* with *PbS* and *MnTe* could produce *n*-type material with obvious band flattening and, thereby, with enhanced m^* (Fig. 10b). Typically, ZT max values of *n*-type *PbTe* are largely boosted from ~1.1 to ~1.4 at 923 K and ~ 1.6 at 723 K through flattening conduction bands [64, 65]. In fact, as suggested by Kane band model, energy dispersion is closely related to bandgap E_g, and broadening E_g will flatten conduction band in *PbTe* following the relationship below [88, 89]:

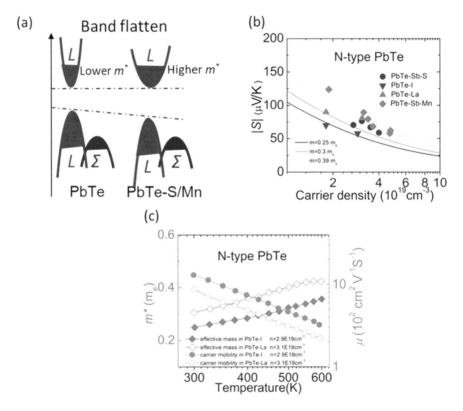

Fig. 10 (**a**) Schematic band flattening after alloying of *PbTe* with *PbS* and *MnTe* in *n*-type material; (**b**) Pisarenko plots for single parabolic band model with different m^* and experimental data for *PbTe* based materials with different doping and alloying; (**c**) comparisons temperature dependences of m^* and μ for *I* and *La*-doped *n*-type *PbTe* [27]. (Reproduced with permission [27]. Copyright 2009, American Chemical Society)

$$\frac{\hbar^2 k^2}{2m_b^*} = E\left(1 + \frac{E}{E_g}\right). \tag{7}$$

It is noteworthy to point out that μ will also be inevitably deteriorated due to the relationship of $\mu \propto 1/m_b^*$ after conduction band flattening. Therefore, it should be careful to balance m^* and μ. Typically, combination of lower m^* and higher μ in Fig. 10c leads to high ZT max value in *I* doped *PbTe* (~ 1.4 at 723 K), which is 40% higher than that in *La* doped *PbTe* (~ 1.0 at 723 K) [90].

3.2.2 Deep Impurity Level

Impurity levels caused by adding group IIIA elements *Ga*, *In*, *Tl* in *PbTe* are systematically investigated [91, 92]. At room temperature, *Ga* and *In* impurities can form deep impurity level in bandgap. On the contrary, *Tl* impurity level will enter into valence band to form a resonant level in *p*-type *PbTe*, as we discussed above [23]. Figure 11a describes schematically deep impurity level in *n*-type *PbTe* produced by doping with *Ga* and *In*. Interestingly, *Ga* and *In* (+3) substitutions on *Pb* (+2) sites do not boost concentration of charge carriers in *PbTe*. At low temperature, energy levels produced by *Ga* and *In* impurities could trap free electrons and work as charge reservoir [64, 93]. With rising temperature, trapped electrons will be released from deep impurity levels into conduction band, finally increasing in concentration of charge carriers as shown in Fig. 11b. In *In* doped *n*-type *PbTe*, concentration of charge carriers undergoes an order of magnitude enhancement from ~5 × 10^{18} cm^{-3} at 300 K to ~2.8 × 10^{19} cm^{-3} at 773 K in Fig. 11c, exhibiting a kind of dynamic doping behavior. As for *PbTe* with single conduction band, this anomalous variation in concentration of charge carriers is related to changing charge states of *Ga* and *In*. With increase in temperature, mixed charge states (+1 and +3) of *Ga* and *In* will change to fully +3 states [91]. Consequently, dynamic doping extends ZT max on larger temperature range, resulting in large average ZT (ZT$_{ave}$) [64, 93]. ZT over the entire working temperature range is important, as it determines TE conversion efficiency [93]. Similarly, *Cr* [94], *Fe* [95], *Ti* [96], *Sc* [97] etc. can introduce impurity states in *PbTe*; however, these resonant impurity levels within conduction band are localized and contribute less to transport of charge carriers.

3.2.3 Summary of Flattening Band Structure and Deep Impurity Level

From Fig. 12, we can find that both flattening band structure and deep impurity level have a positive effect toward improving max ZT and ZT$_{ave}$ (pure *PbTe* has ZT max of less than 1). However, the flattened band structure achieves better TE performance. The reason may be that impurities work as scatter centers and thus deteriorate electric conductivity to some extent.

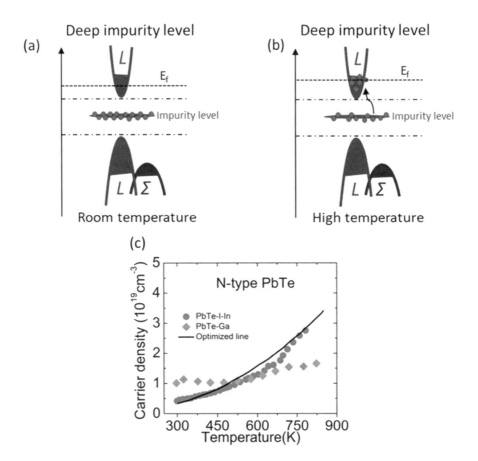

Fig. 11 Band diagram of *n*-type *PbTe* with deep impurity level: (**a**) deep impurity level localizes within bandgap and traps electrons at 300 K; (**b**) deep impurity level releases electrons into conduction band at elevated temperature; (**c**) concentration of charge carriers in both *In* and *Ga* doped *n*-type *PbTe* increases with increase in temperature [27]. (Reproduced with permission [27]. Copyright 2009, American Chemical Society)

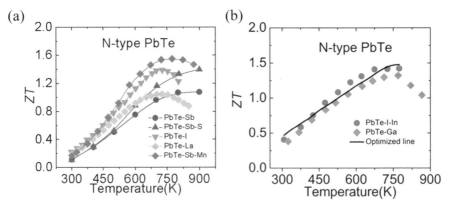

Fig. 12 (**a**) ZT versus temperature in *I* and *La* doped *PbTe*, *n*-type $PbTe_{1-x}S_x$ and $Pb_{1-x}Mn_xTe$ [27]. (**b**) ZT versus temperature in *In* and *Ga*-doped *n*-type *PbTe* [27]. (Reproduced with permission [27]. Copyright 2009, American Chemical Society)

4   Defects Engineering to Lower Lattice Thermal Conductivity

Apart from electrical transport properties, low κ is required to preserve a large temperature gradient, which is vital to realize high TE conversion efficiency. Heat transport is regarded as elementary vibrational motion, and atom oscillation can be quantitatively evaluated using quantum description "phonon." Phonon propagation and dispersion in crystalline materials are closely related to mean free path (MFP), which ranges from atomic-scale, nanoscale to mesoscale. Since MFP of *PbTe* is predominated in nanoscale, nano-structuring is widely applied to reduce κ_L in *PbTe* [6, 98]. When defect sizes are comparable to MFP range, phonon scattering can be dramatically intensified and, therefore, lead to very low κ_L [25, 67]. Based on this idea, artificially designing defects with all-scale hierarchical architectures are put forward to considerably reduce κ_L. In this part, defects are sorted by size.

Theoretical calculations of κ_L are based on modified Callaway's model [99], κ_L is given by formula:

$$\kappa_L = \frac{k_B}{2\pi^2 v}\left(\frac{k_B T}{\hbar}\right)^3 \int_0^{\theta_D/T} \tau_C \frac{e^x}{\left(e^x - 1\right)^2} x^4 dx, \tag{8}$$

where θ_D is Debye temperature, v is average phonon-group velocity, T is working temperature, $x = \dfrac{\hbar\omega}{k_B T}$ is relaxation time due to normal phonon–phonon scattering and τ_C is combined relaxation time. The latter is obtained by integrating relaxation times from various processes. Based on transmission electron microscopy (TEM) studies, for the same frequency, relaxation time depends mainly on scattering from nanoscale precipitates, dislocations, boundaries, and phonon–phonon interactions. The overall relaxation time τ_C is:

$$\frac{1}{\tau_C} = \frac{1}{\tau_U} + \frac{1}{\tau_N} + \frac{1}{\tau_B} + \frac{1}{\tau_S} + \frac{1}{\tau_D} + \frac{1}{\tau_{NP}} + \frac{1}{\tau_{PD}}, \tag{9}$$

where $\tau_U, \tau_N, \tau_B, \tau_S, \tau_D, \tau_{NP}, \tau_{PD}$ are relaxation times corresponding to scattering from Umklapp processes, normal processes, boundaries, strains, dislocations, nanoprecipitates, and point defects. Based on Eq. (9) and parameters obtained from TEM observations (average precipitate size, precipitate density, and dislocation density), one can calculate corresponding relaxation times. It is obvious that the higher τ_C contributes to higher κ_L, which is detrimental to ZT value.

4.1   3D Defects

Here, we sort 3D defects into phase separation and precipitates. Difference between these two are mainly in volume fraction and size of second phase. 3D defects can decrease in κ_L if second phase has lower κ_L than *PbTe*. Meanwhile, 3D defects will certainly bring additional 2D grain boundaries into matrix, which can further decrease in κ_L by scattering the phonons.

4.1.1 Phase Separation

If second phase has large volume fraction and comparable size with *PbTe* grain, we term it phase separation.

H.J. Wu et al. [80] showed that $PbTe_{0.7}S_{0.3}$ phase separates (often referred to as spinodal decomposition) into *PbTe* and *PbS* phases [99], and also showed the coexistence of *PbTe* and *PbS* phases (Fig. 13a). Through powder processing, mesoscale *PbTe* grains (1.2 μm on average, 0.7–1.9 μm in size (Fig. 13b) and *PbS* grains (400 nm on average, size in the range of 100–800 nm) are produced in all samples. Energy-dispersive X-ray spectroscopy (EDS) line scanning and mapping were utilized to further confirm the separation of *PbTe*/*PbS* phases (Fig. 13c–h).

Figure 14 is low magnification bright field image of *PbTe*– $PbSnS_2$ (14 mol. %) sample that shows lamellar microstructure with two types of contrast, dark and bright. This is another example of phase separation in *PbTe* TE material [100].

4.1.2 Precipitates

Fig. 15a, b [86] show typical low magnification TEM images of *PbTe*– *SrTe* samples containing 1 mol. % and 2 mol. % *SrTe*. Both images show numerous regular precipitates having sizes in the range of 5–15 nm with dark (diffraction) contrast. From single-electron diffraction pattern shown in the inset of Fig. 15b we can conclude that *PbTe* matrix and *SrTe* nanocrystals have similar symmetry, structure and lattice parameters, and corresponding crystallographic planes and directions are completely aligned in three dimensions. Indeed, *PbTe* and *SrTe* have similar bulk lattice parameters of 6.453 and 6.660 Å, respectively [101]. Therefore, *SrTe* nanocrystals are endotaxially placed in *PbTe* matrix. Furthermore, STEM investigations reveal high density of small precipitates (1–2 nm). Although it is difficult to quantitatively determine composition of individual precipitates owing to overlap with the matrix, EDS indicates increase in *Sr* signal from precipitates (dark areas in STEM image) compared with the matrix regions. The presence of *SrTe* nanoscale precipitates in *PbTe* matrix was also confirmed by additional STEM analyses of two control samples: *PbTe* with 2 mol. % *SrTe* but no Na_2Te and *PbTe* with 1 mol. % Na_2Te but no *SrTe*. Unfortunately, phase diagram is unknown for *PbTe*–*SrTe* system, so it is not possible to suggest thermodynamic explanation of formation of *SrTe* nanocrystals in *PbTe* matrix at this moment. TEM studies reveal that precipitates number in *PbTe* with 2 mol. % *SrTe* and in *PbTe* with 2 mol. % *SrTe* and 1 mol. % Na_2Te are quite similar, whereas *Sr* free *PbTe*–Na_2Te sample contains very few detectable precipitates.

In order to analyze defects or strain distribution at boundaries between precipitates and *PbTe* matrix, high-resolution TEM (HRTEM) investigations performed on

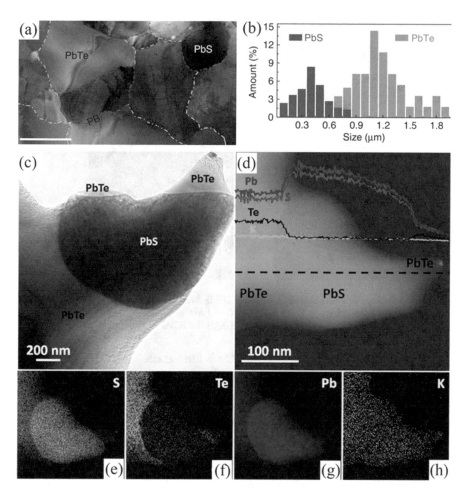

Fig. 13 Microstructures for mesoscale grains and grain/phase boundaries. (**a**) Low-magnification TEM image reveals grains and separated phases. Scale bars, 1 μm; (**b**) grain/phase size distribution histogram; (**c**) Bright-field image showing *PbS* and *PbTe* grains; (**d**) HAADF (*high-angle annular dark-field*) image showing *PbS* and *PbTe* regions with different contrast, with inserted EDS line scanning; (**e–h**) EDS elemental mapping for elements *S, Te, Pb, K*, respectively [80]. (Reproduced with permission [80]. Copyright 2014, Springer Nature Publishing)

PbTe sample containing 2 mol. % *SrTe* and 1 mol. % *Na₂Te*. Figure 15c shows typical phase-contrast HRTEM images containing lattice fringes of several 2–4 nm precipitates with typical interfacial boundary (1 nm dark contrast) between matrix and precipitate. one of precipitate has been enlarged in Fig. 15d. To analyze the presence of elastic and plastic strain, image in Fig. 15c was subjected to geometric phase analysis [102], which is a lattice image-processing method for semi-quantitative spatially distributed strain field analysis. The geometric phase analysis was used to investigate variation in lattice parameter and thus the strain at and around boundaries. To reduce potential artifacts of the strain analysis, it is necessary to obtain

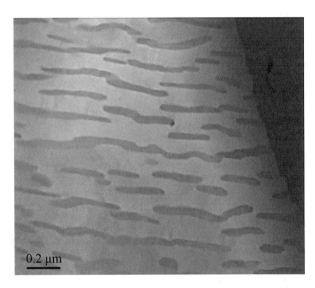

Fig. 14 Low-magnification scanning TEM (STEM) image showing *PbSnS₂* networks (dark regions) extending primarily at grain boundaries of *PbTe* in *PbTe– PbSnS₂* (14 mol. %) [100]. (Reproduced with permission [100]. Copyright 2012, John Wiley & Sons)

high-quality, clear lattice images. Figure 15e is filtered HRTEM image, which indicates at lattice distortion. Figure 15f shows the shear strain map profiles ε_{xy} of precipitates. From strain map distribution in this image, it appears that elastic strain is pervasive in and around all precipitates, although there is additional plastic strain around dislocation cores in particles iii and iv. We can also see that dark contrast in Fig. 15d is not identical to strain region in Fig. 15e. This is due to diffraction contrast. Dislocations (and associated plastic strain) are also observed at interfaces in larger precipitates. Burger's circuit around dislocation core yields a closure failure with projected vector 1/2 *a*[011]. We note that high-density misfit dislocations appear in many larger size precipitates in *PbTe* containing 2 mol. % *SrTe* and 1 mol. % *Na₂Te*, suggesting that density of dislocations is about double density of *SrTe* nanocrystals.

Another interesting feature in *PbTe– PbS* system is that *Na* dopant can control morphology of nanoparticles by tuning ratio of *PbS/Na* [101]. This reduces κ_L and simultaneously lowers Fermi level enough to allow conductivity via two valence band, which enhances power factor [79, 103]. Figure 16a is a low-mag image in which precipitates are homogeneously distributed. Figure 16b-d shows images of faceted nanoparticles, which are some typical shapes in *Na* 1 at. % doped *PbTe– PbS* 12 mol. % sample. Likely, this is entirely new strategy, and possibly a new way to control nanostructure morphology in (pseudo-) binary systems, that is, by partitioning ternary additive to nominally (pseudo-) binary system. Conceptually, *Na* in this system appears to play a role akin to surfactant capping ligand in conventional

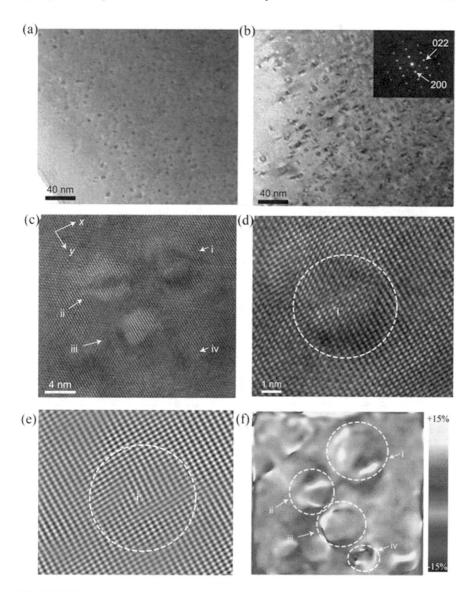

Fig. 15 TEM and strain analysis of *PbTe– SrTe*. (**a**, **b**) Low-magnification TEM images of *PbTe* containing 1 mol. % *SrTe* (**a**) and 2 mol. % *SrTe* (**b**) (both doped with 1 mol. % *Na₂Te*). The inset in (**b**) shows corresponding electron diffraction pattern, which confirms crystallographic alignment of *SrTe* and *PbTe* lattices. (**c**) HRTEM phase-contrast image of several endotaxial nanocrystals of *SrTe* in *PbTe* matrix. (**d**) enlarged area of i in (**c**). (**e**) filtered image of (**d**). (**f**) The shear strain distribution of nanoscale inclusions showing elastic strains at and around all precipitates and the presence of plastic strain at and around dislocation cores in precipitates iii and iv. The color bar indicates 15 to −15% strain [86]. (Reproduced with permission [86]. Copyright 2011, Springer Nature Publishing)

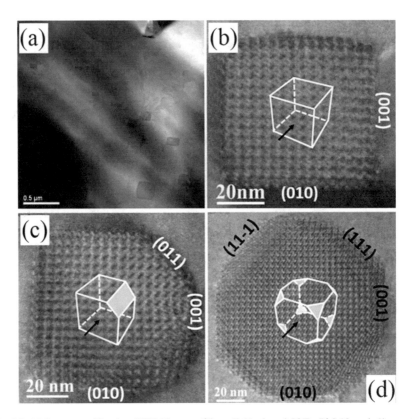

Fig. 16 (**a**) Low-magnification STEM image of 2 at. % *Na* doped *PbTe–PbS* 12 mol. % samples showing a regular shape *PbS* precipitates. (**b–d**) Intermediate magnification TEM images of 1 at. % *Na* doped *PbTe–PbS* 12 mol. % sample, which display three typical projections of *PbS* precipitates. Inserts are schematic of corresponding precipitate morphology in 3D space, and black arrowheads are TEM view directions [104]. (Reproduced with permission [104]. Copyright 2013, Elsevier)

nanocrystal formation from organic solutions such as in *CdSe*, *PbSe*, among others. By controlling the nature of capping ligand and by combining more than one capping ligand, it has been possible to control size and morphology of these materials with highly exotic morphologies. In solid-state systems, such as *PbTe– PbS*, use of *Na* and other dopants separately and in combination may open new pathways for similar size and morphology control.

4.2   2D Defects

4.2.1   Grain Boundary

In principle, all grain boundaries, no matter between the same or different phases, should be counted as 2D defects. So, for polycrystal TE materials, 2D defects are common.

Grain boundary phonon scattering has been shown to be important in improving TE performance of *PbTe* TE materials. Boundary scattering in polycrystalline materials with grain size can be estimated from equation [105, 106]:

$$\frac{1}{\tau_B} = \frac{V_g}{d},$$

(10)

where V_g is phonon group velocity, d is grain or microstructure size. It is easy to find $d \propto \tau_B$, which means that the larger crystal size contributes to larger κ_L.

Figure 17 illustrates an example of grains [6]. Figure 17a shows the grain of spark plasma-sintered (SPS) samples *PbTe–SrTe* (4 mol. %) doped with 2 at. % *Na*. The statistics of grain size is 0.1–1.7 μm (Fig. 17b). This grain size can be used to calculate τ_B^{-1} in Eq. (10).

4.2.2 Stacking Fault

Another kind of 2D defect is the stacking fault.

Some grains in *PbTe–MnTe* alloy system show a high density of stacking faults, as shown in Fig. 18a, b [78]. Such faults are nanoscale in width, but microscale in lateral dimensions, and thus could form a strained network for effective scattering

Fig. 17 (**a**) Low-magnification TEM image showing mesoscale grains in SPS 2 mol. % *Na* doped *PbTe–SrTe* (4 mol. %) sample. (**b**) The statistics of grain size. (Reproduced with permission [6]. Copyright 2012, Springer Nature Publishing)

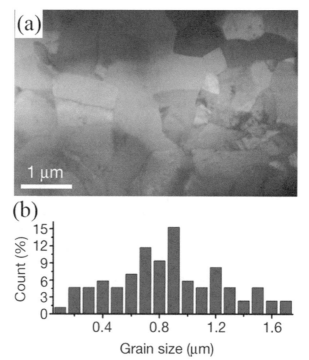

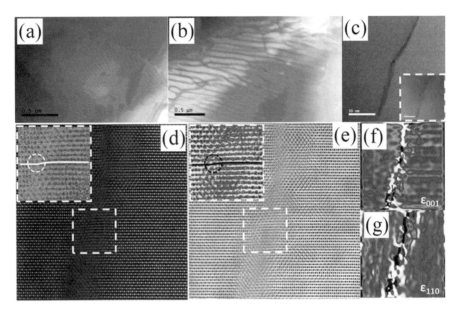

Fig. 18 (**a**) and (**b**) TEM images showing high density of line defects; (**c**) STEM HAADF image of one line defect, with respective STEM ABF image inset; (**d**) and (**e**) simultaneously acquired high-resolution STEM HAADF and ABF images from line defect, with (insets) enlarged images of marked regions; (**f**) and (**g**) GPA strain maps from (**d**) and (**e**) [78]. (Reproduced with permission [78]. Copyright 2018, Royal Society of Chemistry)

phonons from long to medium wavelengths while having little influence on charge carriers transport. Then, we carried out aberration-corrected scanning transmission electron microscopy (Cs-corrected STEM) to look into detailed atomic structure of such defects. Figure 18c and its insets are low-magnification STEM HAADF (high-angle annular dark field) and ABF (annular bright field) images of stacking fault. Simultaneously acquired high-resolution STEM HAADF and ABF images in Fig. 18d, e clearly show the shift in lattice planes across the stacking fault. Geometric phase analysis (GPA) [107], as shown in Fig. 18f, g, shows significant strains associated with the stacking faults.

Stacking fault can also be in second phase. Figure 19a is HRTEM image with electron beam parallel to [110] direction of *PbTe*, which shows one typical feature in *PbSnS₂* as pointed by vertical arrowhead [100]. Dotted boxed area is shown in Fig. 19b. It reveals three parts that are separated by two dashed lines. Careful analysis showed that right and left parts are along [001] direction of *PbSnS₂*; however, middle narrow belt region with width of 3 nm is along [010] direction of *PbSnS₂*. The image simulation based on further analysis inserted in boxed area in Fig. 19b well matches to experimental image. The mismatch between [001] direction and [010] direction of *PbSnS₂* is around 5%, which also can release the strain by formations of some dislocations in *PbSnS₂*.

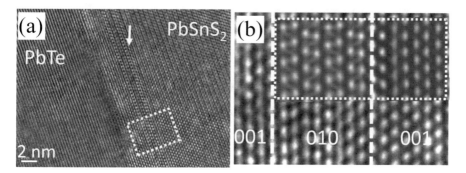

Fig. 19 (**a**) One lattice image of *PbTe/PbSnS₂* boundary along [001] direction of *PbTe* clearly showing one small region different from another regions along [010] direction in *PbSnS₂*. (**b**) Enlarged image in highlighted region of the image (Reproduced with permission [100]. Copyright 2012, John Wiley & Sons)

4.2.3 2D Precipitates

Figure 20 is representative HRTEM image of dark line-like projections in sample *PbTe– K* (1.25 at. %)– *Na* (0.8 at. %), obtained with electron beam parallel to [001], [010] and [111]. As Fig. 20a shows, precipitates that appear as two perpendicular dark lines can clearly be observed along [001] direction. The inserted image taken along [010] direction shows two precipitates with different shapes: one that has dark-line shape, while the other has square shape. Figure 20b, c shows typical HRTEM images of two samples along [011] and [111] directions, respectively. The random shape precipitates show crystallographic coherent relationship with *PbTe* matrix. To analyze the possible strain at or close to precipitate/matrix interface, some regular circular-shaped precipitates were studied by geometric phase-analysis (GPA) [102], which is lattice image-processing method for strain field analysis. Most precipitates show zero or very little strain at the boundaries. Misfit dislocations at the interfaces were observed for very few precipitates.

4.3 1D Dislocations

The scattering mechanism from dislocations can be divided into scattering from dislocation cores and that from dislocation strain [82]; hence, mechanisms have separate phonon relaxation times τ_{DC} [109–112] and τ_{DS} [109, 110, 112, 113], respectively (see Eqs. (11), (12), and (13)). Because array of dislocations at grain boundaries can be physically treated as collection of single dislocations inside grain [111], we include τ_{DC} and τ_{DS} to calculate τ_D [82]:

$$\frac{1}{\tau_D} = \frac{1}{\tau_{DC}} + \frac{1}{\tau_{DS}},\qquad(11)$$

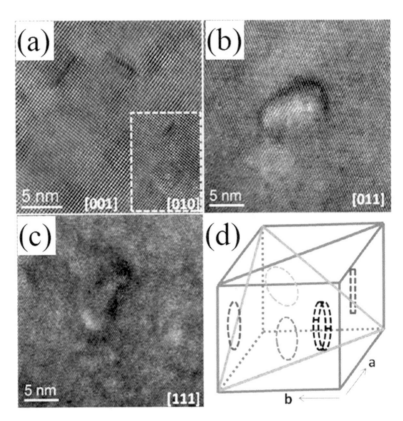

Fig. 20 (**a**) High-magnification lattice image along [001] direction of *PbTe– K* (1.25 at. %)– *Na* (0.8 at. %) depicting two platelet-like precipitates (dark lines). The insert corresponds to analysis along [010] direction and shows both dark line-like and square precipitate. (**b**) and (**c**) Random-shaped precipitates along [011] direction of *PbTe– K* (1.25 at. %)– *Na* (0.6 at. %) and [111] direction of *PbTe– K* (1.25 at. %)– *Na* (0.4 at. %), respectively. (**d**) Schematic diagram describing observed shape of platelet-like precipitate as projected along [100] (red), [110] (green), and [111] (blue) directions [111]. (Reproduced with permission [108]. Copyright 2012, American Chemical Society)

$$\frac{1}{\tau_{DC}} = N_D \frac{\overline{v}^{-4/3}}{v^2} \omega^3, \tag{12}$$

$$\frac{1}{\tau_{DS}} = 0.6 \times B_D^2 N_D (\gamma + \gamma_1)^2 \omega \left\{ \frac{1}{2} + \frac{1}{24} \left(\frac{1-2r}{1-r} \right)^2 \times \left[1 + \sqrt{2} \left(\frac{v_L}{v_T} \right)^2 \right]^2 \right\}, \tag{13}$$

where N_D, B_D, γ, γ_1, r, v_L, and v_T are dislocation density, effective Burger's vector, Grüneisen parameter, change in Grüneisen parameter, Poisson's ratio, longitudinal phonon velocity, and transverse phonon velocity, respectively. The scatter of dislocation cores requires N_D only.

In K doped $PbTe_{0.7}S_{0.3}$ system, dislocations can be found at the boundary between PbS/PbS grains and $PbS/PbTe$ phases. Figure 21a is HRTEM image of boundary between PbS/PbS grains (marked with yellow dashed line) along [001] zone axis, and inserted electron diffraction (ED) pattern indicates its small-angle feature. Array of edge dislocations marked with green dashed circles can be seen on the boundary (tilt grain boundary in this case) between two slightly misaligned grains. The enlarged original and Fourier-transformed images of the same dislocation core (marked with bright dashed square) are inserted in Fig. 21a. To investigate the strain variation around dislocation cores, high-quality HRTEM image was analyzed through geometric phase analysis [102, 110]. The profile of the strain map of dislocation core is also inserted in Fig. 21a.

TEM observations also reveal the presence of misfit dislocations and incompletely relaxed strains at boundaries of $PbTe/PbS$ phases, marked with red dashed line in Fig. 21b because of large lattice mismatch (about 6%) between $PbTe$ and PbS [79]. ED pattern at phase boundary (inset in Fig. 21b) reflects lattice alignment between $PbTe$ and PbS regions. Figure 21c, d are schematic figures showing

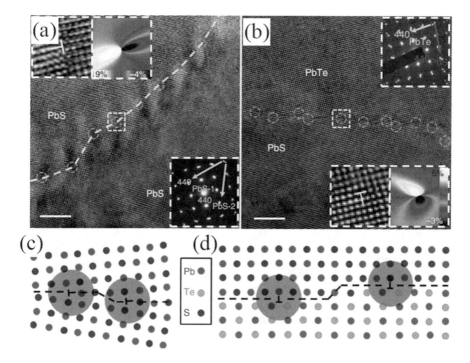

Fig. 21 Microstructures for mesoscale grains and grain/phase boundaries. (**a**) and (**b**) HRTEM images of boundary between PbS/PbS grains (**a**) and $PbTe/PbS$ phases (**b**), marked with yellow and red dashed line, respectively. Edge dislocations along boundaries are marked with green dashed circles. ED pattern, enlarged dislocations and GPA analysis of images are inserted. Scale bars, 5 nm. (**c**) and (**d**) Schematics indicating grain and phase boundaries, respectively [50]. (Reproduced with permission [50]. Copyright 2013, American Chemical Society)

dislocations and strain on boundaries between *PbS/PbS* grains and *PbS/PbTe* phases. Edge dislocations at the grain boundary are formed to relieve the strain between misaligned grains and thus those extra semi-atomic planes are along the boundary. On the contrary, edge dislocations at the phase boundary are formed to offset the lattice mismatch between two distinct phases and thus those extra semi-atomic planes are perpendicular to the boundary. Despite the difference between these two types of boundaries, both help scatter long wavelength phonons. Meanwhile, atomic-scale distortions at grain/phase boundaries can act as effective scattering centers for short wavelength phonons.

Another example is *Se*-doped *PbTe* bulk material [114]. Figure 22a shows HRTEM image with nano-precipitates and Fig. 22b gives inverse fast Fourier transform (IFFT) of yellow dotted square zone showing dislocations by ⊥ symbol.

4.4   0D Point Defects

Scattering by point defects arises from both mass and strain contrast within crystal lattice [115]. In simple case of alloying on single crystallographic site, τ_{PD} is given by:

$$\frac{1}{\tau_{PD}} = \frac{v\omega^4}{4\pi v_p^2 v_g}\left\{\sum_i f_i\left(1-\frac{m_i}{\bar{m}}\right)^2 + \sum_i f_i\left(1-\frac{r_i}{\bar{r}}\right)^2\right\}, \quad (14)$$

here, f_i is fraction of atoms with mass m_i and radius r_i that reside on site with average mass $\bar{m}$ and radius $\bar{r}$ respectively [109, 116]:

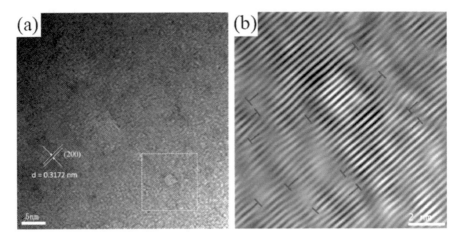

Fig. 22 (a) HRTEM image showing nanoprecipitates and (b), IFFT of yellow dotted square zone showing dislocations by ⊥ symbol. (For interpretation of the references to color in this figure legend, see [114]). (Reproduced with permission [114]. Copyright 2017, Elsevier)

Here, we found that point defects can improve τ_{PD}^{-1} and then decrease in κ_L. The larger the mass and radius differences, the more remarkable the scattering effect.

4.4.1 Random Distributed Doped Atoms in Solid Solution

For any kinds of doping, there will be a certain percentage of doped atoms in solid solution. Unfortunately, randomly distributed atoms in solid solution cannot be imaged in TEM/STEM due to the lack of atomic column.

4.4.2 Ordered Point Defects in the Matrix

Dopant atoms ordered in *PbTe* matrix forming atom columns can be imaged [109]. Two kinds of ordered point defects are reported so far. Figure 23 a1–d1 show crystal structure of *MnTe* doped *PbTe*, while Fig. 23 a2–d2 show corresponding GPA analysis. HAADF image is taken from [110] direction. Some ordered interstitial atom columns have been circled by black rings (Fig. 23a1–d1). GPA analysis is showing that there are intensely stress fields in this local region.

Another example is in *Cu₂Te* doped *PbTe* system [118]. Similar to *MnTe* doped *PbTe*, ordered atom columns are observed in [110] zone axis. Ordered *Cu* atoms are circled by green rings, Fig. 24a. Figure 24b shows cluster of ordered *Cu* atoms. GPA result shows that *Cu* ordered atoms produce local stress field alone (002) and (220) crystal plane. At lower magnification, high density of *Cu* interstitial clusters are observed (Fig. 24c). The enlarged area (Fig. 24d) shows high density of these *Cu* interstitial clusters. GPA strain analysis of Fig. 24c using (002) and (220) reflections

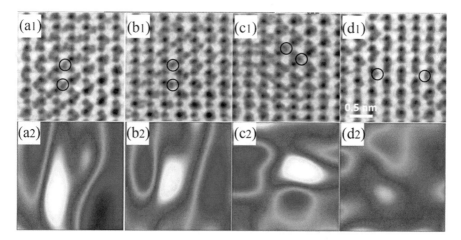

Fig. 23 (**a1**)–(**d1**) Enlarged STEM ABF images of interstitial clusters, where interstitials locate in different positions and (**a2**)–(**d2**) GPA strain analysis of (**a1**)–(**d1**) [117]. (Reproduced with permission [117]. Copyright 2018, Royal Society of Chemistry)

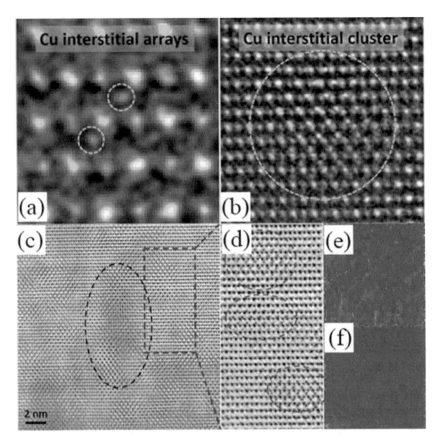

Fig. 24 Microstructures of atomic vacancies and interstitials. (**a**) Enlarged image (colorized) showing *Cu* interstitial arrays. (**b**) Enlarged image (colorized) showing *Cu* interstitial clusters. (**c**) ABF image shows high density of *Cu* interstitial clusters. (**d**) Further enlarged image of blue-square marked region showing three *Cu* interstitial clusters. (**e**) GPA strain analysis of (**c**) using (002) and (220) reflections as primitive vectors. (**f**) GPA strain analysis of (**c**) using two (111) reflections as primitive vectors. The mechanism of ordered point defect may different from the random one, which needs a further calculation to illustrate [118]. (Reproduced with permission [118]. Copyright 2017, American Chemical Society)

as primitive vectors are shown in Fig. 24e. GPA strain analysis (Fig. 24c) using two (111) reflections as primitive vectors is shown in Fig. 24f. One can easily find that the stress field is along (002) and (220) planes.

4.5 All Scale Hierarchical Architectures [104]

Varied contributions to phonon scattering by structures at all length scales (i.e. atomic-, nano- and mesoscale), to various mean free paths have been calculated for *PbTe* [119, 120], *PbTe*$_{1-x}$*Se*$_x$ [119], and *Si* [121], shown in Fig. 25. For these

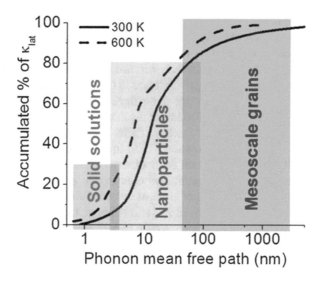

Fig. 25 Contributions of phonons with varied mean free paths to accumulated κ_L value for *PbTe* at different temperatures. Phonons with short, medium, and long mean free paths can be effectively scattered by atomic-scale point defects, nanoscale precipitates, and mesoscale grain boundaries, respectively [104]. (Reproduced with permission [104]. Copyright 2013, American Chemical Society)

specific systems, it indicates that around 25% of κ_L value of *PbTe* is contributed by phonon modes with mean free paths of less than 5 nm, which can be primarily attributed to scattering by combination of atomic-scale alloying and/or dislocations, and about 55% of one is given by phonon modes with mean free paths between 5 and 100 nm, which can be scattered by nanoscale particles embedded in *PbTe* and associated interphase interfaces and spatially distributed strain [120] (Fig. 25). The remaining 20% of κ_L in *PbTe*, however, is contributed by phonon modes with mean free paths of 0.1–1 mm. The mesoscale grain structure is comparable in size to mean free path and thus can scatter notable fraction of these additional phonons. Therefore, structures at all length-scales in one bulk material can strongly scatter broader spectrum of heat carrying phonons [104].

5 Conclusion and Perspectives

As a classic thermoelectric material, *PbTe* system has been intensively studied in the recent 20 years. Here, we summarized three main strategies that were successfully applied in *PbTe* based TE materials: (1) engineering of concentration of charge carriers to peak ZT in *PbTe*; (2) band engineering to optimize σ; and (3) defects engineering to lower κ_L. These three methods are examined by experiment and theory and proved to be effective. Synergistically manipulating these three methods in one material is a promising way to improve TE performance. However, due to the correlation between concentration of charge carriers, band structure and crystal structure, simply adjusting one parameter will usually deteriorate the others, which is the key challenge to apply multiple strategies in one material. Besides, theoretical and experiential breakthrough toward these three methods is still available. For example, due to the complexity, formation mechanism of defects in *PbTe* system such as 0D point clusters and 2D stacking faults have not been fully understood using STEM, although atomic resolution STEM images have been offered. In addition, precise calculations for thermal conductivity by phonons have not been fulfilled, and this restricts direct quantitative comparison of contribution to lattice thermal conductivity among different kinds of defects. With these problems solved, design of TE material will be more rational and efficient. Considering the complexity of TE materials, future development would benefit greatly from close collaborations between chemists, physicists, and material scientists. With increasing shortage of conventional fossil fuels, TE material including *PbTe* system will play more and more important role in energy conversion engineering, which will promote more investment and accelerate the development in this area.

References

1. J.R. Sootsman et al., Large enhancements in the thermoelectric power factor of bulk PbTe at high temperature by synergistic nanostructuring. Angew. Chem. **120**, 8746–8750 (2008)
2. J.R. Sootsman, D.Y. Chung, M.G. Kanatzidis, New and old concepts in thermoelectric materials. Angew. Chemie - Int. Ed. Engl **48**, 8616–8639 (2009)
3. Y. Gelbstein, Z. Dashevsky, M.P. Dariel, High performance n-type PbTe-based materials for thermoelectric applications. Phys. B Condens. Matter **363**, 196–205 (2005)
4. J. Vazquez-Arenas et al., Theoretical and experimental studies of highly active graphene nanosheets to determine catalytic nitrogen sites responsible for the oxygen reduction reaction in alkaline media. J. Mater. Chem. A **4**, 976–990 (2016)
5. K. Kishimoto, T. Koyanagi, Preparation of sintered degenerate n-type PbTe with a small grain size and its thermoelectric properties. J. Appl. Phys. **92**, 2544–2549 (2002)
6. K. Biswas et al., High-performance bulk thermoelectrics with all-scale hierarchical architectures. Nature **489**, 414–418 (2012)
7. R.J. Korkosz et al., High ZT in p-Type (PbTe) 1–2 x (PbSe) x (PbS) x thermoelectric materials. J. Am. Chem. Soc. **136**, 3225–3237 (2014)
8. M. Ohta et al., Enhancement of thermoelectric figure of merit by the insertion of MgTe nanostructures in p-type PbTe doped with Na2Te. Adv. Energy Mater. **2**, 1117–1123 (2012)
9. X. Ji, B. Zhang, T.M. Tritt, J.W. Kolis, A. Kumbhar, Solution-chemical syntheses of nanostructured Bi2Te3 and PbTe thermoelectric materials. J. Electron. Mater. **36**, 721–726 (2007)
10. Y. Takagiwa, Y. Pei, G. Pomrehn, G.J. Snyder, Dopants effect on the band structure of PbTe thermoelectric material. Appl. Phys. Lett. **101**, 092102 (2012)
11. Y. Pei, J. Lensch-Falk, … E. T.-A. F. & 2011, undefined. High thermoelectric performance in PbTe due to large nanoscale Ag2Te precipitates and La doping. Wiley Online Libr
12. Y. Takagiwa, Y. Pei, G. P.-A. P. Letters& 2012, undefined. Dopants effect on the band structure of PbTe thermoelectric material. *aip.scitation.org*

13. Y. Cao, T. Zhu, Physics, X. Z.-J. of P. D. A. & 2008, undefined. Low thermal conductivity and improved figure of merit in fine-grained binary PbTe thermoelectric alloys. *iopscience.iop.org*
14. W. Li et al., Promoting SnTe as an eco-friendly solution for p-PbTe thermoelectric via band convergence and interstitial defects. Wiley Online Libr
15. Y. Gelbstein, Z. Dashevsky, Matter, M. D.-P. B. C. & 2005, undefined. High Performance n-type PbTe-based Materials for Thermoelectric Applications. Elsevier
16. T. Harman, D. Spears, Materials, M. M.-J. of E. & 1996, undefined. High Thermoelectric Figures of Merit in PbTe Quantum Wells. Springer
17. T.C. Harman, D.L. Spears, M.J. Manfra, High thermoelectric figures of merit in PbTe quantum wells. J. Electron. Mater. **25**, 1121–1127 (1996)
18. Y.Q. Cao, T.J. Zhu, X.B. Zhao, Low thermal conductivity and improved figure of merit in fine-grained binary PbTe thermoelectric alloys. J. Phys. D. Appl. Phys. **42**, 015406 (2009)
19. Y. Pei, J. Lensch-Falk, E.S. Toberer, D.L. Medlin, G.J. Snyder, High thermoelectric performance in PbTe due to large Nanoscale Ag2Te precipitates and La doping. Adv. Funct. Mater. **21**, 241–249 (2011)
20. W. Li et al., Promoting SnTe as an eco-friendly solution for p-PbTe thermoelectric via band convergence and interstitial defects. Adv. Mater. **29**, 1605887 (2017)
21. X. Ji et al. Solution-chemical syntheses of nano-structured Bi2Te3 and PbTe thermoelectric materials. Springer
22. T. Zhu, Y. Liu, Bulletin, X. Z.-M. R. & 2008, undefined. Synthesis of PbTe Thermoelectric Materials by Alkaline Reducing Chemical Routes. Elsevier
23. J. Heremans, V. Jovovic, E. Toberer, … A. S.- & 2008, undefined. Enhancement of thermoelectric efficiency in PbTe by distortion of the electronic density of states. *science.sciencemag.org*
24. Y. Pei, A. LaLonde, S. Iwanaga, G. S.-E. Environmental & 2011, undefined. High thermoelectric figure of merit in heavy hole dominated PbTe. *pubs.rsc.org*
25. L. Fu et al., Large enhancement of thermoelectric properties in n-type PbTe via dual-site point defects. Energy Environ. Sci. **10**, 2030–2040 (2017)
26. G. Tan et al., Non-equilibrium processing leads to record high thermoelectric figure of merit in PbTe–SrTe. Nat. Commun. **7**, 12167 (2016)
27. Y. Xiao, L.D. Zhao, Charge and phonon transport in PbTe-based thermoelectric materials. npj Quantum Mater. **3**, 55 (2018)
28. G. Tan, L.-D. Zhao, M.G. Kanatzidis, Rationally designing high-performance bulk thermoelectric materials. Chem. Rev. **116**, 12123–12149 (2016)
29. M. Cutler, J.F. Leavy, R.L. Fitzpatrick, Electronic transport in semimetallic cerium sulfide. Phys. Rev. **133**, A1143 (1964)
30. C. Gayner, K.K. Kar, Recent advances in thermoelectric materials. Prog. Mater. Sci. **83**, 330–382 (2016)
31. G.J. Snyder, E.S. Toberer, Complex thermoelectric materials, in *Materials For Sustainable Energy: A Collection of Peer-Reviewed Research and Review Articles from Nature Publishing Group*, (World Scientific, Hackensack, 2011), pp. 101–110
32. E.S. Toberer, A.F. May, G.J. Snyder, Zintl chemistry for designing high efficiency thermoelectric materials. Chem. Mater. **22**, 624–634 (2009)
33. R. Franz, G. Wiedemann, Ueber die Wärme-Leitungsfähigkeit der Metalle. Ann. Phys. **165**, 497–531 (1853)
34. I. Kudman, Thermoelectric properties of dilute PbTe-GeTe alloys. Metall. Trans. **2**, 163–168 (1971)
35. A. Guéguen et al., Thermoelectric properties and nanostructuring in the p-Type materials NaPb18− x Sn x MTe20 (M= Sb, Bi). Chem. Mater. **21**, 1683–1694 (2009)
36. J. Androulakis et al., Nanostructuring and high thermoelectric efficiency in p-type Ag (Pb1−ySny) mSbTe2+ m. Adv. Mater. **18**, 1170–1173 (2006)
37. Y. Zhang, L. Wu, J. Zhang, J. Xing, J. Luo, Eutectic microstructures and thermoelectric properties of MnTe-rich precipitates hardened PbTe. Acta Mater. **111**, 202–209 (2016)

38. Q. Zhang et al., Heavy doping and band engineering by potassium to improve the thermoelectric figure of merit in p-type pbte, pbse, and pbte1–y se y. J. Am. Chem. Soc. **134**, 10031–10038 (2012)

39. K. Biswas et al., High thermoelectric figure of merit in nanostructured p-type PbTe–MTe (M= Ca, Ba). Energy Environ. Sci. **4**, 4675–4684 (2011)

40. Y. Pei, A.D. LaLonde, N.A. Heinz, G.J. Snyder, High thermoelectric figure of merit in PbTe alloys demonstrated in PbTe–CdTe. Adv. Energy Mater. **2**, 670–675 (2012)

41. H.J. Wu et al., Broad temperature plateau for thermoelectric figure of merit ZT> 2 in phase-separated PbTe 0.7 S 0.3. Nat. Commun. **5**, 4515 (2014)

42. H. Wang et al., High thermoelectric performance of a heterogeneous PbTe nanocomposite. Chem. Mater. **27**, 944–949 (2015)

43. H. Wang et al., Right sizes of nano-and microstructures for high-performance and rigid bulk thermoelectrics. Proc. Natl. Acad. Sci. **111**, 10949–10954 (2014)

44. Y. Gelbstein, J. Davidow, Highly efficient functional Ge x Pb 1– x Te based thermoelectric alloys. Phys. Chem. Chem. Phys. **16**, 20120–20126 (2014)

45. K. Ahn et al., Enhanced thermoelectric properties of p-type nanostructured PbTe–MTe (M= Cd, Hg) materials. Energy Environ. Sci. **6**, 1529–1537 (2013)

46. D. Ginting et al., Enhancement of thermoelectric performance in Na-doped Pb0. 6Sn0. 4Te0. 95–x Se x S0. 05 via breaking the inversion symmetry, band convergence, and nanostructuring by multiple elements doping. ACS Appl. Mater. Interfaces **10**, 11613–11622 (2018)

47. Z. Jian et al., Significant band engineering effect of YbTe for high performance thermoelectric PbTe. J. Mater. Chem. C **3**, 12410–12417 (2015)

48. Y. Pei et al., Convergence of electronic bands for high performance bulk thermoelectrics. Nature **473**, 66 (2011)

49. Y.-J. Kim, L.-D. Zhao, M.G. Kanatzidis, D.N. Seidman, Analysis of nanoprecipitates in a Na-doped PbTe–SrTe thermoelectric material with a high figure of merit. ACS Appl. Mater. Interfaces **9**, 21791–21797 (2017)

50. L.D. Zhao et al., All-scale hierarchical thermoelectrics: MgTe in PbTe facilitates valence band convergence and suppresses bipolar thermal transport for high performance. Energy Environ. Sci. **6**, 3346–3355 (2013)

51. Z. Chen et al., Lattice dislocations enhancing thermoelectric PbTe in addition to band convergence. Adv. Mater. **29**, 1606768 (2017)

52. K. Biswas et al., High-performance bulk thermoelectrics with all-scale hierarchical architectures. *nature.com*

53. D. Wu et al., Superior thermoelectric performance in PbTe–PbS pseudo-binary: Extremely low thermal conductivity and modulated carrier concentration. Energy Environ. Sci. **8**, 2056–2068 (2015)

54. E.K. Chere et al., Enhancement of thermoelectric performance in n-type PbTe 1– y Se y by doping Cr and tuning Te: Se ratio. Nano Energy **13**, 355–367 (2015)

55. S.N. Girard et al., PbTe–PbSnS 2 thermoelectric composites: Low lattice thermal conductivity from large microstructures. Energy Environ. Sci. **5**, 8716–8725 (2012)

56. I. Cohen, M. Kaller, G. Komisarchik, D. Fuks, Y. Gelbstein, Enhancement of the thermoelectric properties of n-type PbTe by Na and Cl co-doping. J. Mater. Chem. C **3**, 9559–9564 (2015)

57. P.K. Rawat, B. Paul, P. Banerji, Exploration of Zn resonance levels and thermoelectric properties in I-doped PbTe with ZnTe nanostructures. ACS Appl. Mater. Interfaces **6**, 3995–4004 (2014)

58. K. Ahn et al., Exploring resonance levels and nanostructuring in the PbTe– CdTe system and enhancement of the thermoelectric figure of merit. J. Am. Chem. Soc. **132**, 5227–5235 (2010)

59. Y. Xiao et al., Synergistically optimizing thermoelectric transport properties of n-type PbTe via Se and Sn co-alloying. J. Alloys Compd. **724**, 208–221 (2017)

60. P. Jood et al., Enhanced average thermoelectric figure of merit of n-type PbTe 1– x I x–MgTe. J. Mater. Chem. C **3**, 10401–10408 (2015)

61. D. Ginting et al., High thermoelectric performance due to nano-inclusions and randomly distributed interface potentials in N-type (PbTe 0.93− x Se 0.07 Cl x) 0.93 (PbS) 0.07 composites. J. Mater. Chem. A **5**, 13535–13543 (2017)

62. G. Ding, J. Si, S. Yang, G. Wang, H. Wu, High thermoelectric properties of n-type Cd-doped PbTe prepared by melt spinning. Scr. Mater. **122**, 1–4 (2016)

63. Y. Pei et al., Optimum carrier concentration in n-type PbTe thermoelectrics. Adv. Energy Mater. **4**, 1400486 (2014)

64. Q. Zhang et al., Deep defect level engineering: A strategy of optimizing the carrier concentration for high thermoelectric performance. Energy Environ. Sci. **11**, 933–940 (2018)

65. G. Tan et al., Subtle roles of Sb and S in regulating the thermoelectric properties of N-type PbTe to high performance. Adv. Energy Mater. **7**, 1700099 (2017)

66. Y. Xiao et al., Remarkable roles of Cu to synergistically optimize phonon and carrier transport in n-type PbTe-Cu2Te. J. Am. Chem. Soc. **139**, 18732–18738 (2017)

67. J. Zhang et al., Extraordinary thermoelectric performance realized in n-type PbTe through multiphase nanostructure engineering. Adv. Mater. **29**, 1703148 (2017)

68. Z. Dashevsky, S. Shusterman, M.P. Dariel, I. Drabkin, Thermoelectric efficiency in graded indium-doped PbTe crystals. J. Appl. Phys. **92**, 1425–1430 (2002)

69. V.L. Kuznetsov, L.A. Kuznetsova, A.E. Kaliazin, D.M. Rowe, High performance functionally graded and segmented Bi2Te3-based materials for thermoelectric power generation. J. Mater. Sci. **37**, 2893–2897 (2002)

70. Y. Pei, A.F. May, G.J. Snyder, Self-tuning the carrier concentration of PbTe/Ag2Te composites with excess Ag for high thermoelectric performance. Adv. Energy Mater. **1**, 291–296 (2011)

71. S.A. Yamini et al., Rational design of p-type thermoelectric PbTe: Temperature dependent sodium solubility. J. Mater. Chem. A **1**, 8725–8730 (2013)

72. K. Bergum, T. Ikeda, G.J. Snyder, Solubility and microstructure in the pseudo-binary PbTe–Ag2Te system. J. Solid State Chem. **184**, 2543–2552 (2011)

73. R.F. Brebrick, R.S. Allgaier, Composition limits of stability of PbTe. J. Chem. Phys. **32**, 1826–1831 (1960)

74. Z. Dashevsky, Y. Gelbstein, I. Edry, I. Drabkin, M. Dariel, Twenty-second international conference on thermoelectrics, 421–424 (2003)

75. S.A. Yamini et al., Heterogeneous distribution of sodium for high thermoelectric performance of p-type multiphase lead-chalcogenides. Adv. Energy Mater. **5**, 1501047 (2015)

76. Y. Pei et al., Stabilizing the optimal carrier concentration for high thermoelectric efficiency. Adv. Mater. **23**, 5674–5678 (2011)

77. I. Yu, B.A.E. Ravich, I.A. Smirnov, *Semiconducting Lead Chalcogenides* (Plenum Press, New York, 1970)

78. Y. Xiao et al., Realizing high performance n-type PbTe by synergistically optimizing effective mass and carrier mobility and suppressing bipolar thermal conductivity. Energy Environ. Sci. **11**, 2486–2495 (2018)

79. S.N. Girard et al., High performance Na-doped PbTe–PbS thermoelectric materials: Electronic density of states modification and shape-controlled nanostructures. J. Am. Chem. Soc. **133**, 16588–16597 (2011)

80. H.J. Wu et al., Broad temperature plateau for thermoelectric figure of merit ZT>2 in phase-separated PbTe0.7S0.3. Nat. Commun. **5**, 4515 (2014)

81. J.P. Heremans, B. Wiendlocha, A.M. Chamoire, Resonant levels in bulk thermoelectric semiconductors. Energy Environ. Sci. **5**, 5510–5530 (2012)

82. S.I. Kim et al., Dense dislocation arrays embedded in grain boundaries for high-performance bulk thermoelectrics. Science (80-.) **348**, 109–114 (2015)

83. W.J. Parker, R.J. Jenkins, C.P. Butler, G.L. Abbott, Flash method of determining thermal diffusivity, heat capacity, and thermal conductivity. J. Appl. Phys. **32**, 1679–1684 (1961)

84. J.P.C. Baena et al., Highly efficient planar perovskite solar cells through band alignment engineering. Energy Environ. Sci. **8**, 2928–2934 (2015)

85. C.-H.M. Chuang, P.R. Brown, V. Bulović, M.G. Bawendi, Improved performance and stability in quantum dot solar cells through band alignment engineering. Nat. Mater. **13**, 796 (2014)
86. K. Biswas et al., Strained endotaxial nanostructures with high thermoelectric figure of merit. Nat. Chem. **3**, 160–166 (2011)
87. C.M. Bhandari, D.M. Rowe, *CRC Handbook of Thermoelectrics*, vol 49 (CRC Press, Boca Raton, 1995)
88. Y. Pei, H. Wang, G.J. Snyder, Band engineering of thermoelectric materials. Adv. Mater. **24**, 6125–6135 (2012)
89. E.O. Kane, Band structure of indium antimonide. J. Phys. Chem. Solids **1**, 249–261 (1957)
90. Y. Pei, A.D. LaLonde, H. Wang, G.J. Snyder, Low effective mass leading to high thermoelectric performance. Energy Environ. Sci. **5**, 7963–7969 (2012)
91. B.A. Volkov, L.I. Ryabova, D.R. Khokhlov, Mixed-valence impurities in lead telluride-based solid solutions. Physics-Uspekhi **45**, 819–846 (2002)
92. V.I. Kaidanov, Y.I. Ravich, Deep and resonance states in AIV BVI semiconductors. Physics-Uspekhi **28**, 31–53 (1985)
93. X. Su et al., Weak electron phonon coupling and deep level impurity for high thermoelectric performance Pb1− xGaxTe. Adv. Energy Mater. **8**(21), 1800659 (2018)
94. M.D. Nielsen, E.M. Levin, C.M. Jaworski, K. Schmidt-Rohr, J.P. Heremans, Chromium as resonant donor impurity in PbTe. Phys. Rev. B **85**, 45210 (2012)
95. E.P. Skipetrov, O.V. Kruleveckaya, L.A. Skipetrova, E.I. Slynko, V.E. Slynko, Fermi level pinning in Fe-doped PbTe under pressure. Appl. Phys. Lett. **105**, 22101 (2014)
96. B. Wiendlocha, Localization and magnetism of the resonant impurity states in Ti doped PbTe. Appl. Phys. Lett. **105**, 133901 (2014)
97. E.P. Skipetrov, L.A. Skipetrova, A.V. Knotko, E.I. Slynko, V.E. Slynko, Scandium resonant impurity level in PbTe. J. Appl. Phys. **115**, 133702 (2014)
98. K.F. Hsu et al., Cubic AgPbmSbTe2+ m: bulk thermoelectric materials with high figure of merit. Science (80-.) **303**, 818–821 (2004)
99. J. Callaway, H.C. von Baeyer, Effect of point imperfections on lattice thermal conductivity. Phys. Rev. **120**, 1149 (1960)
100. J. He et al., Strong phonon scattering by layer structured PbSnS2in PbTe based thermoelectric materials. Adv. Mater. **24**, 4440–4444 (2012)
101. D.L. Partin, C.M. Thrush, B.M. Clemens, Lead strontium telluride and lead barium telluride grown by molecular-beam epitaxy. J. Vac. Sci. Technol. B Microelectron. Process. Phenom. **5**, 686–689 (1987)
102. M.J. Hÿtch, E. Snoeck, R. Kilaas, Quantitative measurement of displacement and strain fields from HREM micrographs. Ultramicroscopy **74**, 131–146 (1998)
103. J. He et al., Morphology control of nanostructures: Na-doped PbTe–PbS system. Nano Lett. **12**, 5979–5984 (2012)
104. J. He, M.G. Kanatzidis, V.P. Dravid, High performance bulk thermoelectrics via a panoscopic approach. Mater. Today **16**, 166–176 (2013)
105. H.J. Goldsmid, A.W. Penn, Boundary scattering of phonons in solid solutions. Phys. Lett. A **27**, 523–524 (1968)
106. G. Yang, Q. Ramasse, R.F. Klie, Direct measurement of charge transfer in thermoelectric Ca3 Co4 O9. Phys. Rev. B - Condens. Matter Mater. Phys. **78**,15 (2008)
107. H. Wu et al., Advanced electron microscopy for thermoelectric materials. Nano Energy **13**, 626–650 (2015)
108. J. He, J. Androulakis, M.G. Kanatzidis, V.P. Dravid, Seeing is believing: Weak phonon scattering from nanostructures in alkali metal-doped lead telluride. Nano Lett. **12**, 343–347 (2012)
109. P.G. Klemens, The scattering of low-frequency lattice waves by static imperfections. Proc. Phys. Soc. Sect. A **68**, 1113 (1955)
110. J. He, S.N. Girard, M.G. Kanatzidis, V.P. Dravid, Microstructure-lattice thermal conductivity correlation in nanostructured PbTe0. 7S0. 3 thermoelectric materials. Adv. Funct. Mater. **20**, 764–772 (2010)

111. J.M. Ziman, *Electrons and Phonons: The Theory of Transport Phenomena in Solids* (Oxford University Press: Amen house, London E.C.4, 2001)
112. P.G. Klemens, *Thermal Conductivity*, vol 1 (Academic Press: London/New York, 1969)
113. P.G. Klemens, Thermal conductivity and lattice vibrational modes. Solid State Phys. **7**, 1–98. (Elsevier (1958)
114. K. Zhang, Q. Zhang, L. Wang, W. Jiang, L. Chen, Enhanced thermoelectric performance of Se-doped PbTe bulk materials via nanostructuring and multi-scale hierarchical architecture. J. Alloys Compd. **725**, 563–572 (2017)
115. E.S. Toberer, A. Zevalkink, G.J. Snyder, Phonon engineering through crystal chemistry. J. Mater. Chem. **21**, 15843–15852 (2011)
116. J.D. Chung, A.J.H. McGaughey, M. Kaviany, Role of phonon dispersion in lattice thermal conductivity modeling. J. Heat Transf. **126**, 376–380 (2004)
117. Y. Xiao et al., MnTE – realizing high performance n-type PbTe by synergistically optimizing effective mass and carrier mobility and suppressing bipolar thermal conductivity. Energy Environ. Sci. **11**, 2486–2495 (2018)
118. Y. Xiao et al., Supporting information: Remarkable roles of Cu to synergistically optimize phonon and Experimental details, 1–18
119. Z. Tian, K. Esfarjani, G. Chen, Enhancing phonon transmission across a Si/Ge interface by atomic roughness: First-principles study with the Green's function method. Phys. Rev. B **86**, 235304 (2012)
120. B. Qiu, H. Bao, X. Ruan, G. Zhang, Y. Wu, Molecular dynamics simulations of lattice thermal conductivity and spectral phonon mean free path of PbTe: Bulk and nanostructures, in *ASME 2012 Heat Transfer Summer Conference collocated with the ASME 2012 Fluids Engineering Division Summer Meeting and the ASME 2012 10th International Conference on Nanochannels, Microchannels, and Minichannels*, (American Society of Mechanical Engineers, 2012), pp. 659–670
121. K. Esfarjani, G. Chen, H.T. Stokes, Heat transport in silicon from first-principles calculations. Phys. Rev. B **84**, 85204 (2011)

Fabrication of High-Performance Flexible Thermoelectric Generators by Using Semiconductor Packaging Technologies

Yusufu Ekubaru and Tohru Sugahara

1 Introduction

1.1 Need for Flexible Thermoelectric Generators

In the near future, heralded as Society 5.0 era (supersmart society), it is expected that our living space will be filled with various sensors networked via wireless communication technology (IoT technology) [1, 2]. Following to concept, the big data that is collected from physical space via various IoT technologies and sensing technologies can be optimized to highly actionable data via artificial intelligence (AI) and information and communications technology (ICT) and subsequently fed back to physical space as actionable information (Fig. 1). This cycle, facilitated by Cyber Physical System (CPS), can ensure healthy and safe life for the next generation of humanity.

IoT technology in such CPSs is supported by several sensor systems [2]. Consequently, the sensor network plays a critical role in effective operation of CPS. The system, e.g., can sense human health and mental conditions that change constantly, grasp data accurately via the network, and promptly manage and improve on surrounding environment that poses a risk to human health and safety. Sensor module incorporates not only sensor (device) but also interface, microcomputer, memory, and communication unit, among other parts, as well as power supply system. To facilitate the implementation of IoT network in society, sensor module that is compact and lightweight and has high degree of elasticity and plasticity is required, so that it can be worn at all times (comfortable to wear), transported easily (highly portable), and installed in any environment, as necessary. In addition,

Y. Ekubaru (✉) · T. Sugahara (✉)
Department of Advanced Interconnection Materials at the Institute of Scientific and Industrial Research, Osaka University, Ibaraki, Osaka, Japan
e-mail: ekubaru@eco.sanken.osaka-u.ac.jp; sugahara@sanken.osaka-u.ac.jp

© Springer Nature Switzerland AG 2021
S. Skipidarov, M. Nikitin (eds.), *Thin Film and Flexible Thermoelectric Generators, Devices and Sensors*, https://doi.org/10.1007/978-3-030-45862-1_7

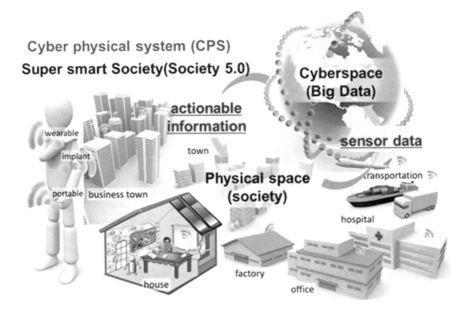

Fig. 1 Conceptual diagram of cyber physical system (CPS) realized via next-generation IoT technology such as AI, ICT, and sensors

because several of these small electronic devices are required to be arranged in all locations of our living space, therefore, power source should be able to generate maintenance-free and on-site power. There is thus a requirement for technology that can efficiently recover unused energy that is wasted currently – that process known as energy harvesting – and use it as power source for various electronic devices that constitute sensor network. Furthermore, usually, in power supply units, rechargeable power supply such as capacitor or battery is mounted. However, because it is difficult to replace secondary battery even in implant device or stationary IoT sensor, it is desirable to incorporate of maintenance-free power source so that the device can be easily employed even under harsh environments.

Considering this background, we introduce our recent achievements pertaining to the development of FlexTEG modules that can be applied to IoT self-power sources employed in severe environment.

1.2   Thermoelectric Generators

Thermoelectric (TE) device can directly generate electricity from thermal radiation energy when temperature difference occurs between sides of TE couple, in accordance with Seebeck effect; inversely, it can produce temperature difference via application of electrical current, in accordance with Peltier effect [3]. Thus, TE devices can be used for both power generation and solid-state refrigeration. As

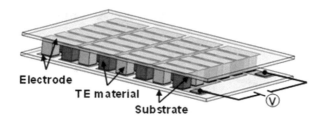

Fig. 2 Schematic of TE module. Both n – and p – type thermoelectric legs forming thermocouples are sandwiched between two electrodes and substrates with high thermal conductivities

shown in Fig. 2, TE devices consist mainly of semiconductor p – and n – type dices (legs) of TE materials, metal electrodes, and insulate substrate. Generally, TEGs are fabricated by bonding p – and n – type legs made of TE materials (TE couples) in π pattern with metal electrodes; subsequently, legs are sandwiched between two insulate substrates with high thermal conductivities. Heat flow occurs from bottom substrate to top, and all TE couples are thus thermally connected in parallel. To improve output voltage, thermocouples are electrically connected in series. When heat is injected into hot side and ejected from cold side, then certain temperature difference is occurred across TEG.

1.2.1   Figure of Merit ZT

The effectiveness of TEG module is represented by dimensionless figure-of-merit ZT:

$$ZT = \frac{S^2 T}{R_{TEG} K_{TEG}}, \tag{1}$$

where S, T, R_{TEG}, K_{TEG} denote total Seebeck coefficient, absolute temperature, total electrical resistance, and total thermal conductance of TE module, respectively [4, 5]. ZT is value characterizing integral quality of TE module and is used for comparing different TE modules.

1.2.2   Temperature Difference Balance in TEG Device

TEG device is a system that includes heat source (heater), thermal interface {1} between hot plate (heater) and hot side substrate of TEG module, TEG module itself, thermal interface {2} between cold side substrate of TEG module and heatsink (radiator/cooling unit), and heatsink (radiator/cooling unit). Efficiency of TEG device to generate power (harvesting) will directly depend on part of maximal temperature difference ΔT_{max} in system that can be provided across TEG module

ΔT_{TEG}. Therefore, TEG device design and high-quality assembling play key role in effective power generation.

General balance of ΔT is as follows:

$$\Delta T = \Delta T_{max} = T_{heater} - T_{heatsink/radiator} \left(T_{ambient}, T_{cooling\,unit} \right),$$ (2)

$$\Delta T_{max} = \Delta T_{heater} + \Delta T_{interface1} + \Delta T_{TEG} + \Delta T_{interface2} + \Delta T_{radiator}.$$ (3)

Here:

- ΔT_{heater} is loss of temperature determined by thermal resistance $R_{th,\,h}$ of hot plate and ability of heater to compensate thermal loss in area of contact between hot plate and hot side substrate of TEG module. In the case of weak source of thermal energy, local overcooling of hot plate in area of thermal contact can take place that can provide adverse effects on efficiency. Obviously, to avoid hot plate overcooling during harvesting, thermal resistance of TEG module $R_{th,\,TEG}$ must be as high as possible.

- $\Delta T_{interface1}$ is drop (loss) of temperature on interface between hot plate (heater) and hot side substrate of TEG module determined by its thermal conductance $R_{th,\,int\,1}$. Bad thermal contact (high $R_{th,\,int\,1}$) between TEG module surface and hot plate surface kills conversion efficiency in thermoelectric power generation. In the case of flexible TEG harvester, there are crucial uncertainties to successful application: unknown value of thermal resistance of interface 1; ΔT_{TEG} can be different even for the same configuration because it depends on the tightness of hot side substrate surface of flexible TEG module and hot plate surface (micro and macro gaps between surfaces can be available).

- ΔT_{TEG} is temperature difference across TEG module (key working parameter for effective power generation). Many adverse effects can seriously lower ΔT_{TEG}: high thermal resistance of interfaces 1 and 2, low potential of thermal energy source (heater), channels of heat leakage from hot side toward cold side (e.g., convection by air or due to high thermal conductivity of another filler), and ineffective removal of heat from outer surface of radiator to environment (e.g., bad or missing of ventilation or not effective cooling unit). $\Delta T_{max} = T_{heater} - T_{heatsink/radiator}$ can be relatively high, but ΔT_{TEG} can be near to zero (thermal isolation of TEG module – Dewar effect) due to high thermal loss in TEG system, e.g., in flexible TEG harvester.

- $\Delta T_{interface2}$ is drop (loss) of temperature on interface between cold side substrate surface of TEG module and radiator/cooling unit surface determined by interface thermal conductance $R_{th,\,int\,2}$. This loss can be minimized (made negligible) by professional attachment.

- $\Delta T_{heatsink/radiator}$ is drop (loss) of temperature on heatsink/radiator due to its thermal conductance $R_{th,\,c}$. This loss depends on design, thermal conductivity of radiator material, and mode of removing heat from radiator surface, all this is very critical for successful operation of flexible TEG harvester. Ineffective removal of heat from radiator surface to environment (e.g., bad or missing of ventilation or not effective cooling unit) leads to increase in $\Delta T_{heatsink/radiator}$ and, therefore, to decrease in ΔT_{TEG}, and humidity of environment air and intensive blowing of radiator with air flow can improve ΔT_{TEG} seriously.

Absolute values of thermal resistance $R_{\text{th, h}}$, $R_{\text{th, int 1}}$, $R_{\text{th, TEG}}$, $R_{\text{th, int 2}}$, $R_{\text{th, c}}$ and especially ratios $R_{\text{th, TEG}}/(R_{\text{th, h}} + R_{\text{th, int 1}}) = R_{\text{th, TEG}}/R_{\text{th, Hot}}$ and $R_{\text{th, TEG}}/(R_{\text{th, c}} + R_{\text{th, int 2}}) = R_{\text{th, TEG}}/R_{\text{th, Cold}}$ are crucial for efficient operation of TEG. When TEG device works in conventional mode with load resistor matched to TEG module electrical resistance $R_{\text{L}} = R_{\text{TEG}}$, two additional thermal sources can affect ΔT_{TEG} – Joule heat on electrical contacts and Peltier heat which can lower seriously ΔT_{TEG} and hence conversion efficiency of TEG device [6]. As usual manufacturer of TEG module takes comprehensive measures to ensure ohmic contacts with the lowest possible contact resistance, so, negative effect of Joule heat is negligible. Because the influence of Peltier effect on ΔT_{TEG} depends on transfer efficiency of heat flows into hot side and away from cold side of TEG module, i.e., $R_{\text{th, Hot}} + R_{\text{th, Cold}}$, it must be considered. When $R_{\text{th, TEG}} \gg R_{\text{th, Hot}} + R_{\text{th, Cold}}$, the influence of Peltier effect is negligible [6].

1.2.3 Output Voltage, Power Output, and Efficiency

1. Output Voltage

According to Seebeck effect, output voltage V_{OUT} of TEG module under a certain temperature difference can be expressed as [7]:

$$V_{\text{OUT}} = N\left(S_{\text{p}} - S_{\text{n}}\right)\Delta T_{\text{TEG}}, \tag{3}$$

where S_{p} (positive value) and S_{n} (negative value) are Seebeck coefficients of p – and n – type semiconductor materials, respectively, and N denotes the number of thermocouples. In the case of low temperature losses on hot and cold sides, i.e., when $R_{\text{th, TEG}} \gg R_{\text{th, Hot}} + R_{\text{th, Cold}}$ (when Peltier effect is negligible), ΔT_{TEG} can be expressed as [8]:

$$\Delta T_{\text{TEG}} = \frac{R_{\text{th,TEG}}}{R_{\text{th,TEG}} + R_{\text{th,Hot}} + R_{\text{th,Cold}}} \Delta T \tag{4}$$

Normally, temperature drop across TEG module ΔT_{TEG} is lower than temperature difference ΔT with environment. This aspect can be mainly attributed to reasons given in paragraph 1.2.2 and is especially important for TEG devices.

2. Output Power

The maximum output power P_{max} of TEG module with matched load $R_{\text{L}} = R_{\text{TEG}}$ and constant Seebeck coefficients in operating temperature range can be expressed as [7]:

$$P_{\max} = \frac{V_{\text{OUT}}^2}{4R_{\text{TEG}}}. \tag{5}$$

3. Conversion Efficiency

The conversion efficiency η can be expressed as [9]:

$$\eta = \frac{P}{Q_{\text{h}}}, \tag{6}$$

where Q_{h} is incoming heat flux into hot side and can be expressed as:

$$Q_{\text{h}} = \frac{\Delta T_{\text{TEG}}}{R_{\text{th,TEG}}}. \tag{7}$$

To enhance performance of TE power generating modules, temperature difference can be increased and ZT of TE materials can be improved. Recent studies on TEGs focused mainly on developing high-performance TE materials. In this context, nanostructure approaches led to dramatic improvements in TE materials, and materials with ZT close to two have been developed successively [10, 11]. However, developments were not focused enough on TEG devices. Recently with increasing demand for IoT development, several $Bi - Te$ based micro-TEGs were developed to allow operation with small temperature difference, and these devices have been applied to novel applications such as IoT, business electronics, and biomedical applications. Because $Bi - Te$ based alloys are used in most commercial TEGs, those represent the best traditional TE materials for applications involving room temperatures.

However, most of these devices have a rigid structure with high cost and complicated fabrication processes, which limits applications, e.g., on nonplanar surfaces. To broaden range of applications, TEG devices must be adapted for heat sources of various shapes (e.g., nonplanar surfaces), allowing powering of portable and wearable devices. Organic TEGs have recently received increasing attention owing to relatively low cost, up-scalability, low-temperature sintering, and high flexibility. However, power output of organic TEGs is often very low, with values of less than 1 µW.

Further, we describe fabrication and characterization of FlexTEGs based on our recent achievements [12].

2 Fabrication of Flexible TEG

The structure of TEG modules can be categorized as vertical, lateral, or hybrid. The vertical structure is traditional and widely adopted in commercial TE products; it is also called the sandwich structure, possibly because its packaging is easier than for other structures [9].

Several techniques for fabrication of TEG modules, based on microthermoelectrics, have been adopted, such as stacking thin film technology, integrated circuit (IC) technology, and nanoscale architecture technology. However, semiconductor packaging is usually employed to fabricate vertical TEG modules, and it is suitable for large-scale structures. Semiconductor packaging technologies were established more than three decades ago, and associated techniques are being refined even now with advancements in semiconductor technology. As shown in Fig. 3, packaging technology includes a variety of steps, such as dicing, plating, patterning, printing, etching, mounting, soldering (bonding), and wiring [13]. Recent research and development in advanced materials and microelectronics has allowed this technology to achieve high performance, thereby contributing considerably to dramatic improvement in the performance of electronic devices. It is expected that using these semiconductor integration packaging technologies, high-performance and highly reliable large-scale flexible TE modules can be fabricated.

Furthermore, output power of TEG module is proportional to squared Seebeck coefficient and temperature difference applied across it. To retain temperature difference, TEGs must realize efficient heat dissipation and absorption; in addition, heat resistance of TE materials and bonding interfaces in various harsh environments should be ensured.

In this study, we fabricated large-scale FlexTEG using described semiconductor packaging technique. Flexible TE module was fabricated on polyimide film (125 μm, Kapton, DuPont) substrate. Fabrication procedure is shown in Fig. 3, and

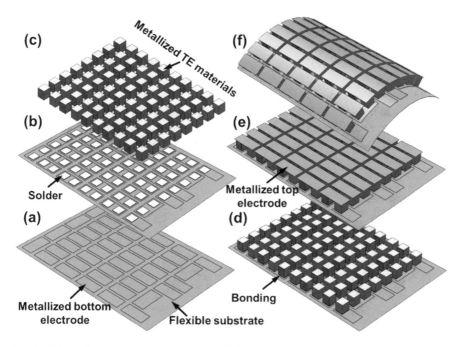

Fig. 3 Schematic manufacturing sequence of FlexTEG

can be described as follows. The first step was formation of substrate: Layer of Au with 100 nm thickness was electroplated onto 18 μm-thick Cu film on flexible plastic substrate. Then, bottom electrode pattern on substrate has been formed using lithography technique and chemical etching. Gaps between bottom electrodes were set as approximately 0.5 mm. The second step was preparation of TE semiconductor dices. Hot-pressed n – and p – type $Bi − Te$ sintering bodies having size of 4 in. were obtained from Toshima Manufacturing Co., Ltd. Each sintering body was sliced into layers with thickness of 1 mm. After non-electroplating 100 nm of Au on 1 μm Ni, slices were cut into dices of approximately 1.4 mm × 1.9 mm × 1 mm. The third step was solder bonding of dices and bottom electrode patterned substrate. Lead-free $Sn − Ag − Cu$ $(Sn − 3.0\%, Ag − 0.5\% − Cu)$ type cream solder was used. Dices were mounted using high-speed chip mounter at rate of 0.1 s/dice on flexible substrate after patterning islands of cream solder through stencil. Then, semi-fabricated module was heated in reflow furnace system at rate of 3 °C/s to 245 °C and set aside for approximately 30 s for solder bonding. After annealing, solder bonded dices on semi-fabricated module were cooled at room temperature at cooling rate of 3 °C/s. The last step was connecting top electrodes, which was realized using Isotropically Conductive Adhesive (ICA) via stencil printing and curing at 150 °C, either with or without elastomer filled by dispenser into gaps between dices.

A comparison between conventional design and design realized in this study is shown in Fig. 4. In conventional design (Fig. 4a), top electrodes at two edges are perpendicular to other top electrodes; therefore, TEG module cannot bend in any direction. However, in proposed design, all top electrodes are in parallel, and thus, TE module is flexible, as shown in Fig. 4b, c.

Using aforementioned steps, 500 dices (250 $p − n$ pairs) were located electrically in series and thermally in parallel on 50 mm × 50 mm flexible plastic substrate in TE module, as shown in Fig. 5a. Compared to TE modules using traditional rigid substrate, fabricated flexible module can be deemed suitable for applications with curved surfaces, as shown in Fig. 5b, c. Further, bending tests demonstrated reliable and stable adhesion of TE module with electrical contacts between dices and flexible substrate. Such flexible TEG modules have several potential applications not only in IoT electronics but also in other portable, wearable, and implantable electronic devices.

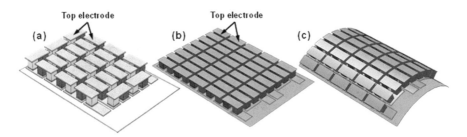

Fig. 4 Comparison of schematics of TE modules (**a**) conventional design (**b**) and (**c**) design realized in this study

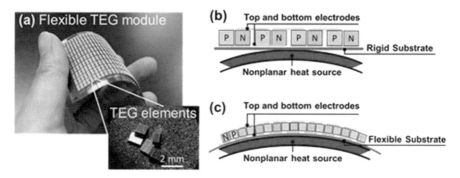

Fig. 5 (**a**) Image of flexible TEG module fabricated in this study. The inset shows n – and p – type dices (elements). Schematic of cross section of (**b**) conventional TEG module on rigid substrate and (**c**) novel TEG module on flexible substrate

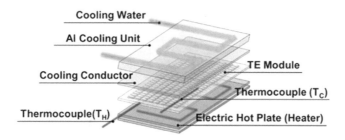

Fig. 6 Schematic of experimental setup for characterization of electrical properties of TEG module

3 Characterization of Flexible TEG

3.1 Electrical Properties of TEG Device

To evaluate power generation characteristics of fabricated flexible TEG module, temperature difference between its two sides must be accurately measured. As seen from the schematic in Fig. 6, flexible TEG module was clamped between hot plate heater at bottom and cooling water block at top with thermal contact sheets (cooling conductor with thermal conductivity $10 \text{ W} \times \text{m}^{-1} \times \text{K}^{-1}$). The temperature on both sides was controlled by using thermocouples. The hot plate temperature varied from 333 to 423 K, while circulation cooling water was maintained at 283 K. Output voltage and output power of TEG module were measured at each temperature difference.

TE conversion efficiency was evaluated with power measurement system as follows: flexible TEG module was mounted between hot plate and cold block. Output power generation characteristics were measured at each temperature difference ΔT across TEG module by using thermally steady-state measurement method. Simultaneously, current–voltage $I - V$ characteristics were measured using computer-controlled multisource meter. Next, output power curves were calculated using values of output current and voltage generated by TEG module at corrected ΔT determined via sandwiched thermocouples.

Figure 7 shows output voltage and power as a function of current at different temperature differences. At $\Delta T = 50$ K, 85 K, and 105 K, maximum open-circuit voltages V_{OC} were 1.4 V, 3.8 V, and 5.0 V, and values of P_{max} generated were approximately 0.6 W, 1.5 W, and 2.1 W, respectively. P_{max} density calculated according to actual area of dices of TEG module was 158 mW/cm^2, and corresponding maximum η was equal to 1.84% at $\Delta T = 105$ K. Here, η was calculated as the ratio of output power P to heat flux into hot side q_h: $\eta_{max} = P/q_h$, and heat flux was calculated as the ratio of temperature difference to thermal resistance between hot side of TEG module and heat reservoir. This efficiency is record value among flexible TEG modules, despite insufficient thermal contact to cooling block.

In principle, all $I - V$ curves at different temperature differences must be linear and parallel to each other. Nonparallel $I - V$ curves between $\Delta T = 85$ K and 105 K indicate that some heat losses may be occurring at hot or cold sides, resulting in lower V_{OUT} at $\Delta T = 105$ K.

3.2   Mechanical Properties of TEG Device

Mechanical properties (bonding strength) of bonded n- and p-type dices, solder, and joints to flexible substrate were evaluated through die shear tests. Head speed of tester was set to 100 μm/s, and fly height was 100 μm from the base tip surface, as shown in Fig. 8. To evaluate stability of electrical contacts and mechanical strength

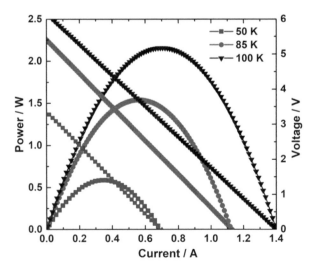

Fig. 7 Output voltage and power as a function of current for FlexTEG module at different temperature differences

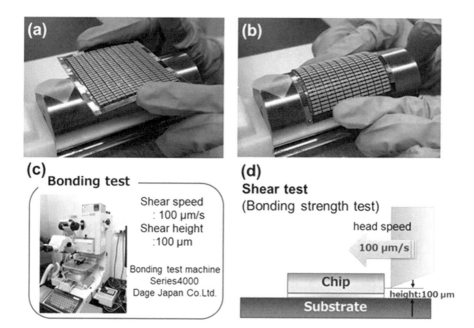

Fig. 8 Bending test and joint strength test of flexible TE module. (**a**) and (**b**) Images of repeated bending tests for each bending radius. (**c**) Shear strength test equipment and (**d**) schematic of test conditions

during bending cycles, electrical resistance and shear strength were measured. The bending was carried out using pipes with radii of 150 mm, 200 mm, and 250 mm (see Fig. 9a, b). After 100, 200, 500, and 1000 bending cycles, electrical resistances were measured and compared to initial value. Figure 9a shows relative resistance ratio of five blocks (each with 50 series $p - n$ pairs of dices) as a function of bending time and cycles. The resistance of each block can be noted to be almost unchanged after various bending cycles, demonstrating that flexible TE module exhibits high reliability during mechanical bending.

4   Conclusions

The findings reported in this chapter demonstrate that traditional semiconductor packaging technique is effective method to fabricate large-scale FlexTEG modules for use in energy harvesting on both planar and nonplanar surfaces. Flexible TEG module with substrate sizes 50×50 mm^2 and $250\,p - n$ pairs (TE couples) was successfully prepared. Further, bending test demonstrated that TE flexible module is reliable and stable during bending and can achieve $P_{\max}$ density of 158 mW/cm^2 at $\Delta T = 105$ K, corresponding to η value of 1.84%, which is comparable to that of

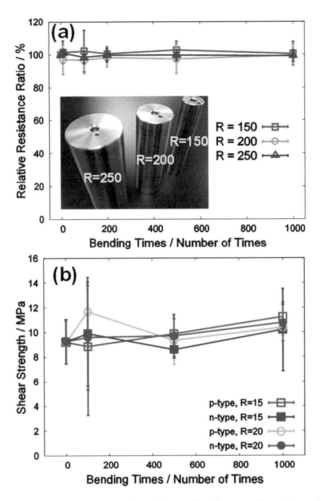

Fig. 9 (**a**) Relative electrical resistance dependence at bending test. Inset shows the pipes with different radii (R (mm)) used in bending test. (**b**) Shear strength of n – and p – type dices during bending test

conventional bulk TEG module. Mechanical tests revealed that flexible module is reliable and stable during bending. These results indicate the excellent potential of FlexTEG devices for application in portable, wearable, or implantable electronic devices. However, some challenges still remain that require further work. Optimization of TEG structure in terms of TE device geometry, dimension, structure, arrangement, etc. is required. In addition, standard technology for characterization of FlexTEGs and wearable TEGs must be developed. Further, suitable materials must be developed for application in higher temperature ranges.

Acknowledgment This work was partially supported by the Programs for "Advancing Strategic Networks to Accelerate the Circulation of Talented Researchers" from Japan Society for the Promotion of Science (JSPS), and the programs of JST CREST (grantnumber: JP MJCR19J1), and "Dynamic Alliance for Open Innovation Bridging Human, Environment and Materials" in "Network Joint Research Center for Materials and Devices" from the Ministry of Education, Culture, Sports, Science and Technology of Japan (MEXT), and "Feasibility Study Program" from New Energy and industrial Technology Development Organization (NEDO). Thanks also to Noriko Kagami, Michio Okajima, Shutaro Nambu, and Professor Katsuaki Suganuma for their help and advice.

References

1. M. Weiser, The computer for the 21st century. Sci. Am. **265**, 94–104 (1991)
2. L. Nummenmaa, E. Glerean, R. Hari, Bodily maps of emotions. PNAS **111**, 646–651 (2014)
3. G.J. Snyder, E.S. Toberer, Complex thermoelectric materials. Nat. Mater. **7**, 105–114 (2008)
4. H.J. Goldsmid, Ch. 3: Conversion efficiency and figure of merit, in *CRC Handbook of Thermoelectrics*, ed. by D. Rowe, (CRC Press, Boca Raton, FL, 1995)
5. M. Cobble, Ch. 3: Calculation of generator performance, in *CRC Handbook of Thermoelectrics*, ed. by D. Rowe, (CRC Press, Boca Raton, FL, 1995)
6. F. Cheng, Ch 19: Calculation methods for thermoelectric generator performance, in *Thermoelectrics for Power Generation - A Look at Trends in the Technology*, ed. by S. Skipidarov, M. Nikitin, (Intech Open, Rieka, Croatia, 2016), pp. 481–506
7. Y. Du, J. Xu, B. Paul, P. Eklund, Flexible thermoelectric materials and devices. Appl. Mater. Today **12**, 366–388 (2018)
8. W. Glatz, S. Muntwyler, C. Hierold, Optimization and fabrication of thick flexible polymer based micro thermoelectric generator. Sens. Actuators A. **132**, 337–345 (2006)
9. Y. Jiabin, L. Xiaoping, Y. Deyang, Y. Chen, Review of micro thermoelectric generator. J. Microelectromech. Syst. **27**, 1–18 (2018)
10. K. Biswas, J. He, I.D. Blum, C.I. Wu, T.P. Hogan, D.N. Seidman, V.P. Dravid, M.G. Kanatzidis, High-performance bulk thermoelectrics with all-scale hierarchical architectures. Nature **489**, 414–418 (2012)
11. L.D. Zhao, S.H. Lo, Y. Zhang, H. Sun, G. Tan, C. Uher, C. Wolverton, V.P. Dravid, M.G. Kanatzidis, Ultralow thermal conductivity and high thermoelectric figure of merit in SnSe crystals. Nature **508**, 373–377 (2014)
12. T. Sugahara, Y. Ekubaru, N.V. Nong, N. Kagami, K. Ohata, L.T. Hung, M. Okajima, S. Nambu, K. Suganuma, Fabrication with semiconductor packaging technologies and characterization of a large - scale flexible thermoelectric module. Adv. Mater. Technol. **4**, 1800556(1)–1800556(5) (2018)
13. T. Zoller, C. Nagel, R. Ehrenpfordt, A. Zimmermann, Packaging of small-scale thermoelectric generators for autonomous sensor nodes. IEEE Trans. Compon. Packag. Manuf. Technol. **7**, 1043–1049 (2017)

Part II
Prospects and Application Features of Wearable TEGs

The Nature of Heat Exchange of Human Body with the Environment and Prospects of Wearable TEGs

Mikhail Nikitin and Sergey Skipidarov

1 Thermal Radiation Energy as Energy Source for Thermoelectric Power Generation and Role of Human Body for That

In accordance with thermodynamic and physical laws, the generation of electromotive force (EMF) by direct conversion of an alternative form of energy into electrical energy is possible in nonequilibrium systems only. The more alternative energy and the greater the deviation from the equilibrium is in the system, the higher is the possible conversion efficiency. There are two main ways of direct conversion of electromagnetic radiation energy into electrical energy: photovoltaic (PV) effect and thermocouple effect (widely used type is Seebeck thermoelectric effect) which is basic for thermoelectric power generating systems.

To convert some sort of thermal radiation energy (heat) to electrical energy via thermoelectricity, two are needed for tango: primary source of thermal radiation energy and thermodynamically nonequilibrium system, i.e., thermoelectric converter with temperature difference $\Delta T \neq 0$ between hot and cold sides.

Due to favorable combination of circumstances, planet Earth is a unique place in the universe where life was born and currently it represented by a wide variety of biological species. Homo sapiens is top species of life on the Earth.

That combination includes:

- presence nearby (approximately 150 million km) of continuous powerful source of electromagnetic radiation (mostly visible (VIS) and infrared (thermal) (IR)) – the Sun with surface temperature about 5800 K and core with temperature in interval of 10–20 million Kelvin;

M. Nikitin (✉) · S. Skipidarov
RusTec LLC, Moscow, Russia

© Springer Nature Switzerland AG 2021
S. Skipidarov, M. Nikitin (eds.), *Thin Film and Flexible Thermoelectric Generators, Devices and Sensors*, https://doi.org/10.1007/978-3-030-45862-1_8

– high temperature of Earth's core estimated 4000–5000 K;
– atmosphere consisting of a mixture of oxygen, nitrogen, and carbon dioxide gases in a certain proportion optimal for biological species living.
– availability of fresh and salt liquid water;
– and rather fast rotation around its axis.

Planet Earth is a unique place in the universe where life was born represented by a wide variety of biological species. *Homo sapiens* is the top species of life on the Earth. The average annual temperature of the Earth's surface and, hence, near surface layer of atmosphere equals to approximately 14 °C. It is enough to live for biological species. But temperature on the Earth is very nonuniform. In hot places, human would like to have a lower ambient temperature and in colder places – a higher temperature. Production, transportation means, amenities, home, and personal equipment and accessories require energy to function as well. Sustainable energy is a modern challenge. Where to get more and more energy that is the question? *Homo sapiens* creates problems for himself and solves those himself.

It would seem that energy sources on the Earth are well known:

Fire (heating) – thermal power (coal), thermal power (oil), thermal power (LNG), thermal power (wood, biological by-products, chemical and biochemical reactions), thermal power (solar radiation heating), geothermal power, and thermal power (hot production processes)
Controlled nuclear reaction – nuclear power
Wind – wind power
Water – small-, medium-, and large-scale hydroelectric power, wave power, and ocean thermal power
Sun – photovoltaic (PV) power
Thermoelectric – (TE) power

However, fire (including nuclear reaction), water, and wind power generation techniques are indirect. And only two techniques – PV and TE – are direct.

PV power generating cell is based on *quantum* effect, i.e., on absorption of electromagnetic radiation photons by electron subsystem of semiconductor, and heating of lattice material is not required, and, generally, growth of total cell temperature is not desirable and has detrimental effect on conversion efficiency of PV cell. All *thermal* generation techniques are based on absorption of electromagnetic radiation photons by whole volume of nonselective absorber (solid, liquid, or gaseous) with immediate conversion of photon energy to vibration energy (phonons) in solid phase, to vibration and rotation energy in liquid phase, or into kinetic energy of particles in gaseous phase. Temperature of absorber increases, and heat flow or kinetic energy (pressure) transfers to parts of energy converting systems.

The primary form of energy of all abovementioned energy sources on the Earth is electromagnetic radiation which originates from chemical, biochemical, or nuclear reactions. During chemical and biochemical reactions, there is a huge number of electron transitions between energy levels of atoms and molecules involved in the reactions which lead to the initiation of many acts of radiative and nonradiative recombination. Nonradiative recombination is usually a multistage process of transferring energy of excited electrons to the lattice (0D, 1D, 2D, or 3D atomic or molecular structure) of solid phase or to liquid phase with weak bonds between atoms, ions, molecules, and radicals or to gaseous phase consisting of practically free moving atoms or molecules. Analogous situation is with nuclear reactions, but in this case electromagnetic radiation is a product of transitions of nucleons between energy levels of contacting nuclei.

When source of energy (the Sun) locates far beyond the Earth (has no direct contact with soil, water, and atmosphere), then we get on the Earth pure electromagnetic radiation energy flux from black body with temperature approximately 5800 K located at distance about 150 million km.

When sources of electromagnetic radiation energy (chemical, biochemical, or nuclear reactions) locate on the Earth or in atmosphere and have direct contact with environment (envelopes), then *part of that energy* can be slowly, fast, or extremely fast transferred to adjacent environment – solid, liquid, or gas phases – up to explosion accompanied by blast wave. Rise in temperature of adjacent environment occurs and can be very high as well.

Therefore, in the mind of ordinary people, energy of the Sun is only electromagnetic radiation; the energy of sources on the Earth is multiform: light, warm air, hot water including geothermal, operating internal combustion engine, molten volcanic lava, and catastrophic damage caused by explosions, be it an explosion of falling meteorite in atmosphere like Tunguska event or nuclear disaster or gas-air mixture blast.

So, all objects and medias on the surface of the Earth or in atmosphere with temperature $T > 0$ emit own equilibrium electromagnetic (mostly) thermal radiation. The emission of thermal radiation occurs under any conditions, without any irradiation of objects from the outside. The intensity of radiation following Stefan-Boltzmann law is proportional to $\varepsilon \sigma T^4$, where σ is Stefan-Boltzmann constant, T is absolute temperature, and ε is emissivity equals from 0 to 1 ($\varepsilon = 1$ for absolute black body source of radiation) [1–3]. Respectively, thermal radiation energy flux from hot to cold substance will be proportional to $\varepsilon \sigma (T_h^4 - T_c^4)$.

The primary source of energy of any object with $T > 0$ is electromagnetic (mostly thermal) radiation energy. Then this energy is absorbed by absorber of a different nature and converts to heat inducing rise in temperature. Generated heat can be transferred from zone of absorption within complex system by different mechanisms of transfer: reradiation, thermal conductivity, convection, evaporation, condensation, etc. Energy in nonequilibrium system inevitably dissipates, often without any benefit to humans.

Note, that it is very difficult (on practice impossible) to catch and use for intended purposes electromagnetic radiation flux totally (in our case to heat hot side of thermoelectric generator (TEG)). Losses are inevitable due to not the whole absorption of electromagnetic radiation flux by absorber, reflection, diffuse scattering, reradiation, etc.

Therefore, to get the highest possible conversion efficiency of thermoelectric generators, all forms of transferring thermal energy from thermal radiation source to TEG module hot side and removing arriving thermal energy from TEG module cold side to environment should be involved. Most often direct (tight) coupling of radiating surface with TEG hot side and surface of high-effective heatsink with TEG module cold side through interfaces with minimal thermal resistance is used. Therefore, development of high-performance wearable TEG is a nontrivial task.

The result of the indomitable consumption of fossil fuels is gigajoules of low-potential waste heat and huge amount of greenhouse gases. Waste heat energy is estimated to be from 60% to 70% of primary energy produced from the burning of fossil fuels. Energy of electromagnetic radiation emitted by the Sun is absorbed by soil, water, atmosphere, flora, fauna, and humans on the Earth and contributes to determination of temperature and its distribution. Converting at least small part of waste heat back to electrical energy is a great challenge for mankind.

So, mankind should solve both problems of the deficit of electricity and the recovery of waste heat simultaneously. It is possible in principle. And thermoelectric power generators can help here. TEG can convert low-potential heat from any sources with $\Delta T \neq 0$ between hot side and heatsink of TEG. It is especially important that TEG can convert thermal radiation into electrical energy at any time of the day, including in dark and nighttime and in bad weather conditions.

Due to long time evolution and adaptation to conditions on the Earth and its atmosphere, human and other endothermic organisms became life species with constant temperature interval of 30–40 °C with own systems of thermoregulation. So, there are above 7.7 billion people – low-potential thermal radiation energy sources for generation electric power with wearable TEGs.

A new era in thermoelectric power generation is becoming a reality as a result of the synergistic effect of recent years' success both in thermoelectricity and related areas of science and technology, such as materials science and technology, microelectronics technology, and design and manufacturing of electronic devices. Impressive advances in microelectronics and electronic devices technology have led to a radical reduction in energy consumption and cost of individual wearable sensors, information, storage, and other devices that could make such devices affordable for all segments of the population.

Motivation for TEG development and production was born resulting in great potential market niche on personal wearable TEGs for self-powering of personal electronic devices. Note, that standard temperature of healthy human body $T_{body} \approx 36.6$ °C is higher than average temperature of the Earth's surface (14 °C) and, hence, temperature in near surface atmosphere T_{amb}. Therefore, almost always, the basic condition for TEG operation $T_{body} - T_{amb} = \Delta T = \Delta T_{max.} \neq 0$ is satisfied.

Further we consider the nature and features of heat production and heat exchange of human body with environment and make estimation in concern of possible amount of electric power which would be generated using human body's thermal radiation power.

2 Features of Heat Production and Heat Exchange of Human Body with Environment

Human is child of nature, and his physiology is fully adapted to living conditions on the Earth. Living healthy human is self-sufficient to interact with environment – sun, soil, water, and air. The primary powerful inexhaustible energy source on the Earth is electromagnetic radiation of the Sun which consists mostly of visible (VIS) and infrared (IR) (thermal) radiation.

Following Stephan-Boltzmann law, the maximum power density of thermal radiation emitted by healthy human body occurs in spectrum range of 8–14 μm (Fig. 1), where the atmosphere has an "optical window" of transparency (Fig. 2) [1].

As seen in Fig. 1, the maximum radiation power density of a healthy human body lays in interval of 9.3–9.5 μm, that is, in the middle of the optical window of atmospheric transparency 8–14 μm. Thus, the release of thermal radiation from human body to environment occurs free under natural conditions, not being absorbed in the adjacent atmosphere.

A healthy human's body with constant average temperature $\approx$ 36.6 °C and emissivity of the skin of almost 1 (0.98) (absolute black body with temperature $\approx$ 36.6 °C) emits freely thermal (IR) radiation through environment, and human is easily observed using thermal imaging device when ambient/background/underly-

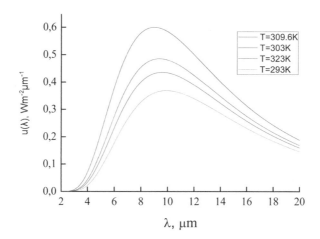

Fig. 1 Spectral power density of thermal radiation of absolute black body source ($\varepsilon = 1$) in wavelength range of 2–20 microns. The temperature of absolute black body source equals to 293 K (20 °C), 303 K (30 °C), 309.6 K (36.6 °C is normal average body temperature of a healthy human, $\varepsilon \approx 0.98$), and 323 K (50 °C)

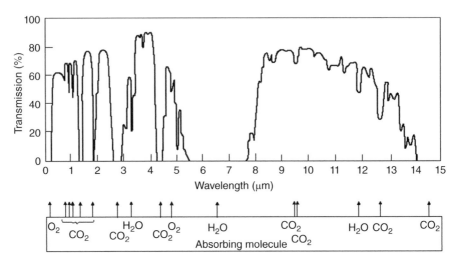

Fig. 2 Normalized transmission of the atmosphere versus wavelength measured along 2 km horizontal path at sea level

Fig. 3 The thermal image of a human figure on the underlying surface with temperature of 4.4 °C (the measuring point of the background temperature is marked with a cross). Positive thermal image: high temperature, white shades; low temperature, dark shades. On the right of the picture is a scale of temperature gradations (thermographic scale) by color shades

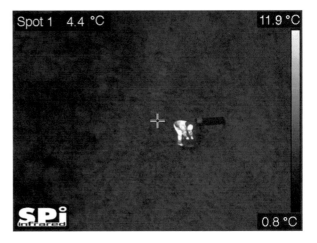

ing surface temperature is below temperature of open skin areas (32–34, 5 °C) or outer layers of light clothing (24–28 °C) (Fig. 3). The human body shines like a lamp, only in infrared spectrum, and, hence, acts as heater.

From the point of view of engineering, human is smart biochemical reactor consuming food, water, and air and converting those to energy of different types, thermal, electrical, and mechanical. Human's organism is a complex system with many sensors, actuators, and neurophysiological controllers [4].

In order to provide good functional (efficient) state of human for a long time, normal physiological functioning of the human body must be ensured in various living conditions on and off work and in different situations. The most important physiological function is thermoregulation which allows to provide a constant average human body temperature of $\approx 36.6\ °C$.

Thermoregulation is the process of regulating balance between heat production and release to maintain constant average temperature of the human body [1]. Thermoregulation allows to keep temperature of internal human's organs constant (36.5–37.2 °C) and whole organism is involved for that. Reaction to cool or to heat is governed by neurophysiological system, which involves specific functional organs, ensuring that constant temperature is maintained in the most efficient and economical way.

Temperature of human body T_{body} is indicator of thermoregulation process. The value of T_{body} depends on rates of heat production and loss, which, in turn, depends on the temperature and humidity of ambient air, air blowing intensity, availability of external thermal irradiation, and heat-protective properties of clothing. Even slight deviations from normal temperature toward hot or cold lead to feeling of unwell. When the body temperature rises by 1 °C and more, the state of health begins unwell, lethargy and irritability appear, pulse and breathing become more frequent, attentiveness decreases, and probability of erroneous actions increases. At T_{body} (internal organs) of 39 °C, human cannot work practically, breakdown is observed, and human can fall into unconsciousness.

Thus, large temperature differences between the skin and inner side of the cloth layer or environment are unacceptable. At long time temperature difference of even 2–3 °C with temperature gradient toward the skin of the body, a gradual increase in average temperature of the body (internal organs) will begin, followed by the onset of unwell and even fatal outcome. In the case of temperature difference greater than 10–15 °C with gradient from the body to cold environment, rapid cooling of the body will begin, and human will freeze followed by the onset of unwell and even fatal outcome.

The human body is of very delicate constitution, energy (heat) production and thermoregulation abilities are not unlimited, and both processes must be carefully considered to optimize living and working conditions, physical activity and, obviously, when "stealing" of energy for thermoelectric generation purpose.

Waste heat power removal by human body P_{body} (heat exchange with the environment) occurs through:

- Emitting thermal radiation of human body into environment, $P_{th.rad}$
- Sweating and evaporation of sweat from the skin P_{sweat}
- Breathing P_{breath}
- Thermal conductivity of clothing and equipment, P_{eq}
- Convection by air blowing human body, P_{conv}

So, human body total waste heat consists of a few components:

$$P_{body} = P_{th.rad} + P_{sweat} + P_{breath} + P_{eq} + P_{conv}. \tag{1}$$

Experiments proved that optimal body metabolism and, hence, the maximum human performance take place at the following balance of heat removal processes:

$$P_{\text{th.rad}} = 50-55\%; P_{\text{sweat}} = 15-20\%; P_{\text{breath}} = 5\%; P_{\text{eq}} + P_{\text{conv}} = 25\%;$$

This balance characterizes relaxed state of thermoregulation system.

Thermoregulation system of a healthy man is very stable and even with heavy physical work does not let increase in body temperature by more than 0.3–0.5 °C. Because healthy man's body temperature is almost always constant, including during heavy physical work, the absolute value of thermal radiation power emitted by the body remains constant, and its share in body heat balance decreases during heavy physical work (hyperactivity), while the share of heat removed by sweating increases, respectively.

The heavier is the physical work, the more heat production increases. The average values of heat production of an adult human depending on the ambient temperature and of physical activity are given in Table 1.

The component of heat production in the form of thermal (IR) radiation emitted by human skin is determined by average temperature of the body (skin) and emissivity of the skin, which is practically equal to 1 in IR spectral range. Integrating spectral distribution of radiation power density emitted by absolute black body with temperature of 36.6 °C (Fig. 1) in spectral range of 2–30 μm, and with accounting for average skin area of adult body of the order of 1.4–1.6 m² (Fig. 4), results in $P_{\text{th.rad}}$ component of body heat production approximately equals to 50 ± 5 W.

$P_{\text{th.rad}}$ is the single source of power for harvesting (conversion) by TEG. Even in rest state, $P_{\text{th.rad}}$ is less than 50% of total human body power P_{body}, and its contribution to P_{body} decreases down to 10–20% with growing physical activity. Excess heat power (over $P_{\text{th.rad}}$) produced by human body in active state is dissipated by other mechanisms, sweating, sweat evaporation, air convection, etc. that are useless for harvesting. Fact is that value of electrical power P_{har} that can be generated by wear-

Table 1 Total heat production of adult human depending on the ambient temperature and physical activity

Ambient air temperature, °C	Heat production, J/s (W)	Ambient air temperature, °C	Heat production, J/s (W)
Rest state		Mid-physical activity	
10	103.7	10	332.0
18	103.7	18	334.1
28	112.1	28	354.3
35	116.2	35	359.1
45	119.7	45	354.3
Light physical activity		Hard and hyperphysical activity	
10	179.6	0	735.0
18	179.6	22	650.1
22	176.8	32	500.4
35	197.0	45	696.0
45	204.6	–	–

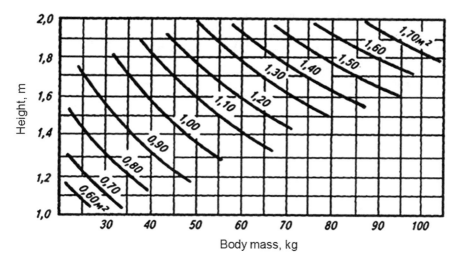

Fig. 4 Graphs for determining the body skin area depending on human mass and height

able TEG is independent on state of human body, being the same at rest, light-, mid-, hard-, or hyperactive state.

Statement is easy to prove by experiment with TEG in climate chamber. Let's place human with attached wearable TEG in climate chamber and start to increase temperature of chamber environment ($T_{\text{chamber inv}}$) higher than ambient, e.g., RT (room temperature). We will observe decrease in generated electric power, and when $T_{\text{chamber inv}}$ reaches T_{body}, i.e., at $T_{\text{chamber inv}} = T_{\text{body}} = \Delta T = 0$, a surprise awaits us – no electrical power generation at all. In hot chamber ($T_{\text{chamber inv}} = T_{\text{body}} \approx 35.5$–$36.6\,°C$), due to thermal regulation in human body, all heat removal processes listed will work: hundreds of watts of heat produced by the human body (Table 1) will be dissipated by sweating, thermal conduction, vasodilation, air convection, hard breathing, etc., but there will be no power generation at all (ZERO), because due to Stephan-Boltzmann law, thermal radiation energy fluxes out and to body will be equal to each other (equilibrium system cannot generate energy). With further rising in $T_{\text{chamber inv}}$ higher T_{body}, TEG will generate power again but with opposite sign of polarity. In this case TEG will operate in reverse mode, i.e., human body will be heatsink, and environment – thermal radiation energy source. But this situation is not good because human cannot stay in the environment with temperature higher than temperature of the human body for a long time; he will simply receive heat stroke.

Therefore, TEG will convert only $P_{\text{th.rad}}$ of the human body (constant part of P_{body}), and harvested power P_{har} will be proportional to area of harvester A_{har} (small part of human body skin area A_{skin}) and conversion efficiency of harvester η_{har} that depends on achievable ΔT_{TEG} (Eq. 2). ΔT_{TEG} depends on optimization (minimization) of parasitic thermal resistances and successful transferring of heat from the skin through better tightness of hot side of wearable TEG to the skin surface (micro and macro gaps between the surfaces can be available); skin moisture (sweat (salted water) has a higher thermal conductivity than air) can assist in better thermal contact at interface and increasing ΔT_{TEG}; muscle activity of the body can improve tightness and enlarging ΔT_{TEG} as well; humidity of environment air and blowing intensity of heatsink with air flow (ventilation) can increase ΔT_{TEG} seriously.

$$P_{\mathrm{har}} = P_{\mathrm{th.rad}} \times \left(A_{\mathrm{har}} / A_{\mathrm{skin}} \right) \times \eta_{\mathrm{har}} \tag{2}$$

Efficiency of TEG power generation (harvesting) will depend on part of ΔT_{max} that can be provided across TEG module (ΔT_{TEG}). This part can be negligible.

The general balance of ΔT is as follows:

$$T_{\mathrm{body}} - T_{\mathrm{amb}} = \Delta T = \Delta T_{\mathrm{max}} = \Delta T_{\mathrm{skin}} + \Delta T_{\mathrm{interface1}} + \Delta T_{\mathrm{TEG}} + \Delta T_{\mathrm{interface2}} + \Delta T_{\mathrm{heatsink}}. \tag{3}$$

Here:

- ΔT_{skin} determines thermal resistance of the skin and ability of human body to compensate thermal loss (lowering skin temperature) in area of contact between body skin and hot side of TEG module. Human body is a very weak source of thermal energy, and, therefore, local overcooling and discomfort can be adverse effects. Obviously, to avoid skin overcooling during harvesting, thermal resistance of TEG must be as high as possible.

- $\Delta T_{\mathrm{interface1}}$ is drop (loss) of temperature on interface between body (skin) and hot side of TEG. Bad thermal contact (high thermal resistance) between TEG module surface and the skin (human body) kills conversion efficiency in thermoelectric power generation. In the case of wearable TEG harvester, there are crucial uncertainties to successful application: unknown value of thermal resistance of interface 1; ΔT_{TEG} could be different even for the same person because it depends on tightness of hot side of wearable generator and the skin surface (micro and macro gaps between the surfaces can be available); skin moisture (sweat (salted water) has a higher thermal conductivity than air) can assist better thermal contact at interface and lowering loss; muscle activity of the body can improve tightness and decreasing loss; humidity of environment air and blowing intensity of heatsink with air flow can improve ΔT_{TEG} seriously.

- ΔT_{TEG} is drop (benefit) of temperature across TEG module. Many adverse effects can seriously lower ΔT_{TEG}: high thermal resistance of interfaces 1 and 2, low potential of thermal energy source (body), channels of heat leakage (dissipation) from hot side toward cold side (e.g., convection by air or due to high thermal conductivity of another filler), and ineffective sink of heat from heatsink surface to environment (e.g., bad or missing of ventilation) make the situation worse. $T_{\mathrm{body}} - T_{\mathrm{amb}} = \Delta T = \Delta T_{\mathrm{max}}$ can be relatively high but ΔT_{TEG} can be near to zero (thermal isolation of TEG – Dewar effect) due to high thermal loss in generating system, e.g., in wearable TEG harvester.

- $\Delta T_{\text{interface2}}$ – is drop (loss) of temperature on interface between cold side of TEG module and heatsink surface. This loss can be minimized (made negligible) by professional attachment.
- $\Delta T_{\text{heatsink}}$ – is drop (loss) of temperature on heatsink. This loss depends on design, thermal conductivity of heatsink material, and mode of removing heat from heatsink surface, all this is very critical for successful operation of wearable TEG harvester. Ineffective sink of heat from heatsink surface to environment (e.g., bad or missing of ventilation) makes the situation worse, i.e., leads to increase in $\Delta T_{\text{heatsink}}$ and, therefore, to decrease in ΔT_{TEG}.

Muscle activity, increased moisture of the skin in walking, better tightness between wearable TEG harvester, and more intensive removing of heat from heatsink surface (enforced ventilation) when person is walking (running) enlarge ΔT_{TEG} that results in higher open circuit voltage and output power compared to sitting (rest) body situation of same TEG at the same ambient temperature.

The challenge is how much of $P_{\text{th.rad}}$ can be harvested by wearable TEG to power different sensors and other personal equipment subsystems. The human body is IR lamp of approximately 50 W. Averaged $P_{\text{th.rad}}$ density of an adult healthy man equals to 3–4 mW/cm^2. However, the distribution diagram of thermal radiation emission is specific. Emitted IR radiation density is not uniform on the skin of a human body (Fig. 5). Energy flux density is low (low-potential heat). It is extremely difficult to convert effectively low-potential heat which is nonuniformly distributed over the

Fig. 5 Author M. Nikitin's thermal signatures in profile and full face. Thermal radiation emission pattern is nonuniform. Positive thermal images: higher temperature, white shades; lower temperature, dark shades

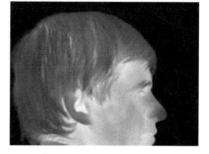

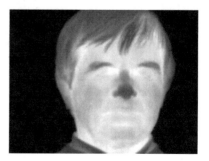

surface of the human body into electrical energy. Style of donning and time duration of wearing TEG on open areas of the skin should not cause discomfort or undesirable consequences. For this, skin areas of the forehead and wrist can be considered.

3 Wearable TEG Harvester Output Power Density Estimate

As noted above, due to uncertainties in thermal energy transfer from the human body skin to wearable TEG and from heatsink to environment, there is no reason to hope getting optimistic high figures of ΔT_{TEG} in wearable TEG at time-to-time donning.

Let's take expected ΔT_{TEG} = 1, 3, 6 K, average $P_{th.rad}$ = 50 W, typical skin area 1.5 m^2, and normal average body temperature of healthy human T_{body} = 309.6 K (36.6 °C). Respectively, harvester optimistic (Carnot) conversion efficiency $\eta_{har} = \eta_{Carnot} \sim \Delta T_{TEG}/T_{body} \approx 0.0033; 0.01; 0.02 \approx 0.33\%, 1\%, 2\%$.

Using Eq. (2) we get max estimate for wearable TEG harvester output power density (per cm^2) as:

$$P_{har} \approx 11 \ \mu W / cm^2. \ \Delta T_{TEG} = 1 \ K\left(^\circ C\right);$$

$$P_{har} \approx 33 \ \mu W / cm^2. \ \Delta T_{TEG} = 3 \ K\left(^\circ C\right);$$

$$P_{har} \approx 66 \ \mu W / cm^2. \ \Delta T_{TEG} = 6 \ K\left(^\circ C\right).$$

Due to inevitable losses caused by current low level of thermoelectric figure-of-merit ZT ≤1, real conversion efficiency and, hence, P_{har} will be lower, maybe five to ten times lower.

4 Conclusions

Due to rapid development of electronic device manufacturing technology resulting in radical decrease in energy consumption, weight, and sizes and significant increase in functionality, the need grows for development of autonomous low-power generators providing constant operational readiness of state-of-the-art future devices. Therefore, development of high-performance wearable TEGs converting low-potential waste heat to electricity is highly topical task now.

To solve the task successfully, researchers and engineers have to do the following:

- Develop innovative lightweight inorganic, organic, or composite low-temperature thermoelectric materials with high thermoelectric performance characterized by ZT >1 for n-type and p-type thermoelectric materials. Breakthrough in TE materials up to ZT = 3–5 could revolutionize application of wearable devices.
- Develop optimal design of wearable TEG.
- Develop innovative construction materials and advanced assembling techniques of wearable TEG to minimize thermal losses in generating system.
- Take comprehensive measures to ensure safe and comfortable use of wearable TEGs by humans. Wearable TEG must be friendly to human.

References

1. G. Gaussorgues, S. Chomet, *Infrared Thermography* (Springer, 1994). https://doi.org/10.1007/978-94-011-0711-2
2. M.A. Bramson, Infrared radiation of heated bodies, Nauka, Moscow, 1964 (in Russian)
3. M.A. Bramson, Reference tables for infrared radiation of heated bodies, Nauka, Moscow, 1964 (in Russian)
4. J.E. Hall, Unit VIII, Metabolism and temperature regulation, Ch. 73, in *Guyton and Hall Textbook of Medical Physiology*, 12th edn., (Elsevier, 2011), Cambridge, MA, pp. 867–880
5. Health and safety. Textbook, ed. by S.V. Belov, Vyschaya shkola, (Moscow, 1999) (in Russian)
6. http://ohrana-bgd.narod.ru/proizv_68.html

Heatsinks and Airflow Configurations for Wearable Thermoelectric Generators

Beomjin Kwon and Jin-Sang Kim

1 Introduction

In the last two decades, there have been efforts to develop TEGs as charging units for wearable electronics. Table 1 lists power density P' and other measurement parameters of previously reported TEGs tested for wearable applications. Note, the usage condition for wearable TEG is $\Delta T_{ext} = T_s - T_a > 0$, where T_s is the body skin temperature and T_a is the ambient air temperature. The difference between ambient (atmospheric) temperature and skin temperature derives heat from body to air, while fraction of heat can be harvested by TEGs. In previous works, power density P per square of applied temperature gradient (P') equaled to merely a few hundreds of nW/(cm²×K²) in stationary air environment and increased to a few μW/(cm²×K²) when air flowed at walking speed (~ 1.4 m/s). Power densities of wereable TEGs have been still less than 1% of the theoretical limit. Unless surface areas of heatsink A_{hs} and TEG–skin contact A_{sc} were uncomfortably large.

Wearable TEG harvester is a system including heat source (human body); thermal interface SG between skin surface and hot side surface of TEG module; TEG module itself; thermal interface GHS between cold side surface of TEG module and adjacent surface of heatsink; heatsink. The efficiency of wearable TEG harvester to generate power will directly depend on part of external temperature difference $\Delta T_{ext} = \Delta T_{max}$ that can be provided across TEG module: $\Delta T_{int} = \Delta T_{TEG}$. Therefore, for effective power generation by wearable TEGs, optimal device design, high-quality

B. Kwon (✉)
Arizona State University, Department of Mechanical Engineering, Tempe, AZ, USA
e-mail: kwon@asu.edu

J.-S. Kim (✉)
Korea Institute of Science and Technology, Center for Electronic Materials,
Seoul, Republic of Korea
e-mail: jskim@kist.re.kr

© Springer Nature Switzerland AG 2021
S. Skipidarov, M. Nikitin (eds.), *Thin Film and Flexible Thermoelectric Generators, Devices and Sensors*, https://doi.org/10.1007/978-3-030-45862-1_9

Table 1 List of previously reported TEGs for wearable applications

Ref	P^r $\mu W/(cm^2 \times K^2)$	A_{sc} (cm^2)	A_{hs} (cm^2)	ΔT_{ext} (K)	Test condition
[1]	0.35	1	~6	8	Sitting
[2]	0.21	6	~30	8	Sitting
	0.78				Walking (1.1 m/s)
[3]	0.47	5	~223	8	Sitting
[4]	0.45	10	~27	8	Sitting
[5]	0.01	5	NA	8	Sitting
[6]	0.38	0.97	~3.5	7.4	Sitting
	2.26				Airflow (0.9 m/s)
[7]	0.15	16	16	5.7	Sitting
[8]	0.18	1.45	~6.4	6	Sitting
	0.56				Airflow (1 m/s)
[9]	0.10	40	40	20	Sitting
					Initial heatsink temperature = 10 °C

assembling, attachment to human body and effective heat removal from external surface of heatsink play a key role.

The general balance of ΔT in TEG system is as follows:

$$\Delta T_{ext} = \Delta T_{max} = T_s - T_a, \tag{1}$$

$$\Delta T_{ext} = \Delta T_{max} = \Delta T_s + \Delta T_{sg} + \Delta T_{TEG} + \Delta T_{ghs} + \Delta T_{hs} \tag{2}$$

Limited power densities generated by wearable TEG result inherently from small temperature difference across TEG module $\Delta T_{int} = \Delta T_{TEG}$, low thermoelectric efficiency of TE materials, and parasitic factors involved in TEG fabrication and application. A simple model for TEG predicts that output power P of TEG device with load resistor matched to TEG module electrical resistance $R_L = R_g$ and constant Seebeck coefficients S_n of n – type and S_p of p – type materials of TEG legs in operating temperature range can be expressed as:

$$P \sim \frac{\left[N\left(S_p - S_n\right)\Delta T_{TEG} \right]^2}{4R_g} = \frac{\left[S_g \Delta T_{TEG} \right]^2}{4R_g} = \frac{V_{OUT}^2}{4R_g}, \tag{3}$$

where S_p is positive value and S_n is negative value, S_g is total Seebeck coefficient of TE module, N denotes number of $p - n$ pairs/thermocouples, and V_{OUT} is output voltage. Human body is very weak source of thermal radiation energy (heat) and, therefore, local overcooling and discomfort can have adverse effects at skin–TEG contact area. Overcooling degree will be determined by thermal resistance of skin $R_{th, s}$ and ability of human body to compensate thermal loss in the area of contact between body and hot side of TEG module. Obviously, to avoid skin overcooling during harvesting, thermal resistance of TEG module $R_{th, g}$ must be as high as possible.

$\Delta T_{int} = \Delta T_{TEG}$ is a key working parameter for effective power generation. Many adverse effects can seriously lower ΔT_{TEG}: high thermal resistance $R_{th, sg}$ of interface SG between rough skin surface and hot side surface of TEG module and thermal resistance $R_{th, ghs}$ of interface GHS between cold side surface of TEG module and adjacent surface of heatsink, low potential of thermal energy source – human body, channels of heat leakage from hot side toward cold side (e.g., convection by air or due to high thermal conductivity of another filler), high thermal resistance of heatsink $R_{th, hs}$ due to ineffective removal of heat from external surface of heatsink (radiator) to environment (e.g., bad or missing of ventilation). Value $\Delta T_{ext} = \Delta T_{max} = T_s - T_a$ can be relatively high but $\Delta T_{int} = \Delta T_{TEG}$ can be near to zero (thermal isolation of TEG module) due to high thermal loss in wearable TEG harvester. Poor thermal contact (high $R_{th, sg}$) between TEG module surface and rough skin surface significantly deteriorates conversion efficiency in thermoelectric power generation. In the case of wearable TEG harvester, there are crucial uncertainties to its successful application: unknown value $R_{th, sg}$; as a consequence, $R_{th, sg}$ and, hence, $\Delta T_{int} = \Delta T_{TEG}$ can be different even for the same configuration because it depends on tightness of hot side surface of wearable TEG and rough skin surface (micro and macro gaps between surfaces can be available) and skin moisture. Absolute values of $R_{th, s}$, $R_{th, sg}$, $R_{th, g}$, $R_{th, ghs}$, $R_{th, hs}$ and especially ratios $R_{th, g}/(R_{th, s} + R_{th, sg}) = R_{th, g}/R_{th, h}$ and $R_{th, g}/(R_{th, hs} + R_{th, ghs}) = R_{th, g}/R_{th, c}$ are crucial for efficient operation of wearable TEG harvester.

When TEG device operates with a load resistor matched to TEG module electrical resistance $R_L = R_g$, two additional thermal sources – Joule heat on electrical contacts and Peltier heat – can reduce seriously ΔT_{TEG} and hence conversion efficiency of TEG device. As usual, manufacturer of TEG module takes comprehensive measures to ensure ohmic contacts with the lowest possible contact resistance for the purpose of minimizing negative effect of Joule heat but, in general case, contact resistance should be considered. Because the influence of Peltier effect on ΔT_{TEG} depends on transfer efficiency of heat flows into hot side and away from cold side of TEG module, i.e., on $R_{th, h} + R_{th, c}$, it must be considered as well.

In order to determine optimal sizes and types of heatsinks, $R_{th, c}$ and $R_{th, g}$ should be carefully considered. $R_{th, g}$ changes with fill factor (F), while $R_{th, c}$ varies with geometry and material of heatsink and conditions of removal of heat from the surface. Not only TEG device structure but also convective boundary condition affect $R_{th, c}$. The ratio of thermal resistances is important for designing heatsink and can be classified into three regimes: (1) perfect heatsink condition $R_{th, g} \gg R_{th, h}, R_{th, c}$, (2) practical heatsink condition $R_{th, c} \geq R_{th, g}$, and (3) wearing condition $R_{th, h}, R_{th, c} \geq R_{th, g}$. For wearable TEGs, due to limited sizes of heatsink and skin roughness, external resistances usually overwhelm $R_{th, g}$. Such unbalanced thermal resistances cause tiny ΔT_{TEG}, and limit power generation, which, therefore, necessitates increase in $R_{th, g}$ and improved designs of heatsinks.

The decrease in ΔT_{TEG} due to Peltier effect is negligible when $R_{\text{th, g}} \gg R_{\text{th, h}} + R_{\text{th, c}}$.

The temperature drop across TEG module ΔT_{TEG} is always lower (may be seriously) than temperature difference with the environment ΔT_{ext}.

Thus, to reduce in $R_{\text{th, c}}$, equivalently to improve in $\Delta T_{\text{TEG}}/\Delta T_{\text{ext}}$, TEG modules have been combined with bulky heatsinks. Furthermore, TE figure-of-merit ZT of bulk materials has been limited below 2 [10], which also restricted actual conversion efficiency less than the half of Carnot efficiency. To improve such low efficiency near those of mechanical heat engines, ZT should reach at least 4 [11].

In order to improve ΔT_{TEG}, there have been efforts to optimize $R_{\text{th, g}}$, and reduce $R_{\text{th, c}}$ through incorporating various types of heatsinks. The most prevalent type of heatsinks for wearable TEGs have been metal fins due to large thermal conductivity of metal >100 W/(m×K) and commercial availability. An array of pin fins, plate fins, or metal plate with grooves could enlarge surface area of heat transfer between cold side surface of TEG and ambient air, leading to reduction in $R_{\text{th, c}}$. Metal fins could enhance in TEG performance more than 50% [8]. When used with wearable TEGs, the most critical design constrictions for heatsinks have been sizes. Although large sizes of heatsinks ensure significant reductions in $R_{\text{th, c}}$, it is essential to minimize volume and weight of heatsinks for wearable applications. Thus, characteristic dimension of heatsink, usually height, has been a few millimeters. Shapes and gaps between fins are also important for heatsink performance. By increasing in fin density, surface area of heatsinks can be extended. However, extremely large fin density would result in reduced airflow between the fins, indicating that it is necessary to optimize fin density. Unusual material, cotton glued with carbon fabric, was also employed to construct heatsink for wearable TEGs [3]. When TEGs are attached inside clothes, flexible and thermally conductive material can be used to spread heat across large cloth area. Although thermal conductivity of clothes is small <1 W/(m×K), heat spreading using carbon fabric demonstrated power generation enhancement of about 30%.

Recently, heat capacitors have also been employed to maintain small ΔT_{TEG} over a long period. Thin layers of silica gel mixed with hydrogel [7] or water absorbing polymer particles [9] have been attached on TEG substrates. Due to large heat capacity, TEG substrate could maintain at or below ambient temperature over a few hours. While ΔT_{TEG} was kept around 20 K, TEG with water absorbing polymer generated ~4.7× larger power than TEG with compact metal fins.

2   Theory

For designing TEG module or heatsink, theoretical model is necessary that calculates ΔT_{TEG} when one of TEG substrates is exposed to air and another substrate is placed on the human body. The human body and air are considered as thermal reservoirs such that temperatures T_{s} and T_{a} are constants. Assuming that thermal resistance along direction from skin to air is dominantly large, heat flow is one-dimensional. The resultant ΔT_{TEG} generates Seebeck voltage V_{OUT}.

Figure 1a shows the schematic for heat transfer model. Upon ΔT_{ext}, associated thermal resistances $R_{\mathrm{th,\,h}}$, $R_{\mathrm{th,\,g}}$, $R_{\mathrm{th,\,c}}$ and TE phenomena such as Peltier effect and Joule heating determine temperature at cold side T_{c} and at hot side T_{h} of TEG module. Figure 1b shows the schematics for TEG device where heatsink is incorporated at TEG–air interface.

Particularly, in this chapter, plate fin heatsink is attached on cold side surface of TEG which undergoes small pressure drop and is simple to design and fabricate. Fin density D is important for flow rate, flow bypass, and convection heat transfer coefficient. Fin density is defined as $D = tN_{\mathrm{f}}/L$, where t is fin thickness, N_{f} is number of fins, and L is footprint length of heatsink. Although fin height is critical to fin performance, it is limited to few mm for wearable applications. For simplicity, lumped heat transfer coefficient is defined at skin–TEG interface h_{s} and at heatsink–air interface h_{a}. However, TEG–skin contact area A_{sc} is difficult to estimate and plate fin heatsink geometry complicates heat transfer analysis. Thus, we may further simplify interfacial heat transfer by employing effective heat transfer coefficient for skin–TEG interface $h_{\mathrm{eff,\,s}}$ and TEG–air interface $h_{\mathrm{eff,\,a}}$ assuming that interfacial area is merely to substrate area A_{g}.

Our heat transfer model accounts for relevant TE phenomena and effects of filler material and electrical contact resistance to obtain explicit solution for ΔT_{TEG}. The energy balance equation at TEG hot side [12] is:

$$S_{\mathrm{g}}T_{\mathrm{h}}I - 0.5I^{2}R_{\mathrm{g}} + K_{\mathrm{g}}\left(T_{\mathrm{h}} - T_{\mathrm{c}}\right) = K_{\mathrm{sg}}\left(T_{\mathrm{s}} - T_{\mathrm{h}}\right), \qquad (4)$$

where I is electrical current, $K = 1/R_{\mathrm{th}}$ is thermal conductance and subscript sg indicates skin–TEG interface. An energy balance equation at TEG cold side [12] is:

$$S_{\mathrm{g}}T_{\mathrm{c}}I + 0.5I^{2}R_{\mathrm{g}} + K_{\mathrm{g}}\left(T_{\mathrm{h}} - T_{\mathrm{c}}\right) = K_{\mathrm{ga}}\left(T_{\mathrm{c}} - T_{\mathrm{a}}\right), \qquad (5)$$

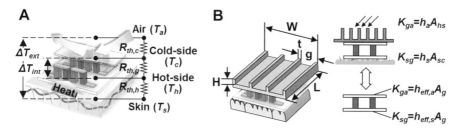

Fig. 1 (**a**) Schematic for heat transfer model of wearable TEG device. (**b**) Schematics for TEG module integrated with heatsink that defines heatsink parameters and heat transfer coefficients used at TEG module interfaces

where the subscript *ga* corresponds to TEG–air interface. Here, K_g combines thermal conductance of TE material (denoted as leg) and filler material:

$$K_g = \left[0.5\left(\kappa_p + \kappa_n\right)A_g F + \kappa_f A_g \left(1-F\right)\right]/l, \tag{6}$$

where κ_p, κ_n and κ_f are thermal conductivities of *p* - and *n* - type materials of TEG legs and filler material, respectively, F is fill factor, l is leg length. The electrical resistance of TEG module R_g is given by:

$$R_g = N\left(\rho_p l + \rho_n l + 4\rho_{cont}\right)/w^2, \tag{7}$$

where N is number of $p - n$ pairs, ρ is electrical resistivity, w is leg width, and ρ_{cont} is specific contact resistivity. Now, Eqs. (4) and (5) can be combined into a matrix form to solve for T_h and T_c as:

$$\begin{pmatrix} S_g I + K_g + K_{sg} & -K_g \\ K_g & S_g I - K_g - K_{ga} \end{pmatrix}\begin{pmatrix} T_h \\ T_c \end{pmatrix} = \\ = \begin{pmatrix} 0.5 I^2 R_g & K_{sg} T_s \\ -0.5 I^2 R_g & -K_{ga} T_a \end{pmatrix}, \tag{8}$$

For explicitly obtaining $\Delta T_{TEG} = T_h - T_c$, I should be replaced with:

$$I = S_g \left(T_h - T_c\right)/\left(1+m\right)R_{TEG}, \tag{9}$$

where $mR_g = R_L$ is external load resistance. This method gives a cubic equation for ΔT_{TEG} with coefficients in terms of TEG module physical properties, T_s and T_a. A feasible root for cubic equation provides ΔT_{TEG}.

$$T_h = \frac{\left(K_{sg} T_s + 0.5 I^2 R_g\right)\left(K_{ga} - S_g I\right) + K_g \left(K_{sg} T_s + K_{ga} T_a + I^2 R_g\right)}{\left(K_{sg} + S_g I\right)\left(K_{ga} - S_g I\right) + K_g \left(K_{sg} + K_{ga}\right)}, \tag{10}$$

$$T_c = \frac{\left(K_{ga} T_a + 0.5 I^2 R_g\right)\left(K_{sg} + S_g I\right) + K_g \left(K_{sg} T_s + K_{ga} T_a + I^2 R_g\right)}{\left(K_{sg} + S_g I\right)\left(K_{ga} - S_g I\right) + K_g \left(K_{sg} + K_{ga}\right)}, \tag{11}$$

To eliminate I in the above equations, I is substituted by other parameters from Eq. (9). For simplicity, we may define $x = \Delta T_{TEG} = T_h - T_c$. Then, the above equations can be combined as:

$$\frac{S_g^4 m}{R_g^2 \left(1+m\right)^3} x^3 + \frac{S_g^2 \left(K_{sg} - K_{ga}\right)\left(0.5+m\right)}{\left(1+m\right)^2 R_g} x^2 \\ - \left[\frac{S_g^2 \left(K_{sg} T_s + K_{ga} T_a\right)}{\left(1+m\right)R_{TEG}} + K_{sg} K_{ga} + K_g \left(K_{sg} + K_{ga}\right)\right]x, \\ + K_{sg} K_{ga} \left(T_s - T_a\right) = 0 \tag{12}$$

The above equation is simply a cubic equation. For expressing the explicit solution for x, six coefficients are defined as follows:

$$c_1 = \frac{S_g^4 m}{R_g^2 (1+m)^3 K_{sg} K_{ga}},$$

$$c_2 = \frac{S_g^2 (1/K_{ga} - 1/K_{sg})(0.5 + m)}{(1+m)^2 R_g},$$

$$c_3 = -\frac{S_g^2 (T_s/K_{ga} + T_a/K_{sg})}{(1+m) R_g} - 1 - K_g (1/K_{ga} + 1/K_{sg}),$$

$$c_4 = T_s - T_a,$$

$$c_5 = -27 c_1^2 c_4 + 9 c_1 c_2 c_3 - 2 c_2^3,$$

$$c_6 = 3 c_1 c_3 - c_2^2. \tag{13}$$

A feasible root for cubic equation is then expressed as:

$$x = \frac{\left(1 + i\sqrt{3}\right) \sqrt[3]{\sqrt{c_5^2 + 4 c_6^3} + c_5}}{6 \sqrt[3]{2} c_1} + \frac{\left(1 - i\sqrt{3}\right) c_6}{3 \times 2^{2/3} c_1 \sqrt[3]{\sqrt{c_5^2 + 4 c_6^3} + c_5}} - \frac{c_2}{3 c_1}, \tag{14}$$

Effective heat transfer coefficients at TEG interfaces are critical factors for calculation of ΔT_{TEG}. Heat transfer at skin–TEG interface depends on actual contact area, applied pressure, and skin moisture. Previous works have employed 15 W/(m²×K) $\leq h_{eff, s} \leq$ 100 W/(m²×K) [2, 3, 6]. Heat transfer at TEG–air interface relies on efficiency and size of heatsink and airflow induced by body motions. Generally, heatsink made of highly conductive material with large surface area is advantageous to obtain large $h_{eff, a}$. When air velocity varies from 0 to 1.4 m/s (walking velocity), an order of magnitude of $h_{eff, a}$ may improve. However, air velocity distribution around heatsink is difficult to predict such that theoretical estimation of h_a is also challenging. Empirically, $h_{eff, a}$ is between 5 and 20 W/(m²×K) under natural convection, and it is between 100 and 300 W/(m²×K) under forced convection [2, 4, 6].

3 TEG Design

Based on the theory, $\Delta T_{int} = \Delta T_{TEG}$ and, in turn, P' can be predicted in order to design critical TEG parameters such as fill factor, length, and width of individual TE leg. To optimize such parameters, it is important to consider convection condition for TEG, which, however, depends largely on human activities. Depending on if a person is motionless, walks, or runs, airflow surrounding TEG as well as convective heat transfer varies significantly. In addition, contact between skin and TEG varies well with applied pressure and interfacial material.

Thus, this chapter considers four different convection configurations: (1) natural convection under motionless situation $h_{\text{eff, a}} = 10$ W/(m²×K), (2) moderate forced convection under walking situation $h_{\text{eff, a}} = 100$ W/(m²×K), (3) enhanced forced convection under running situation $h_{\text{eff, a}} = 500$ W/(m²×K), and (4) enhanced body contact $h_{\text{eff, a}} = 100$ W/(m²×K), $h_{\text{eff, s}} = 500$ W/(m²×K). Except configuration for enhanced body contact, $h_{\text{eff, s}}$ is a constant as 100 W/(m²×K).

Figure 2 shows predicted P'' multiplied by squared applied temperature gradient or P and $\Delta T_{\text{int}} = \Delta T_{\text{TEG}}$ as a function of fill factor and individual leg length. These two parameters are dominant factors for P and $\Delta T_{\text{int}} = \Delta T_{\text{TEG}}$, as both determine TEG thermal conductance.

Table 2 lists all calculation parameters. As reported earlier, maximum P'' is possible when internal and external thermal resistances become similar such that $\Delta T_{\text{TEG}}/T_{\text{ext}} \sim 0.5$. When TE legs lengthen, then optimum fill factor F_{opt} also linearly increases to meet thermal matching condition (TMC). If TE legs are longer beyond TMC, excessive electrical resistance cuts down P'' in spite of increased ΔT_{TEG}. As compared to natural convection, forced convection improves P and F_{opt} by about an order of magnitude, and ΔT_{TEG} by about a factor of two. However, for forced convection configurations, enhancement in $h_{\text{eff, a}}$ and $h_{\text{eff, s}}$ does not linearly improve P,

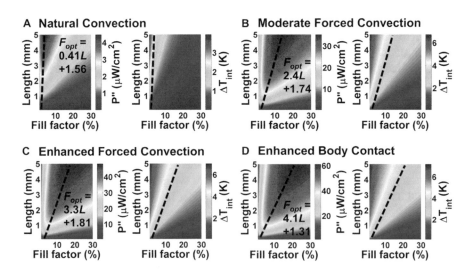

Fig. 2 (**a–d**) Predicted P and $\Delta T_{\text{int}} = \Delta T_{\text{TEG}}$ as a function of fill factor and length of TE legs. Datasets (**a–c**) correspond to convection conditions with typical contact condition with skin $h_{\text{eff, s}}$ = 100 W/(m²×K): (**a**) natural convection $h_{\text{eff, a}} = 10$ W/(m²×K), (**b**) moderate forced convection $h_{\text{eff, a}} = 100$ W/(m²×K), and (**c**) enhanced forced convection $h_{\text{eff, a}} = 200$ W/(m²×K). Dataset (**d**) is for enhanced contact with body $h_{\text{eff, s}} = 500$ W/(m²×K) with moderate forced convection $h_{\text{eff, a}}$ = 100 W/(m²×K). The calculation assumes $\Delta T_{\text{ext}} = 8$ K and $\rho_{\text{cont}} = 20$ μOhm×cm²

Table 2 Calculation parameters

Parameter	Value
Seebeck coefficient of TE materials, $S_g/2N$	±195 μV/K
Thermal conductivity of TE material, κ_p, κ_n	1.4 W/(m×K)
Electrical resistivity of TE material, ρ_p, ρ_n	0.9 mOhm×cm
Leg-electrode specific contact resistivity, ρ_{cont}	20 μOhm×cm²
Thermal conductivity of air, κ_f	0.027 W/(m×K)
Effective heat transfer coefficient for skin–TEG interface, $h_{eff, s}$	100 W/(m²×K) (natural/moderate forced/ enhanced forced convection) 500 W/(m²×K) (enhanced body contact)
Effective heat transfer coefficient for TEG–air interface, $h_{eff, a}$	10 W/(m²×K) (natural convection) 100 W/(m²×K) (moderate forced convection/ enhanced body contact) 200 W/(m²×K) (enhanced forced convection)
TEG area, A_g	1 cm²
Individual leg width, w	0.1–1.5 mm
Individual leg length, l	0.1–5 mm
Fill factor, F	0.01–0.3
Air temperature, T_a	22 °C
Skin temperature, T_s	30 °C
External load resistance, mR_g	R_g ($m = 1$)

although it helps to increase in F_{opt} slightly. Therefore, for a desirable leg length and for common convection configuration, F should be chosen properly.

For wearable TEG, thickness and output voltage V_{OUT} are important factors. Figure 3a, b show predicted P and $\Delta T_{int} = \Delta T_{TEG}$ with $F = F_{opt}$ as a function of leg length. As l increases, P ($F = F_{opt}$, denoted as P_{opt}) and $\Delta T_{int} = \Delta T_{TEG}$ drastically rise when $l \leq 1$ mm, and level off with larger l. Especially, under forced convection configurations, P_{opt} at $l \sim 1.6$ mm reaches nearly 80% of P_{opt} at $l = 5$ mm. Thus, the loss of P due to leg length constraints for wearable applications would not be significant as long as legs are 1–2 mm long at least. Figure 3c, d show generated voltage V and N as a function of leg width when $F = F_{opt}$ for $l = 3$ mm. For constant F, reduction in leg width results in increase in N as well as V. Practically, $Bi - Te$ alloys (common type of TE material) are mechanically brittle such that leg width should be larger than a few hundreds of μm if $l \geq 1$ mm. Commonly, commercially available TEGs include legs with $w \sim 1.5$ mm, $N \sim 2/$cm², and $l \sim 1.5$ mm. If F is optimized near 10%, such TEGs would produce $P \sim 20$ μW/cm² and $V \sim 4$ mV/cm² under forced convection.

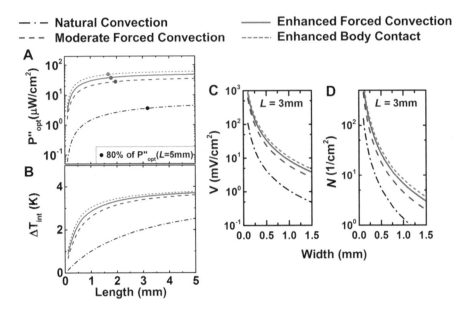

Fig. 3 Predicted (**a**) P'', (**b**) ΔT_{TEG}, (**c**) voltage V, and (**d**) number of TE couples N when F is optimized for each convection condition. TE leg length l is 3 mm for calculating V and N. The calculation assumes $\Delta T_{ext} = 8$ K and $\rho_{cont} = 20$ μOhm×cm²

4   Heatsinks and Airflow Configurations

This section shows and discusses experimental results obtained with various heatsinks and airflow configurations. Under different external conditions, internal thermal resistance of TEG module needs to be optimized in order to maximize output power. This section shows how internal thermal resistance can be engineered through varying fill factor of TEG modules. Three test configurations with different external thermal resistances will be considered: (1) perfect heatsink condition, (2) practical heatsink condition, and (3) wearing condition.

4.1   TEG Modules with Different Fill Factors

In order to vary internal thermal resistance, TEG modules with four different fill factors ($F = 6\%$, 11%, 21%, 33%) were fabricated. Figure 4a shows TEG module with $F = 33\%$. Dimensions and material parameters of TEG modules are identical as provided in Table 3. TE legs are made of $Bi - Te$ alloys, where $S_p = 196$ μV/K, $S_n = -198$ μV/K, $\rho_p = 0.9$ mOhm×cm, $\rho_n = 0.8$ mOhm×cm, $\kappa_p = 1.3$ W/(m×K) and $\kappa_n = 1.8$ W/(m×K) at room temperature. Leg dimensions are based on those widely

Fig. 4 (**a**) Fabricated TEG with $F = 33\%$ ($N = 12$). Lower image is the schematic for TEG configuration. (**b**) Fabricated TEG combined with plate fin heatsink ($D = 0.43$ or $N_f = 6$). Lower image is top-view schematic for heatsink

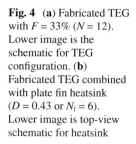

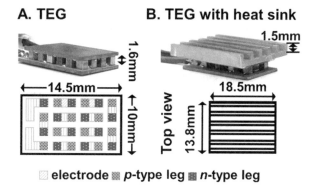

A. TEG

14.5mm — 1.6mm — 10mm

B. TEG with heat sink

1.5mm

18.5mm

Top view 13.8mm

▨ electrode ▨ p-type leg ▧ n-type leg

Table 3 Dimensions and materials parameters of fabricated TEGs and plate fin heatsinks

Parameter	Value
Individual leg width, w	1.4 mm
Individual leg length, l	1.6 mm
TEG substrate material	Cu (thickness = 530 μm)
TEG passivation material for substrates	Epoxy (thickness = 30 μm)
TEG electrode material	Cu (thickness = 35 μm)
Width of heatsink base, W	14.3 mm
Length of heatsink base, L	19.1 mm
Thickness of heatsink base	0.8 mm
Fin thickness, t	0.9 mm
Fin height, H	1.7 mm

adopted for commercial TEGs. To ensure uniform temperature distribution across substrate, TEGs employ Cu substrates. An insulating layer exists between Cu substrate and TEG electrode, which is ~35 μm thick epoxy composite including boron nitride and alumina powders. Such filler particles help to enhance in thermal conductivity of epoxy composite from ~0.2 W/(m×K) to the order of few W/(m×K).

4.2   Test Heatsinks

Among various types of heatsink, plate fin heatsinks are one of commonly used and the most studied. Thus, this section particularly investigates effect of plate fin geometries. For predicting performance of heatsink $h_{\text{eff, a}}$, airflow rate V_f, total heatsink area A_{hs} and heat transfer coefficient h_a are critical parameters. Due to restriction on fin height, optimization of fin density, space between fins g or shape is necessary. For developing laminar forced convection between parallel plates, Nusselt number is $Nu = 7, 55 + 0.024L^* - 1.14/(1 + 0.0358Pr^{0.17}L^{*-0.64})$, where $L^* = L/(2gRe_{2g}Pr)$, $Re_{2g} = V_f(2g)/\upsilon$ is parallel plate Reynolds number, υ is fluid kinematic viscosity, and

Pr is Prandtl number [13]. If *g* reduces, airflow speed and Re_{2g} would increase, leading to enhancement in h_a. Moreover, the smaller the fin space is, the larger the fin number and A_{hs} are. However, extremely small fin space would increase flow bypass and even interrupt airflow due to boundary layer effect and increased friction. Thus, there exists an optimum fin density.

Three types of plate fin heatsinks were fabricated and combined with TEG modules. Heatsink material was copper and fin density *D* was varied as 0.14 ($N_f = 2$), 0.29 ($N_f = 4$), and 0.43 ($N_f = 6$). Figure 4b shows especially TEG with heatsink (*D* = 0.43). Table 3 lists dimensions and materials for heatsinks. For easily soldering heatsinks to TEGs, base area of heatsink $A_{b, hs}$ is larger than TEG substrate (X1.8). As a reference, plain copper plate is also soldered to TEG. Generally, as flow rate becomes smaller, flow bypass easily increases with *D*, and optimum *D* decreases.

4.3   Perfect Heatsink Condition

With perfect heatsink, hot side and cold side of TEG will have the same temperature as thermal reservoirs ($T_h = T_s$ & $T_c = T_a$). This condition is made when external thermal resistances are negligible as compared to internal thermal resistance ($R_{th, h}$, $R_{th, c}$ $\ll R_{th, g}$). Figure 5a shows the experimental setup for evaluating TEG with perfect heatsink. TEG was sandwiched by heated and chilled *Cu* blocks that maintain ΔT_{ext}. To reduce in external thermal resistances, the entire system was compressed with dummy mass (~ 300 kPa). And to eliminate convective heat transfer, the entire setup was within vacuum chamber (~ 10^{-3} torr). V_{OUT} was recorded using load resistance (with $m \sim \sqrt{1 + ZT_g}$, where ZT_g is TE figure-of-merit of tested TEG) when variation of ΔT_{ext} reached below 0.3 K.

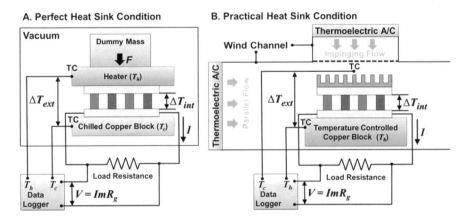

Fig. 5 Schematics for TEG evaluation setup under perfect heatsink condition (**a**), and practical heatsink condition (**b**). For practical heatsink condition, air flows in either parallel or perpendicular direction to TEG. For each flow direction, different wind channel was employed

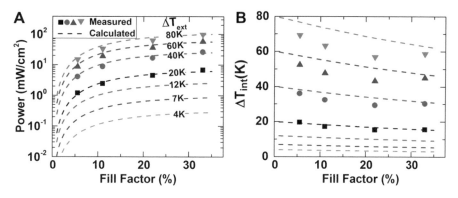

Fig. 6 (**a**) Generated power density, (**b**) internal temperature differential as a function of fill factor under perfect heatsink condition

Figure 6 shows measured data with varied ΔT_{ext} (T_c = 25 °C). Measured power monotonically improves as F increases. Here, ΔT_{int} = ΔT_{TEG} decreased ~15% with increased F. The model fit to estimated ΔT_{int} = ΔT_{TEG} provides $h_{eff, s}$ (= $h_{eff, a}$) of 2700 W/(m^2×K), indicating that $R_{th, g}/R_{th, h}$ (= $R_{th, g}/R_{th, c}$) ranges from 8 to 35. Therefore, when TEG is tightly connected with thermal reservoirs, high-density array of legs within large footprint is desirable.

4.4   Practical Heatsink Condition

With practical heatsink, external thermal resistances become typically larger than internal thermal resistance of TEG. The ratio $R_{th, c}/R_{th, g}$ may exceed 1000 for thin film TEG. Thermal resistance at TEG-air interface will reduce with integrated heatsink or airflow induced by walking or body movements.

Figure 5b shows the experimental setup for evaluating TEG under practical heatsink condition. TEG was placed on heated Cu block (T_h ~ 30 °C) while its top side is subjected to either natural convection (no airflow), parallel or impinging airflow (1.4 m/s). Two different wind channels and TE air conditioners (A/Cs) were used for each airflow configuration. TE A/C controlled air temperature at 18, 23 or 34 °C to make ΔT_{ext} = 4, 7, 12 K. For parallel flow configuration, however, ΔT_{ext} = 12 K could not be measured due to limited performance of dedicated TE A/C. For natural convection configuration, the setup for parallel flow configuration was used and enclosure made of aluminum foil covered TEG to eliminate of airflow around TEG. To mimic tight contact between skin and TEG, test TEG was attached to Cu block using adhesive tape. V_{OUT} was measured using load resistance with $m \sim \sqrt{1 + ZT_g}$, when variation of ΔT_{ext} maintained below 0.07 K.

Figure 7 shows the measured data of TEGs (F = 11%) with heatsinks and as a function of ΔT_{ext}. Due to large base area of heatsinks ($A_{b, hs}/A_g$ = 1.8), the model fit

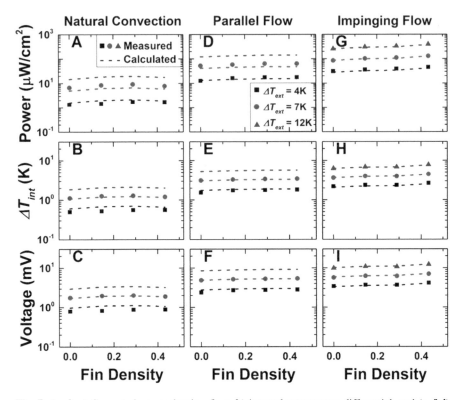

Fig. 7 (**a, d, g**) Generated power density, (**b, e, h**) internal temperature differential, and (**c, f, i**) output voltage of TEGs ($F = 11\%$) combined with plate fin heatsinks as a function of fin density under practical heatsink condition. Data were obtained when TEGs were subjected to (**a–c**) natural convection, (**d–f**) parallel flow (1.4 m/s), (**g–i**) impinging flow (1.4 m/s)

to $\Delta T_{int} = \Delta T_{TEG}$ shows that $h_{eff, a}$ is improved as 25 W/(m²×K) (natural convection), 130 W/(m²×K) (parallel flow), and 190 W/(m²×K) (impinging flow) even when there is no fin in heatsink (D or $N_f = 0$). Accordingly, the maximum power density is also enhanced as 7 μW/cm² (natural convection), 47 μW/cm² (parallel flow), and 126 μW/cm² (impinging flow) when $\Delta T_{ext} = 7$ K. Optimal fin density (D_{opt}) varies with convection configuration. D_{opt} is ~0.3 with natural convection, between 0.3 and 0.4 with parallel flow, and larger than 0.4 with impinging flow. When D increases from 0 to D_{opt} (about two-fold increase in surface area), effective heat transfer coefficient $h_{eff, a}$ and cold side thermal resistance $R_{th, c}$ improved 16% (natural convection), 18% (parallel flow), and 84% (impinging flow). Although D_{opt} may change with different fin heights, heatsink base area, and heatsink material, this result implies that excessively large D does not improve TEG–air thermal resistance as it will cause flow bypass and poor flow rate due to the overlap of fluid boundary layers.

4.5 Wearing Condition

Under wearing condition, external thermal resistances are typically larger than internal resistance $R_{th, h}$, $R_{th, c} \geq R_{th, g}$. Figure 8a, b shows the experimental setup for evaluating TEG under this situation. TEG with heatsink is on the wrist which is fastened by a rubber band. To enhance in skin–TEG contact, a piece of 0.5-mm thick thermal pad with nominal thermal conductivity of 12 W/(m×K) was inserted. An electric fan provides airflow either from the front (parallel flow) or from the top (impinging flow) of TEG. Anemometer with temperature sensor confirmed air velocity (1.4 m/s) and air temperature (~ 26 °C). The wrist temperature measured by K-type thermocouple was ~32 °C, thus $\Delta T_{int} = \Delta T_{TEG}$ was 6 K. The output voltage was acquired using load resistance with $m \sim \sqrt{1 + ZT_g}$ when variation of $\Delta T_{int} = \Delta T_{TEG}$ was below 0.5 K.

Figure 9 shows test data of TEG ($F = 11\%$) with heatsink ($N_f = 6$) measured under wearing condition. Initially, power density, $\Delta T_{int} = \Delta T_{TEG}$, and output voltage are large. When TEG is placed on the skin, then skin–TEG interface temperature increases rapidly, while TEG–air interface temperature requires few minutes to be saturated. Thermal pad is useful to instantly raise in TEG temperature, as it helps in thermal diffusion between the skin and TEG. During the first 10 s, power density and output voltage are similar to values predicted under perfect heatsink condition (Fig. 6). Such large transient response of TEG is advantageous when used together with voltage boost converter. Several voltage boost converters require few hundreds of millivolts to start operating. After a few minutes, TEG reaches thermal equilibrium. Under the steady state, power density, $\Delta T_{int} = \Delta T_{TEG}$ and output voltage approach ~15–100% of values obtained under practical heatsink condition (Fig. 7). Another important aspect of using thermal pad is that thermal time constant during the transient regime is ~50% smaller as compared to the cases without thermal pad, which is due to rapid thermal diffusion within thermal pad.

Under wearing condition, power density reduced to ~6.5 μW/cm^2 with natural convection and ~20 μW/cm^2 with parallel and impinging flows. The natural variation of $R_{th, h}$ and skin temperature and uneven skin–TEG contact are responsible for

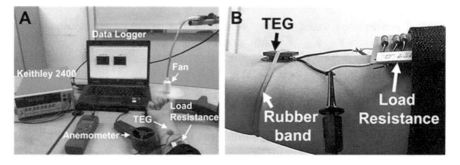

Fig. 8 (**a**) Experimental setup for wearing condition measurement. (**b**) TEG ($F = 11\%$) with plate fin heatsink ($N_f = 6$) on wrist for the measurement under wearing condition

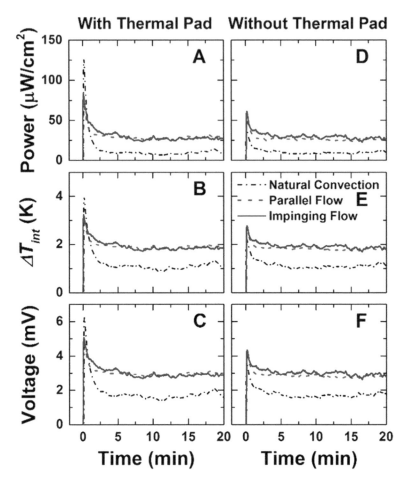

Fig. 9 (**a**, **d**) Generated power density, (**b**, **e**) internal temperature differential, and (**c**, **f**) output voltage of TEG ($F = 11\%$) combined with plate fin heatsink ($N_f = 6$) under wearing condition ($\Delta T_{int} = \Delta T_{TEG} \sim 6$ K). Data were obtained with (**a**–**c**) or without (**d**–**f**) thermal pad at the skin–TEG interface

reduced performance under wearing condition. The model fit to impinging flow condition suggests that $h_{eff, s}$ for skin contact significantly deteriorates to ~60 W/($m^2 \times$K), which is merely 12% of $h_{eff, s}$ for flat surface contact (contact with heated Cu block). Reduced $h_{eff, s}$ implies that skin–TEG interface should improve by applying larger pressure or properly filling interface gaps. Even with reduced power density, total TEG power generation P is expected to be ≥ 500 μW when the total substrate area becomes 20 cm² (approximate area for watchbands). P may be further increased to the order of mW, if the following scenarios can happen: (1) internal thermal resistance of TEG becomes greater by increasing in length of TE legs or (2) skin–TEG interfacial thermal resistance reduces by improving interfacial contact.

References

1. M. Kishi, H. Nemoto, T. Hamao et al., Micro-thermoelectric modules and their application to wristwatches as an energy source. Paper presented at 18th International Conference on Thermoelectrics, Baltimore, August 1999
2. V. Leonov, T. Torfs, P. Fiorini, et al., Thermoelectric converters of human warmth for self-powered wireless sensor nodes. IEEE Sensors J. **7**(5), 650–657 (2007)
3. V. Leonov, Thermoelectric energy harvesting of human body heat for wearable sensors. IEEE Sensors J. **13**(6), 2284–2291 (2013)
4. K.T. Settaluri, H. Lo, R.J. Ram, Thin thermoelectric generator system for body energy harvesting. J. Electron. Mater. **41**(6), 984–988 (2012)
5. S.J. Kim, J.H. We, B.J. Cho, A wearable thermoelectric generator fabricated on a glass fabric. Energy Environ. Sci. **7**(6), 1959 (2014)
6. F. Suarez, A. Nozariasbmarz, D. Vashaee, et al., Designing thermoelectric generators for self-powered wearable electronics. Energy Environ. Sci. **9**(6), 2099–2113 (2016)
7. H. Park, D. Kim, Y. Eom, et al., Mat-like flexible thermoelectric system based on rigid inorganic bulk materials. J. Phys. D: Appl. Phys., **50**, 494006 (2017)
8. Y.G. Lee, J. Kim, M.S. Kang, et al., Design and experimental investigation of thermoelectric generators for wearable applications. Adv. Mater. Technol. **2**(7), 1600292 (2017)
9. C.S. Kim, H.M. Yang, J. Lee, et al., Self-powered wearable electrocardiography using a wearable thermoelectric powergenerator, ACS Energy Letters, **3**, 501–507 (2018)
10. J.P. Heremans, M.S. Dresselhaus, L.E. Bell, et al., When thermoelectrics reached the nanoscale. Nat. Nanotechnol. **8**, 7 (2013)
11. C.B. Vining, An inconvenient truth about thermoelectrics. Nat. Mater. **8**(2), 83–85 (2009)
12. L. Chen, J. Gong, F. Sun, et al., Effect of heat transfer on the performance of thermoelectric generators. Int. J. Therm. Sci. **41**(1), 95–99 (2002)
13. R.A. Wirtz, W. Chen, R. Zhou, Effect of flow bypass on the performance of longitudinal fin heatsinks. J. Electron. Packag. **116**(3), 206 (1994)

Part III
Modeling of Thermoelectric Materials Properties

Modeling of Organic Thermoelectric Material Properties

Daniel B. Cooke and Zhiting Tian

1 Introduction

Thermoelectric devices are capable of direct conversion between thermal and electrical energy. In choosing a material for applications in TE devices, there are three important properties that govern its energy conversion efficiency. These are Seebeck coefficient S, thermal conductivity κ, and electrical conductivity σ. Seebeck coefficient, referred to also as thermopower, thermoelectric power, or thermoelectric sensitivity, is a measure of the magnitude of induced voltage as a result of temperature difference between hot and cold sides of thermoelement. Thermal conductivity describes the ability of material to conduct heat and a small value is needed to create a large temeprature gradient. Electrical conductivity quantifies how well a material facilitates the flow of electric current. These three separate properties are commonly combined into one non-dimensional term called the dimensionless figure of merit ZT which is determined by

$$ZT = \frac{S^2 \sigma}{\kappa} T, \tag{1}$$

T is absolute temperature. To reach a higher ZT value, larger S and σ and smaller κ are desired. However, these three parameters are intricate, and the goal is to optimize overall ZT. The higher the material's ZT value, the more efficient it is to TE applications.

D. B. Cooke
Virginia Polytechnic Institute and State University, Blacksburg, VA, USA

Z. Tian (✉)
Cornell University, Sibley School of Mechanical and Aerospace Engineering,
Ithaca, NY, USA
e-mail: zhiting@cornell.edu

© Springer Nature Switzerland AG 2021
S. Skipidarov, M. Nikitin (eds.), *Thin Film and Flexible Thermoelectric
Generators, Devices and Sensors*, https://doi.org/10.1007/978-3-030-45862-1_10

Until recently, TE materials have largely been composed of inorganic elements, namely bismuth (*Bi*), tellurium (*Te*), antimony (*Sb*), lead (*Pb*), etc. Through the development of bulk nanocomposite materials, values of ZT >1 are common [1–3]. However, these elements are toxic, rare, and costly to process. Having this in mind, organic materials pose to be a promising alternative. Compared to inorganics, organics are lightweight, flexible, abundant, and easy to manufacture.

However, ZT values of organic TE materials are inferior to inorganic counterparts. The highest reported ZT values for organic TE materials are around 0.4 [4], while most other organic materials possess much lower ZT values [3, 5–7]. It is hoped that through a deeper understanding to guide the material design, the performance of organic TE materials can be improved. Ultimately, what is desired is to understand thermal and charge carriers transport characteristics of organic TE materials. By using theoretical modeling, it can help better design and predict TE characteristics.

This chapter seeks to serve as an introduction to modeling material properties of organic TE materials. Section 1 provides an overview of what organic materials are and introduces the fundamentals of thermal and charge carriers transport. Section 2 covers in detail the important theories and concepts which are currently in use. Finally, in Section 3, specific research cases are given that show the application of the developed theories and in some cases how these models compare to experimental data.

1.1 Review of Transport Phenomena in Organic Semiconductors

In designing TE materials, two main mechanisms are important: thermal and charge carriers transport of the material. Currently, research is focused on how to model these mechanisms at nanoscale. In the present view of solid materials, the atomic structure is composed of free electrons and atoms which are arranged in a periodic arrangement called the lattice. Thermal transport in a material is due to the migration of free electrons and to the vibration of the lattice. The quantum of vibration of the lattice is called a phonon. Phonons exhibit both wave and particle-like behaviors. In metals, thermal transport is largely dominated by the transport of free electrons. In organic semiconductors, phonons are the major mechanism of thermal transport [8].

Thermal conductivity κ can be determined from

$$\kappa = \frac{1}{3}C_v v\lambda, \tag{2}$$

where C_v is volumetric heat capacity, v is phonon group velocity, and λ is mean free path. Group velocity is velocity at which phonon propagates through the material [9]. The mean free path $\lambda = v\tau$ is traveling distance of phonon before being scattered

Crystalline   Amorphous

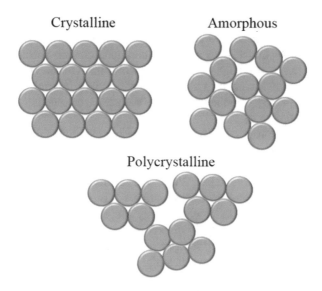

Polycrystalline

Fig. 1 Atoms or molecules of organic material can have crystalline, amorphous, or polycrystalline structures

by an obstacle. An obstacle could be an imperfection in crystalline structure or another particle (phonon or electron). It can be seen from Eq. (2) that, in general, the larger the mean free path, the higher the thermal conductivity. Crystalline materials are composed of ordered structures. Amorphous materials lack long-range order. Polycrystalline materials have multi-domains that are crystalline. An example of each structure is shown in Fig. 1.

In most cases, organic semiconductors have a disordered structure [10]. Crystalline structures allow phonons to travel farther, giving rise to higher κ. Conversely, amorphous materials restrict phonon movement and generally have low κ.

The nature of charge carriers transport in materials is largely dependent on electron–electron and electron–phonon interactions. For traditional inorganic semiconductors, electron–electron interactions mainly contribute to charge carriers transport while electron–phonon interactions play a smaller role [11]. For organic semiconductors, electron–phonon interactions are more dominant. The two main sources of electron–phonon interaction are the electron site energy and the electronic coupling. The electron site energy is influenced by both intra-molecular (internal) and inter-molecular (external) vibrations [11]. These dependences lead to interactions with phonons. Electron–phonon interactions arise also due to the dependence of electronic coupling on spacing and orientation of molecules [11]. In organic TE materials, electron–phonon interactions form a new particle called polarons [12].

The majority of studied organic TE materials are based on conductive polymers. Polymers usually form intermolecular covalent σ-bonds [3] and localize electrons in low-energy orbitals, which makes most polymers electrical insulators [13]. However, conductive polymers form additional π-bonds within the molecule [3]. Because of the mixture of σ- and π-bonded atoms, π band is separated into filled and empty bands which are separated by a gap in electronic excitation spectrum [14]. This energy gap between the Highest Occupied Molecular Orbital (HOMO) and the Lowest Unoccupied Molecular Orbital (LUMO) makes the polymer resemble a semiconductor [14, 15]. Conductive polymer's electrical conductivity can be further enhanced by doping.

Charge carriers transport mechanism for organic materials falls into three main categories: hopping transport, band-like transport, and multiple trapping and release (MTR) theory [16]. Hopping transport occurs when localized charge carriers jump from one localized site to another. When charge carrier hops to a site with lower energy, a phonon is emitted to compensate for the energy difference. Likewise, when charge carrier moves to a higher energy site, a phonon is absorbed. Hopping transport generally occurs in highly disordered materials [11]. In ordered crystals with delocalized charge carriers, such as pentacene [17], there is band-like transport. Because charge carriers move as a highly delocalized plane wave in a broad energy band [16], charge carriers show properties like metals with higher mobility. Some organic materials form polycrystalline films, which are different from amorphous and crystalline structures [18]. For these materials, charge carriers transport cannot be explained by hopping transport or band-like transport. In this case, MTR theory is used. In MTR theory, it is assumed that charge carriers are trapped in the localized states and then transport occurs in extended states [19]. The energy of the localized state and that of the mobility edge are separated [16]. If the energy of localized state is just below that of the mobility edge, the localized state traps charge carrier. By absorbing thermal energy, charge carrier can be released.

1.2   Review of Organic Materials

Organic semiconductors are largely composed of carbon (C), hydrogen (H), and oxygen (O). They can be broadly sorted into two main categories: small molecules or oligomers, such as pentacene, and conductive polymers, such as poly(3,4-ethylenedioxythiophene) (PEDOT) and polypyrrole (PPy) [11]. Although most organic TE materials to date are based on conductive polymers, small molecules are also attractive because those are easier to purify and crystallize [3].

The chemical structures of several organic TE materials are displayed in Fig. 2. Pentacene is one of the best semiconductors in the field of organic semiconductors [11]. Special attention is given due to its well-defined crystalline structure [20]. Because of this, pentacene is useful for studying crystal structure–transport relationships [17]. Rubrene is a derivative of the same family of small molecules. Bis-dithienothiophene (BDT) molecules form a crystal structure through $\pi - \pi$ stacking [21]. Normally, the stacking is coplanar and has high mobility [22]. PPy is an electroactive polymer with relatively high electrical conductivity and low thermal conductivity, making it a good match for TE materials. The undoped crystalline structure

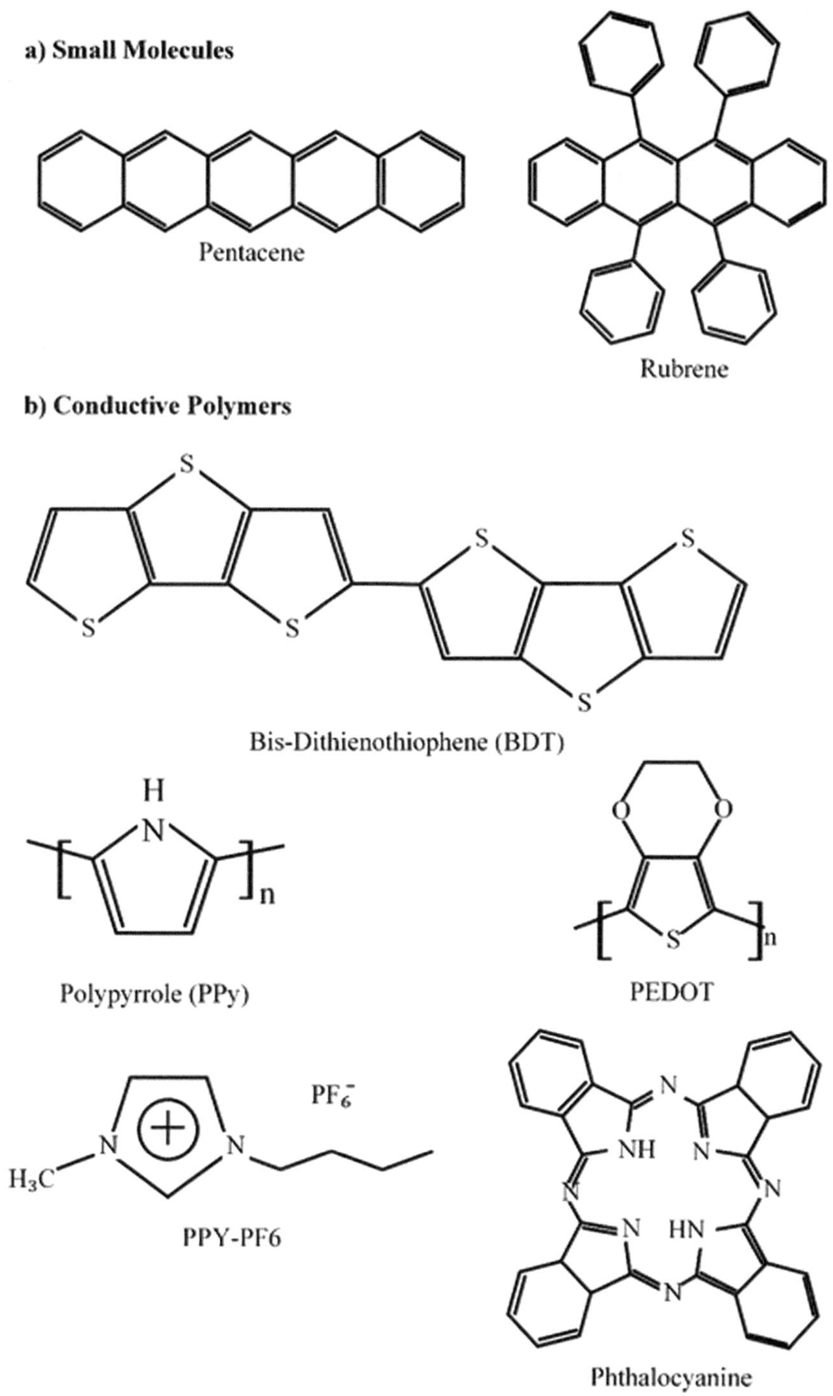
a) Small Molecules
Pentacene
Rubrene
b) Conductive Polymers
S
S
S
S
S
S
Bis-Dithienothiophene (BDT)
H
N
n
Polypyrrole (PPy)
O
O
S
n
PEDOT
PF6-
N
N
H3C
PPY-PF6
N
NH
N
N
N
N
HN
N
Phthalocyanine

Fig. 2 Structures of (**a**) small molecules and (**b**) conductive polymers

of PPy contains a periodic structure of tetragonal unit cell [5]. PEDOT is another organic semiconductor that has been widely studied due to its high electrical conductivity. Crystalline nanowires have been developed using PEDOT with ZT values as high as 0.25 [23]. This is achieved through control of the doping level. Phthalocyanine is 2D disc-like molecule. In the discotic phase, this molecule forms quasi-1D columns with an amorphous structure, which provides pathways for electron and/or hole transport [11].

2   Current Progress

As previously stated, two main mechanisms which are important to TE performance are thermal and charge carries transport of the material. To improve transport, several techniques have been developed to model these mechanisms. The first section introduces methods used to model phonon transport, while the following section focuses on charge carriers transport.

2.1   *Phonon Transport*

2.1.1   Phonon Boltzmann Transport Equation

At small distances which are of note in modeling phonon transport, traditional methods of determining heat transfer such as Fourier's law and energy conservation equations are inadequate [24]. This is because the length scales are approaching mean free paths of the phonons. In order to describe heat transfer at these scales, phonon Boltzmann transport equation (BTE) is used often. Using phonon BTE, the energy distribution carried by phonon f can be expressed as follows:

$$-\frac{f - f_0(T)}{\tau} = v \times \nabla f + \frac{\partial f}{\partial t},$$

(3)

where f_0 is equilibrium distribution, T is temperature, τ is relaxation time, v is group velocity, and t is time [24]. In this model, relaxation time approximation has been applied. Progress has been made in solving phonon BTE numerically and iteratively. This eliminates the need for relaxation time approximation [24].

2.1.2   Molecular Dynamics Simulation

Molecular dynamics (MD) simulations are another tool used to describe phonon transport. MD allows the simulation of material behavior at the atomic scale and can track the approximate trajectory of individual particles. MD requires the knowledge of potentials that describe the interaction between atoms. By solving Newton's equation of motion for every atom in the system of interest numerically, detailed information on the entire microscopic system can be obtained. MD simulation does not require a priori knowledge of heat conduction. Additionally, MD simulation can determine input parameters which are needed by BTE.

MD simulations can be divided into two categories: equilibrium and non-equilibrium. Equilibrium MD method uses Green–Kubo formalism. It works by observing small fluctuations of heat transfer in the material at thermal equilibrium. The phonon transport in the material is derived from linear response of these fluctuations [26]. Non-equilibrium MD method is also known as "the direct method" because it is directly analogous to the way thermal conductivity is measured experimentally in macroscopic materials [26, 27]. Two different approaches have been taken to create non-equilibrium heat conduction conditions: impose a temperature difference to calculate the heat flux or impose a heat flux to calculate the resulting temperature distributions [28].

2.1.3   Monte Carlo Simulation of Phonon Transport

Monte Carlo (MC) simulation is another method for describing phonon transport in an organic system. MC refers to a broad class of numerical algorithms which use random numbers to model systems [29]. The advantage of MC method is that it avoids directly solving high-dimensional BTE, which makes the calculation simpler [30]. The application of MC simulation to particle transport involves the free flight of phonons which encounter scattering [31]. The scattering and duration events are determined through random number sampling. The most common MC simulations used are the ensemble MC method and the phonon-tracing MC method. The ensemble MC simulation functions by tracking the path of all phonons in the simulation [32, 33]. Phonon-tracing MC works by tracking individual phonons independent of each other [34, 35]. Because of this, phonon tracing takes less computational power than ensemble method [36]. The process of phonon-tracing MC simulation is explained by Hua and Cao as follows [36]:

1. The number of phonons and those various properties are initialized.
2. Each phonon bundle is given properties such as position, traveling angle, polarization, etc., which are determined through a random process.
3. The phonon bundle is tracked until it is scattered.
4. If phonon bundle encounters a boundary, it is reflected into the domain. Type of scattering event (specular or diffusive) is determined randomly.
5. When phonon bundle does not encounter a boundary, it is set to a new position and the process starts at step 3.
6. Once the phonon bundle is absorbed at a boundary, the tracing process is complete for that specific phonon bundle.

This process is repeated for each phonon bundle which is to be simulated.

2.2 Charge Carriers Transport

2.2.1 Electron Boltzmann Transport Equation

The earliest concepts for charge carriers transport were based on the method developed by L. Friedmann which applied BTE to organic crystal semiconductors [37]. This equation uses analytical methods to statistically describe distribution of charge carriers in a system. From this, macroscopic property of the system may be determined. BTE is used to describe time evolution of charge carriers' distribution function, external fields, and thermal gradient [38]. When solving BTE, an assumption about relaxation time for energy band is normally made. In Friedman's model, anisotropic energy band is taken to have a constant relaxation time τ. This constant can be determined from experimental mobility data [39]. BTE can be written as follows:

$$\frac{f - f_0}{\tau} = -v_k \times \nabla f + \frac{e}{\hbar}\left(E + \frac{v_k \times H}{c}\right) \times \nabla f, \tag{4}$$

where f_0 is equilibrium distribution function, E and H are external electric and magnetic fields, and $v_k = \hbar^{-1}\nabla E_k$ is velocity of electron in Bloch state. Eq. (4) assumes that DC electric and magnetic fields are applied and that a constant temperature gradient is present. From this, it can be shown that the absolute value of S along ith crystallographic direction is as follows:

$$S^i = -\frac{1}{eT}\left(\frac{K_2^i}{K_1^i} - E_f\right), \tag{5}$$

where E_f is Fermi level and e is electron charge. The use of BTE in this manner is to calculate the order of magnitude of S for organic semiconductors which have a crystalline structure [37].

While this is a good foundation for TE transport, the main shortcoming of BTE presented is the treatment of scattering and relaxation time. The assumptions made by Friedman may be entirely incorrect under certain situations, and, therefore, BTE under constant relaxation time only provides a semi-classical answer. Recent developments on BTE included rigorous treatment of electron–phonon scattering and frequency-dependent relaxation times, yet the application to organic semiconductors remains unexplored [40, 41].

2.2.2 First-Principles Calculations of Charge Carriers Transport

In response to the shortcomings of empirical BTE, a new technique was developed. This method combines first-principles calculations with BTE to simulate TE effect. Because it is less dependent upon empirical models or fitting parameters for electronic band structure, this method is more reliable. Density functional theory (DFT) is one of the most widely used methods in first-principles calculations [5, 42]. DFT is a method to determine electronic structure of atoms and molecules [43]. The advantage of DFT is its computational efficiency which allows it to simulate more realistic systems compared to Quantum MC-based methods [43].

2.2.3 Monte Carlo Simulation of Charge Carriers Transport

As with phonon transport, MC simulation is also used in charge carriers transport. In order to utilize analytical models, a large number of free parameters must be provided to complete the calculation. MC simulation eliminates the need for inputting so many parameters and creates a universal method to describe hopping transport. This provides a more valid result. MC simulation for charge carriers transport generally follows six steps which are described by Ihnatsenka et al. [44]:

1. The first step is to initialize the random energy at site i derived from Gaussian of exponential density of states (DOS).
2. Initialize the placement of charge carriers using Fermi–Dirac occupation probabilities.
3. Choose hopping events. This is based on Miller–Abraham transition rate.
4. Calculate waiting time by drawing a random number from exponential waiting time distribution.
5. Calculate electric current density after a number of predefined jumps have occurred.
6. The final step is to calculate S, which is given by the following equation:

$$S(T) = \frac{E_f - E_{trans}}{|e|T},$$ (6)

where E_{trans} is transport energy.

Although MC technique provides a direct modeling of the hopping transport in organic TE materials and, therefore, gives the most accurate description of electronic conductivity, its disadvantage is that it demands extensive computational resources. This makes it difficult to use this technique to analyze and fit experimental data. A comparison between MC simulation and semi-analytical methods is displayed in Fig. 3. From the results, it is shown that two curves are qualitatively similar.

2.3 General Conclusion on Models

Of the theories presented above, the use of first-principles method coupled with BTE is currently the best way to describe charge carriers transport in organic materials. This method begins by using particular chemical and geometrical structure of the

Fig. 3 MC and semi-
analytical calculations for
Seebeck coefficient for
localization length $\alpha = a$
[44]. (Reprinted figure
with permission from [44].
Copyright (2015) by the
American Physical
Society)

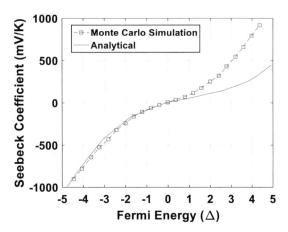

3 Application

system [45]. Therefore, it does not need to make assumptions like other models. For studying phonon transport in disordered (amorphous) structures, MD or MC simulations are used. When comparing these two, MD simulation contains more details but is more computationally expensive than the phonon-tracing MC simulation [36].

To demonstrate the mentioned theories, selection of various studies is presented. These studies look at multiple organic semiconductor materials and utilize several methodologies. In choosing these studies, preference is given to those with TE application as opposed to studying organic semiconductors in general.

The first study is presented by Schmechel and looks at the effect of doping levels on σ and S of p-type doped zinc-phthalocyanine [46]. The analysis uses Miller–Abraham formalism for hopping transport in Gaussian-disordered system. By using this method, σ, mobility μ, and S were found. The results of simulation are compared to measurements made by Maennig et al. [47]. Figure 4 shows the results for S. From this study, it is shown that there are differences between theory and experiment, especially for the polycrystalline sample. This may be due to the model, which considers only the effect of charge carriers transport on S and neglects the effect of phonon movement [46]. It is also important to point out that experimental data were used to fit the model to σ.

In a study by Wang et al., the properties of rubrene and pentacene crystals were modeled [48]. Because of the ordered structure, the model used first-principle calculations and Boltzmann transport theory. It was assumed that relaxation time was constant, and the bands were rigid. The model was used to determine dependency of S on temperature and density of charge carriers. The results of modeling are compared to experimental measurements of two organic TE materials. The comparison of experimental and calculated results for S vs density of charge carriers is shown in

Fig. 4 Simulated
temperature dependences
of S. Comparison of
analytical model by
Schmechel [46] to
measured results by
Maennig et al. [47] for
p-type doped zinc-
phthalocyanine.
(Reproduced from [46]
with the permission of AIP
Publishing)

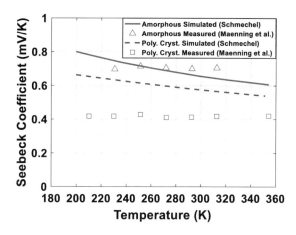

Fig. 5 Dependence of S
on density of charge
carriers. Comparison of
analytical to experimental
results for rubrene and
pentacene crystals.
(Reproduced from [48],
with the permission of AIP
Publishing)

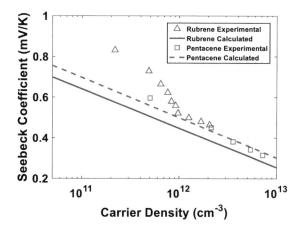

Fig. 5. The analytical results of both rubrene and pentacene show qualitative agreement with the experimental results. The discrepancy between rubrene results is due to the difference in density of state (DOS) of calculations and measurements. Pentacene results show especially good quantitative agreement. For pentacene crystal, ZT value of 0.8–1.1 was found, showing that organic TE materials have the potential to serve as replacement for inorganic materials.

A study by Kim and Pipe used hopping transport with Gaussian DOS model [4]. The model is used to show dependency of S on normalized density of charge carriers. The materials used are pentacene and PEDOT. The results are compared to experimental measurements made for the same materials [49, 50]. The results show good agreement with experimental measurements.

Mendels and Tessler used MC simulations with Gaussian disorder model to determine TE properties of disordered organic semiconductors [51]. Mainly, dependence of S and transport energy on temperature and concentration of charge carriers were modeled (Fig. 6). While not directly compared to experimental results, the

Fig. 6 Dependence of S
on temperature at
concentration of charge
carriers of $n = 10^{17}$ cm^{-3}.
(Reproduced with
permission from [51].
Copyright (2014) by the
American Chemical
Society)

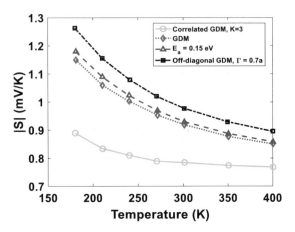

results qualitatively matched other experimental results [47, 52, 53]. Ultimately, this study provides a proof of concept for simulating material properties of disorder systems, and also sheds new light on mechanisms behind charge carriers transport characteristics.

Mi et al. used first-principle calculations along with Boltzmann transport theory to model the properties of BDT molecular crystal [21]. The results are used to find optimal doping which gives the highest ZT value at room temperature. The dependence of ZT at this doping value is then simulated over a range of temperatures (Fig. 7). Deformation potential (DP) theory was used to estimate relaxation time and MD simulation was used to calculate phonon thermal conductivity.

S and σ of PEDOT were calculated in a study by Ihnatsenka et al. [44]. The calculations were performed using both semi-analytical techniques, namely Boltzmann transport theory and MC simulations. Semi-analytical approach is used to fit to experimental data, which is difficult to perform with MC simulations. Two methods are compared and show qualitative agreement. However, semi-analytical approach was not modeling temperature dependence of κ well. In some cases, the models even show qualitative agreement. The results are compared to experimental results from Bubnova et al. and show good agreement [12, 23] (Fig. 8).

Lu et al. developed general analytical model based on hopping transport theory to simulate κ of disordered organic semiconductor PEDOT [10]. Simulation results are compared to measurements made by Duda et al. [54] (Fig. 9). Calculated results show good agreement. Furthermore, the results predict that phonons contribute more to κ than electrons do.

In a study by Li et al., first-principle calculations are used along BTE to study the properties of polypyrrole (PPy) and hexafluorophosphate $\left(PF_6^- \right)$-doped PPy (PPy–PF_6^-) [5]. MD simulations are used to determine κ. It was shown that crystalline PPy–PF_6^- has high σ while amorphous phase has high S and low κ. Despite this, ZT value of the crystalline phase is higher than that of the amorphous. Calculated results are compared to several experimental measurements [55–57]. Calculated results show good qualitative agreement with experimental results (Fig. 10). Overall, the results provide insight into the nature of organic TE material.

Fig. 7 Temperature dependence of ZT value at optimal doping. (Reprinted (adapted) with permission from [21]. Copyright (2015) American Chemical Society)

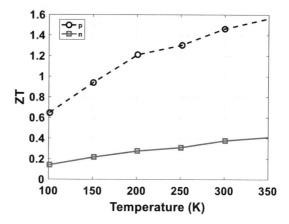

Fig. 8 Comparison of experimental results by Bubnova et al. [12, 23] vs calculated results from Ihnatsenka [44]. (Reprinted figure with permission from [44]. Copyright (2015) by the American Physical Society)

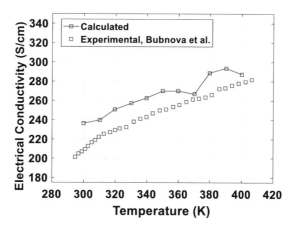

4   Conclusions and Future Work

This chapter has presented an overview of simulating, modeling, and designing organic TE material properties. The various prevailing models are shown. These include BTE, first-principles calculations combined with BTE, MD, and MC simulations. First-principles calculations are more suited to studying ordered (crystalline) organic materials. However, current state of theoretical research on transport in disordered organic semiconductors is still lacking [58]. Several studies which applied the mentioned models were shown. The proposed models act as guides to understanding thermal and charge carriers transport of organic semiconductors. This will lead to better design of organic thermoelectric devices in the future.

Fig. 9 Temperature
dependence of thermal
conductivity: (**a**)
contribution from phonons
and electrons and (**b**)
comparison to
experimental results by
Duda et al. [54].
(Reprinted from [10], with
permission from Elsevier)

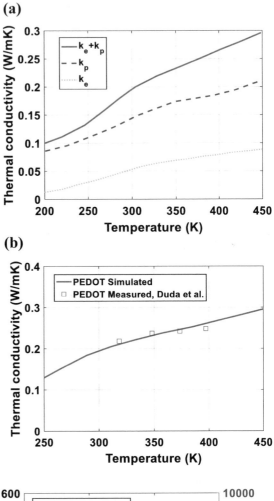

Fig. 10 Temperature
dependences of σ.
Comparison of
experimental work by Sato
et al., Yoon et al., and Lee
et al. [55–57] to simulation
by Li et al. [5]. (Reprinted
from [5], with permission
from Elsevier)

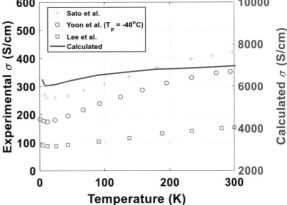

References

1. J.P. Heremans, M.S. Dresselhaus, L.E. Bell, D.T. Morelli, When thermoelectrics reached the nanoscale. Nat. Nanotechnol. **8**(7), 471–473 (2013). https://doi.org/10.1038/nnano.2013.129
2. Z. Tian, S. Lee, G. Chen, Heat transfer in thermoelectric materials and devices. J. Heat Transfer **135**(6), 061605–061615 (2013). https://doi.org/10.1115/1.4023585
3. Q. Zhang, Y. Sun, W. Xu, D. Zhu, Organic thermoelectric materials: Emerging green energy materials converting heat to electricity directly and efficiently. Adv. Mater. **26**(40), 6829–6851 (2014). https://doi.org/10.1002/adma.201305371
4. G. Kim, K.P. Pipe, Thermoelectric model to characterize carrier transport in organic semiconductors. Phys. Rev. B **86**(8) (2012). https://doi.org/10.1103/PhysRevB.86.085208
5. C. Li, H. Ma, Z. Tian, Thermoelectric properties of crystalline and amorphous polypyrrole: A computational study. Appl. Therm. Eng. **111**, 1441–1447 (2017). https://doi.org/10.1016/j.applthermaleng.2016.08.154
6. Q. Wei, M. Mukaida, K. Kirihara, Y. Naitoh, T. Ishida, Recent progress on PEDOT-based thermoelectric materials. Materials (1996–1944) **8**(2), 732 (2015)
7. C. Wan, X. Gu, F. Dang, T. Itoh, Y. Wang, H. Sasaki, et al., Flexible n-type thermoelectric materials by organic intercalation of layered transition metal dichalcogenide TiS_2. Nat. Mater. **14**(6), 622–627 (2015). https://doi.org/10.1038/nmat4251
8. T. Bergman, A. Lavine, F.P. Incropera, D. Dewitt, Introduction to conduction, in *Fundmentals of Heat and Mass Transfer*, 7th edn., (John Wiley, Hoboken, 2011), pp. 71–72
9. L. Brillouin, *Wave Propagation and Group Velocity* (Academic Press, New York, 1960)
10. N. Lu, L. Li, N. Gao, M. Liu, A unified description of thermal transport performance in disordered organic semiconductors. Org. Electron. **41**, 294–300 (2017). https://doi.org/10.1016/j.orgel.2016.11.019
11. V. Coropceanu, J. Cornil, D.A. da Silva Filho, Y. Olivier, R. Silbey, J.-L. Brédas, Charge transport in organic semiconductors. Chem. Rev. **107**(4), 926–952 (2007). https://doi.org/10.1021/cr050140x
12. M. Pope, C.E. Swenberg, *Electronic Processes in Organic Crystals and Polymers*, vol 2 (Oxford University Press, New York, 1999)
13. M. Bharti, A. Singh, S. Samanta, D.K. Aswal, Conductive polymers for thermoelectric power generation. Prog. Mater. Sci. **93**, 270–310 (2018). https://doi.org/10.1016/j.pmatsci.2017.09.004
14. O. Bubnova, Thermoelectric properties of conducting polymers, Ph. D Thesis, Linkoping University, Sweden, 2013
15. O. Bubnova, X. Crispin, Towards polymer-based organic thermoelectric generators. Energy Environ. Sci. **5**(11), 9345–9362 (2012). https://doi.org/10.1039/C2EE22777K
16. N. Lu, L. Li, D. Geng, M. Liu, A review for polaron dependent charge transport in organic semiconductor. Org. Electron. (2018). https://doi.org/10.1016/j.orgel.2018.05.053
17. C.C. Mattheus, G.A. de Wijs, R.A. de Groot, T.T.M. Palstra, Modeling the polymorphism of pentacene. J. Am. Chem. Soc. **125**(20), 6323–6330 (2003). https://doi.org/10.1021/ja0211499
18. G. Horowitz, R. Hajlaoui, P. Delannoy, Temperature dependence of the field-effect mobility of sexithiophene. Determination of the density of traps. J. Phys. III **5**(4), 355–371 (1995). https://doi.org/10.1051/jp3:1995132
19. B. Hartenstein, H. Bässler, A. Jakobs, K.W. Kehr, Comparison between multiple trapping and multiple hopping transport in a random medium. Phys. Rev. B **54**(12), 8574–8579 (1996). https://doi.org/10.1103/PhysRevB.54.8574
20. I.M. Rutenberg, O.A. Scherman, R.H. Grubbs, W. Jiang, E. Garfunkel, Z. Bao, Synthesis of polymer dielectric layers for organic thin film transistors via surface-initiated ring-opening metathesis polymerization. J. Am. Chem. Soc. **126**(13), 4062–4063 (2004). https://doi.org/10.1021/ja035773c

21. X.-Y. Mi, X. Yu, K.-L. Yao, X. Huang, N. Yang, J.-T. Lü, Enhancing thermoelectric figure-of-merit by low-dimensional electrical transport in phonon-glass crystals. Nano Lett. **15**(8), 5229–5234 (2015). https://doi.org/10.1021/acs.nanolett.5b01491
22. H. Sirringhaus, R.H. Friend, X.C. Li, S.C. Moratti, A.B. Holmes, N. Feeder, Bis(dithienothiophene) organic field-effect transistors with a high ON/OFF ratio. Appl. Phys. Lett. **71**(26), 3871–3873 (1997). https://doi.org/10.1063/1.120529
23. O. Bubnova, M. Berggren, X. Crispin, Tuning the thermoelectric properties of conducting polymers in an electrochemical transistor. J. Am. Chem. Soc. **134**(40), 16456–16459 (2012). https://doi.org/10.1021/ja305188r
24. J.G. Christenson, R.A. Austin, R.J. Phillips, Comparison of approximate solutions to the phonon Boltzmann transport equation with the relaxation time approximation: Spherical harmonics expansions and the discrete ordinates method. J. Appl. Phys. **123**(17), 174304 (2018). https://doi.org/10.1063/1.5022182
25. A.J. Minnich, Advances in the measurement and computation of thermal phonon transport properties. J. Phys. Condens. Matter **27**(5), 053202 (2015). https://doi.org/10.1088/0953-8984/27/5/053202
26. R.J. Stevens, L.V. Zhigilei, P.M. Norris, Effects of temperature and disorder on thermal boundary conductance at solid–solid interfaces: Nonequilibrium molecular dynamics simulations. Int. J. Heat Mass Transf. **50**(19), 3977–3989 (2007). https://doi.org/10.1016/j.ijheatmasstransfer.2007.01.040
27. Z. Tian, H. Hu, Y. Sun, A molecular dynamics study of effective thermal conductivity in nanocomposites. Int. J. Heat Mass Transf. **61**, 577–582 (2013). https://doi.org/10.1016/j.ijheatmasstransfer.2013.02.023
28. G. Chen, *Nanoscale Energy Transport and Conversion: A Parallel Treatment of Electrons, Molecules, Phonons, and Photons* (Oxford, Oxford; New York, 2005)
29. W. Kaiser, J. Popp, M. Rinderle, T. Albes, A. Gagliardi, Generalized kinetic Monte Carlo framework for organic electronics. Algorithms **11**(4), N.PAG (2018). https://doi.org/10.3390/a11040037
30. X. Ran, Y. Guo, M. Wang, Interfacial phonon transport with frequency-dependent transmissivity by Monte Carlo simulation. Int. J. Heat Mass Transf. **123**, 616–628 (2018). https://doi.org/10.1016/j.ijheatmasstransfer.2018.02.117
31. Z. Aksamija, Full band Monte Carlo simulation of phonon transport in semiconductor nanostructures, in *14th IEEE International Conference on Nanotechnology*, (2014), pp. 37–40. https://doi.org/10.1109/NANO.2014.6968118
32. V. Jean, S. Fumeron, K. Termentzidis, S. Tutashkonko, D. Lacroix, Monte Carlo simulations of phonon transport in nanoporous silicon and germanium. J. Appl. Phys. **115**(2), 024304 (2014). https://doi.org/10.1063/1.4861410
33. R.B. Peterson, Direct simulation of phonon-mediated heat transfer in a Debye crystal. J. Heat Transf. **116**(4), 815–822 (1994). https://doi.org/10.1115/1.2911452
34. T. Klitsner, J.E. VanCleve, H.E. Fischer, R.O. Pohl, Phonon radiative heat transfer and surface scattering. Phys. Rev. B **38**(11), 7576–7594 (1988). https://doi.org/10.1103/PhysRevB.38.7576
35. D.-S. Tang, B.-Y. Cao, Ballistic thermal wave propagation along nanowires modeled using phonon Monte Carlo simulations. Appl. Therm. Eng. **117**, 609–616 (2017). https://doi.org/10.1016/j.applthermaleng.2017.02.078
36. Y.-C. Hua, B.-Y. Cao, An efficient two-step Monte Carlo method for heat conduction in nanostructures. J. Comput. Phys. **342**, 253–266 (2017). https://doi.org/10.1016/j.jcp.2017.04.042
37. L. Friedman, Transport properties of organic semiconductors. Phys. Rev. **133**(6A), A1668–A1679 (1964). https://doi.org/10.1103/PhysRev.133.A1668
38. J. Chen, D. Wang, Z. Shuai, First-principles predictions of thermoelectric figure of merit for organic materials: Deformation potential approximation. J. Chem. Theory Comput. **8**(9), 3338–3347 (2012). https://doi.org/10.1021/ct3004436

39. D. Wang, W. Shi, J. Chen, J. Xi, Z. Shuai, Modeling thermoelectric transport in organic materials. Phys. Chem. Chem. Phys. **14**(48), 16505–16520 (2012). https://doi.org/10.1039/C2CP42710A

40. G.S. Jung, J. Yeo, Z. Tian, Z. Qin, M.J. Buehler, Unusually low and density-insensitive thermal conductivity of three-dimensional gyroid graphene. Nanoscale **9**(36), 13477–13484 (2017). https://doi.org/10.1039/C7NR04455K

41. B. Qiu, Z. Tian, A. Vallabhaneni, B. Liao, J.M. Mendoza, O.D. Restrepo, et al., First-principles simulation of electron mean-free-path spectra and thermoelectric properties in silicon. EPL (Europhysics Letters) **109**(5), 57006 (2015). https://doi.org/10.1209/0295-5075/109/57006

42. J. Yang, Q. Fan, Y. Ding, X. Cheng, Predicting thermoelectric performance of eco-friendly intermetallic compound p-type CaMgSi from first-principles investigation. J. Alloys Compd. **752**, 85–92 (2018). https://doi.org/10.1016/j.jallcom.2018.04.166

43. T. van Mourik, M. Bühl, M.-P. Gaigeot, Introduction: Density functional theory across chemistry, physics and biology. Philos Trans. Math. Phys. Eng. Sci. **372**(2011), 1–5 (2014)

44. S. Ihnatsenka, X. Crispin, I. Zozoulenko, Understanding hopping transport and thermoelectric properties of conducting polymers. Phys. Rev. B: Condens. Matter Mater. Phys. **92** (2015). https://doi.org/10.1103/PhysRevB.92.035201

45. N. Lu, L. Li, M. Liu, A review of carrier thermoelectric-transport theory in organic semiconductors. Phys. Chem. Chem. Phys. **18**(29), 19503–19525 (2016). https://doi.org/10.1039/C6CP02830F

46. R. Schmechel, Hopping transport in doped organic semiconductors: A theoretical approach and its application to *p*-doped zinc-phthalocyanine. J. Appl. Phys. **93**(8), 4653–4660 (2003). https://doi.org/10.1063/1.1560571

47. B. Maennig, M. Pfeiffer, A. Nollau, X. Zhou, K. Leo, P. Simon, Controlled p-type doping of polycrystalline and amorphous organic layers: Self-consistent description of conductivity and field-effect mobility by a microscopic percolation model. Phys. Rev. B **64**(19), 195208 (2001). https://doi.org/10.1103/PhysRevB.64.195208

48. D. Wang, L. Tang, M. Long, Z. Shuai, First-principles investigation of organic semiconductors for thermoelectric applications. J. Chem. Phys. **131**(22), 224704 (2009). https://doi.org/10.1063/1.3270161

49. K. Harada, M. Sumino, C. Adachi, S. Tanaka, K. Miyazaki, Improved thermoelectric performance of organic thin-film elements utilizing a bilayer structure of pentacene and 2,3,5,6-t etrafluoro-7,7,8,8-tetracyanoquinodimethane (F4-TCNQ). Appl. Phys. Lett. **96**(25), 253304 (2010). https://doi.org/10.1063/1.3456394

50. K.P. Pernstich, B. Rössner, B. Batlogg, Field-effect-modulated Seebeck coefficient in organic semiconductors. Nat. Mater. **7**(4), 321–325 (2008). https://doi.org/10.1038/nmat2120

51. D. Mendels, N. Tessler, Thermoelectricity in disordered organic semiconductors under the premise of the Gaussian disorder model and its variants. J. Phys. Chem. Lett. **5**(18), 3247–3253 (2014). https://doi.org/10.1021/jz5016058

52. A. Nollau, M. Pfeiffer, T. Fritz, K. Leo, Controlled n-type doping of a molecular organic semiconductor: Naphthalenetetracarboxylic dianhydride (NTCDA) doped with bis(ethylenedithio)-tetrathiafulvalene (BEDT-TTF). J. Appl. Phys. **87**(9), 4340–4343 (2000). https://doi.org/10.1063/1.373413

53. Y. Xuan, X. Liu, S. Desbief, P. Leclère, M. Fahlman, R. Lazzaroni, et al., Thermoelectric properties of conducting polymers: The case of poly(3-hexylthiophene). Phys. Rev. B **82**(11), 115454 (2010). https://doi.org/10.1103/PhysRevB.82.115454

54. J.C. Duda, P.E. Hopkins, Y. Shen, M.C. Gupta, Thermal transport in organic semiconducting polymers. Appl. Phys. Lett. **102**(25), 251912 (2013). https://doi.org/10.1063/1.4812234

55. W.P. Lee, Y.W. Park, Y.S. Choi, Metallic electrical transport of pf6-doped polypyrrole: Dc conductivity and thermoelectric power. Synth. Met. **84**(1), 841–842 (1997). https://doi.org/10.1016/S0379-6779(96)04174-4

56. K. Sato, M. Yamaura, T. Hagiwara, K. Murata, M. Tokumoto, Study on the electrical conduction mechanism of polypyrrole films. Synth. Met. **40**(1), 35–48 (1991). https://doi.org/10.1016/0379-6779(91)91487-U
57. C.O. Yoon, H.K. Sung, J.H. Kim, E. Barsoukov, J.H. Kim, H. Lee, The effect of low-temperature conditions on the electrochemical polymerization of polypyrrole films with high density, high electrical conductivity and high stability. Synth. Met. **99**(3), 201–212 (1999). https://doi.org/10.1016/S0379-6779(98)01494-5
58. S.D. Baranovskii, Theoretical description of charge transport in disordered organic semiconductors. Phys. Status Solidi B **251**(3), 487–525 (2014). https://doi.org/10.1002/pssb.201350339

High Thermoelectric Properties in Quasi-One-Dimensional Organic Crystals

Ionel Sanduleac and Anatolie Casian

1 Introduction

It is known that the global energy demand increases each year. At the same time, almost 70% of produced energy is lost as waste heat into atmosphere contributing also to global warming. Therefore, opportunity to recover even a small part of the lost energy is considered as very attractive subject of material science and industrial technologies. It is expected that, namely, thermoelectric (TE) power generation as direct conversion technique of waste heat into electricity will become a sustainable solution for that.

Capacity of TE devices, such as generators (TEGs) and coolers (TECs), is directly related to effectiveness of used p- and n-type TE materials. Choice of materials for intended use in TE device is based on dimensionless TE figure of merit ZT (Ioffe parameter [1]), $ZT = \sigma S^2 T/\kappa$, where σ is electrical conductivity, S is Seebeck coefficient, and $\kappa = \kappa^L + \kappa^e$ is thermal conductivity of TE material. Here κ^L and κ^e are contribution of lattice and free charge carriers to κ, and T is temperature. Product $PF = \sigma S^2$ is called power factor. Values of ZT as high as possible are needed. Now the most used TE materials are alloys based on Bi_2Te_3 (low-temperature TE materials up to 300 °C) and $SiGe$ (high-temperature TE materials up to 1000 °C) with ZT ~1. This is low ZT value. To make TEGs commercially competitive with conventional electrical generators, it needs to use TE materials with $ZT > 3$.

In the last decade, organic functional materials attract more and more attention of researches and engineers as materials with more diverse and often unusual properties in comparison with inorganic ones. Due to the abundance of raw materials and low-cost production technology friendly to environment, these materials are

I. Sanduleac (✉) · A. Casian
Technical University of Moldova, Chisinau, Moldova
e-mail: acasian@mail.utm.md

© Springer Nature Switzerland AG 2021
S. Skipidarov, M. Nikitin (eds.), *Thin Film and Flexible Thermoelectric Generators, Devices and Sensors*, https://doi.org/10.1007/978-3-030-45862-1_11

very prospective for different applications. There is already a new generation of electronic devices based on organic functional materials.

We have predicted theoretically that some Q1D organic crystals after optimization of parameters may have improved TE properties with $ZT \sim 20$ at room temperature [2, 3]. In molecular nanowires of conducting polymers, $ZT \sim 15$ at room temperature were predicted [4]. However, these predictions were made strictly for 1D physical models.

Conducting polymers are intensively investigated for TE applications. Recently, significant progress has been achieved on TE polymers based on PEDOT:PSS [5]. In PEDOT:PSS thin films, $ZT \sim 0.42$ at $T = 300$ K was reported [6], and $ZT \sim 0.44$ in PEDOT nanowire/PEDOT composites [7]. Another prospective research direction is the use of mixed organic-inorganic composites [8]. In phenyl acetylene doped with silicon nanoparticles, $ZT = 0.57$ at $T = 300$ K was reported [9]. A more detailed state-of-the-art of polymer nanocomposites for TE applications is presented in [10]. Optimized organic TE material should have simultaneously high σ and high S [11]. Controlling of dopant distribution within conjugated polymer films is important in order to achieve high charge carriers transport properties. It is found [12] that reducing in energetic disorder while increasing in positional disorder can lead to high values of PF. By suitable engineering of PEDOT crystalline structure, semi-metallic behavior has been reported by Crispin et al. [13]. The corresponding electronic structure was referred to as bipolaron network, and both σ and S were positively affected. In [14], the effect of oxidant concentration on PEDOT:TOS oxidation levels, σ and S, has been investigated. Decoupling of σ and S leading to increase in PF is observed [15].

The recent progress in the development of TE materials based on conjugated polymers is analyzed in [16, 17]. It is demonstrated [18] that effective $\pi - \pi$ stacking is more important to engineering materials with enhanced mobility of charge carriers in polymers than the long-range order. A qualitative analysis of morphology-mobility dependence in PEDOT:TOS is provided in [19]. A review of TE properties and κ of conducting polymer nanocomposites is presented in [20]. In the last decades, scientific research of TE materials has been boosted significantly by new discoveries in physics: wave effects in phonon transport, correlated electron physics, and unconventional transport in organic materials [21]. TEG made of p-type optimized PEDOT:TOS and n-type non-optimized TTF-TCNQ/PVC, which generates power of ~ 0.128 μW at $\Delta T = 10$ K and 0.27 μW at $\Delta T = 30$ K, was reported [5]. Polymeric TE device is realized [22] based on highly conductive ($\sigma \approx 2500$ S/cm) structure of ordered film of PEDOT:PSS with S of 20.6 μV/K, in-plane κ of 0.64 W/(m×K), and peak PF of 10^7 μW/(m×K^2) at room temperature. Ability to integrate FS-PEDOT:PSS TE modules with textiles to power wearable electronics by harvesting human body's heat was demonstrated. High thermal stability of organic TE devices is also confirmed. Printing as high-throughput processing of organic thin-film TE devices is described in [23].

The qualitative theoretical understanding of fundamental principles is indispensable while designing high-performance organic TE materials, because it can shed light on the fundamental processes of charge and heat transport. Parameter-free computational methodology was elaborated [24] to predict TE properties of organic materials, which combines density functional theory (DFT) calculations for band structures, Boltzmann transport theory for electrical transport coefficients, deformation potential theory for electron-phonon scatterings, and nonequilibrium molecular dynamics simulations for phonon transport properties.

Another category of promising TE materials is highly conducting quasi-one dimensional (Q1D) organic crystals. The crystals are formed of segregate chains or stacks of molecules so that distance between two nearest molecules along the chain is much smaller than distance between adjacent chains. Because of such internal structure, the overlap of electronic wave functions along chains becomes sufficient to determine band-type charge carriers transport, whereas between chains the overlap is very small and charge carriers transport is of hopping type. This peculiarity determines the pronounced quasi one dimensioanlity of the crystals. In the following, TE properties of Q1D organic crystals of p-type TTT_2I_3 [25–27] and of n-type $TTT(TCNQ)_2$ [28] will be analyzed.

1.1 Features of Electrical Conductivity in Organic Molecular Solids (Molecular Crystals)

Charge carriers transport (electrical conductivity) in organic semiconductor materials (molecular crystals) is generally different from that in atomic crystals, where it is governed by energy difference between Fermi level (E_F) and the highest occupied molecular orbital (HOMO) state (for hole) or the lowest unoccupied molecular orbital (LUMO) state (for electron). Molecular crystals with Q1D structure possessing weakly interacting interfaces show pinning of Fermi level near HOMO or LUMO. In relation to this Fermi level pinning effect, we observe n- or p-type charge carriers transport property without intentional impurity doping, i.e., there is a symmetry between electrons and holes. These effects seem to be determined by the host molecule itself. However, increase in concentration of charge carriers can be achieved by additional doping as well.

Figure 1 shows the evolution of electronic structure from single molecule to molecular solid consisting of polyatomic molecules with energy bands typically appearing in organic solid. Electrons at deeper levels are localized in atomic potential wells (core levels) and thus have feature of those in atomic orbitals because of very high potential barrier between atoms. The upper energy levels, MO (molecular orbitals) levels, involve interatomic interaction to form delocalized molecular orbitals.

Since, in many organic solids, molecules interact only by weak van der Waals interaction, wave functions of occupied valence states and lower unoccupied states are mainly localized on each molecule, yielding a narrow intermolecular energy band usually of width < 0.5 eV generally known as *conduction band*. Thus, electronic structure of organic solid approximately preserves that of a molecule, and the

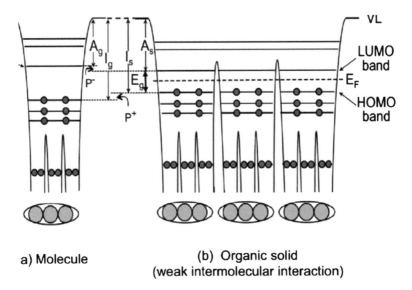

a) Molecule

(b) Organic solid
(weak intermolecular interaction)

Fig. 1 Evolution of electronic structure, from single molecule (**a**) to organic solid (**b**). When intermolecular electronic interaction is weak, width of energy bands is very narrow (**b**). VL is vacuum level, E_F Fermi level, A_g electron affinity of gas phase, A_s electron affinity of solid, I_g ionization energy of gas phase, I_s ionization energy of solid, P⁻ polarization energy for negative ion in solid, P⁺ polarization energy for positive ion in solid, E_g bandgap [29]

validity of usual band theory (atomic crystals with clearly defined conduction and valence bands) is often limited in discussing charge carriers transport in organic solid, which means that such organic solid acts often as one of two options, in some cases as single molecule and in other cases as molecular crystal. This situation in electronic structure of organic solid allows us to simply write the band structure such as HOMO and LUMO levels by using "narrow conduction band" due to very narrow bandwidth as in Fig. 1b which is Q1D case.

1.2 Quasi-One-Dimensional Crystals of TTT₂I₃

Q1D organic crystals of tetrathiotetracene-iodide, TTT_2I_3, have the form of needles of a length of 3–6 mm and dimensions in transversal directions ~30–60 μm, almost ready legs for thermocouples. From structural point of view, the crystals are formed from segregate stacks or chains of TTT molecules and iodine chains. The lattice constants are $a = 18.35$ Å, $b = 4.96$ Å, and $c = 18.46$ Å. Figure 1 shows the structure along **b** axis of TTT_2I_3 crystal (Fig. 2).

TTT_2I_3 is of mixed valence. In stoichiometric crystals, two molecules provide in average of two electrons for three iodine atoms. The iodine chain is formed from I_3^- ions, with strongly localized electronic wave functions. Thus, charge carriers transport along iodine chains can be neglected in comparison to that along TTT chains.

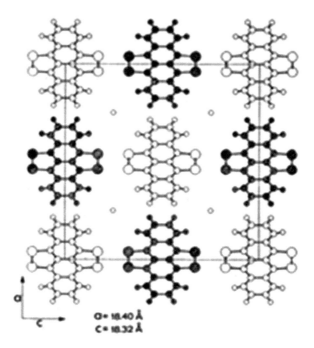

Fig. 2 Structure of TTT$_2$I$_3$ crystal along **b** axis [26]

Electrical conductivity of the crystal is of p - type, provided by TTT chains. Along TTT chains, electrical conductivity σ is almost of three orders of magnitude larger than in transversal to chains direction.

In [30–32] we have studied TE properties of TTT$_2$I$_3$ crystals in 2D physical model and in [33] 3D model. Since interchain interaction is very weak, it is allowed in the first approximation, to apply the band representation for describing charge carriers transport. As a result, Hamiltonian of the crystal in 3D physical model has the form [33]:

$$H = \sum_k E(k)a_k^+ a_k + \sum_q \hbar\omega_q b_q^+ b_q + \sum_{k,q} A(k,q)\left(b_q + b_{-q}^+\right)a_k^+ a_{k-q}$$
$$+ \left(I_l V_{0i} \sum_{l=1}^{N_i} e^{-iq\eta_l} + I_d V_{0d} e^{-E_a/k_0 T} \sum_{m=1}^{N_d} e^{-iqr_m} \right) V^{-1} \sum_{k,q} a_k^+ a_{k-q} \qquad (1)$$

where the first term describes hole energy in tight binding and nearest neighbor approximations, and $E(k)$ is hole energy with quasi-wave vector k and projections (k_x, k_y, k_z), measured from the top of conduction band:

$$E(k) = -2w_1\left(1-\cos k_x b\right) - 2w_2\left(1-\cos k_y a\right) - 2w_3\left(1-\cos k_z c\right), \qquad (2)$$

where w_1, w_2, and w_3 are transfer energies of hole from the given molecule to the nearest ones along lattice vectors b, a, and c; x, y, and z axes are pointing along b, a, and c with axis x along conductive TTT chains. It is considered that w_1 is much greater than w_2 and w_3. The second term in Eq. (1) describes energy of phonons. As it can be shown, charge carriers interact mainly with longitudinal acoustic phonon modes of frequency ω_q, which has the form:

$$\omega_q^2 = \omega_1^2 \sin^2\left(bq_x/2\right) + \omega_2^2 \sin\left(aq_y/2\right) + \omega_3^2 \sin\left(cq_z/2\right), \tag{3}$$

where phonon's quasi-wave vector q has projections (q_x, q_y, q_z), and ω_1, ω_2, and ω_3 are cut-off frequencies in x, y, and z directions, $\omega_1 \gg \omega_2$, ω_2.

The third term in Eq. (1) describes hole-phonon interaction. Two of the most important interactions of holes with acoustic phonons are considered. The first interaction is similar to that of deformation potential with three coupling constants $(w_1', w_2', \text{and } w_3')$ determined by variation of transfer energies with respect to intermolecular distances. The second interaction is like that of polaron and it is caused by variation, due to the same acoustic phonons, of polarization energy of molecules surrounding conduction hole. The coupling constant of this interaction is determined by the mean polarizability of TTT molecule α_0.

The square of the absolute value of matrix element describing hole-phonon interaction has the form:

$$
\begin{aligned}
\left|A(\mathbf{k,q})\right|^2 = 2\hbar / \left(MN\omega_q\right) \Big\{ & w_1'^2 \left[\sin\left(k_x b\right) - \sin\left(\left(k_x - q_x\right)b\right) + \gamma_1 \sin\left(q_x b\right)\right]^2 + \\
& + w_2'^2 \left[\sin\left(k_y a\right) - \sin\left(\left(k_y - q_y\right)a\right) + \gamma_2 \sin\left(q_y a\right)\right]^2 \\
& + w_3'^2 \left[\sin\left(k_y c\right) - \sin\left(\left(k_y - q_y\right)c\right) + \gamma_3 \sin\left(q_y c\right)\right]^2 \Big\}.
\end{aligned}
\tag{4}
$$

Here N is number of molecules constituting volume of the crystal considered and M is mass of TTT molecule. Parameters γ_1, γ_2, and γ_3 have the mean of amplitude ratios of the second interaction to the first one for longitudinal (b) and transversal (a and c) directions, respectively:

$$\gamma_1 = 2e^2\alpha_0 / \left(b^5 w_1'\right), \gamma_2 = 2e^2\alpha_0 / \left(a^5 w_2'\right), \gamma_3 = 2e^2\alpha_0 / \left(c^5 w_3'\right) \tag{5}$$

where e is electric charge of carrier.

The fourth term in Eq. (1) describes hole scattering by point-like and neutral impurities and by thermally activated defects with activation energy E_a, and I_i and I_d are energies (potentials) of hole interaction with point-like impurity and defect, respectively. V_{0i} and V_{0d} are volumes of action of respective potentials. Since conduction band is not very large, we have considered the whole Brillouin zone as a range of values for wave vectors k and q for holes and phonons.

1.3 Charge and Energy Transport

Let us consider that weak electric field and low temperature gradient are applied along x direction of conductive TTT chains and charge and energy transport take place in the same x direction. For nonequilibrium distribution function, we obtain kinetic equation of Boltzmann type. For temperatures $T \sim 1$ K, phonon energy [34] and transversal component of charge carrier's kinetic energy can be neglected in scattering processes, because these energies are much lower than hole longitudinal kinetic energy. The linearized kinetic equation is solved analytically, and electrical conductivity σ_{xx}, Seebeck coefficient S_{xx}, electronic thermal conductivity κ_{xx}^e, and TE figure of merit $(ZT)_{xx}$ can be expressed through transport integrals R_n as follows [32]:

$$\sigma_{xx} = \sigma_0 R_0, \quad S_{xx} = \left(k_0/e\right)\left(2w_1/k_0 T\right) R_1/R_0, \tag{6}$$

$$\kappa_{xx}^e = \left[4w_1^2 \sigma_0/\left(e^2 T\right)\right]\left(R_2 - R_1^2/R_0\right), \quad (ZT)_{xx} = \sigma_{xx} S_{xx}^2 T/\left(\kappa_{xx}^L + \kappa_{xx}^e\right), \tag{7}$$

where

$$\sigma_0 = \left(2e^2 M v_{s1}^2 w_1^3 r\right)/\left(\pi^2 \hbar a b c \left(k_0 T\right)^2 w_1'^2\right), \tag{8}$$

Here $r = 4$ is number of molecular chains per transversal section of elementary cell, κ_{xx}^L is lattice thermal conductivity, v_{s1} is sound velocity along chains, and R_n are transport integrals:

$$R_n = \int_0^2 d\varepsilon \int_0^\pi d\eta \int_0^\pi d\varsigma \left(2-\varepsilon\right) n_{\varepsilon,\eta,\varsigma} \left(1-n_{\varepsilon,\eta,\varsigma}\right) \times$$

$$\frac{\left[\varepsilon + d_1\left(1-\cos\eta\right) + d_2\left(1-\cos\varsigma\right) - \left(1+d_1+d_2\right)\varepsilon_F\right]^n}{\gamma_1^2\left(\varepsilon-\varepsilon_0\right)^2 + D_0 + D_1 e^{-E_a/k_0 T} + \left\{\begin{array}{l} d_1^2\left(1+\gamma_2^2 + 2\sin^2\eta - 2\gamma_2\cos\eta\right) \\ +d_2^2\left(1+\gamma_3^2 + 2\sin^2\varsigma - 2\gamma_3\cos\varsigma\right)\end{array}\right\}/\left(8\varepsilon\left(2-\varepsilon\right)\right)} \tag{9}$$

New variables were introduced: dimensionless energy along TTT chains $\varepsilon = E(k_x)/2w_1 = (1 - \cos k_x b)$, $\eta = k_y a$ and $\varsigma = k_z c$, and $n_{\varepsilon,\eta,\varsigma}$ is Fermi distribution function in these new variables. $\varepsilon_0 = (\gamma_1 - 1)/\gamma_1$ is dimensionless resonance energy, $d_1 = w_2/w_1 = w_2'/w_1'$, $d_2 = w_3/w_1 = w_3'/w_1'$, and $\varepsilon_F = E_F/2w_1$ is dimensionless 1D Fermi energy in unities of $2w_1$. 3D Fermi energy will be $2w_1(1 + d_1 + d_2)\varepsilon_F$. The parameters D_0 and D_1 describe hole scattering on point-like impurities and on defects, respectively. D_0 and D_1 are proportional to impurity and defect concentrations and can be made very small in pure and perfect crystals.

Before analyzing TE properties, let us consider the peculiarities of relaxation time as a function of carrier energy $\tau(\varepsilon)$, $\varepsilon = [0, 2]$ in 1D approximation:

$$\tau(\varepsilon) \sim \frac{\left[\varepsilon(2-\varepsilon)\right]^{1/2}}{\gamma_1^2(\varepsilon - \varepsilon_0)^2 + D_0 + D_1 e^{-E_a/k_0 T}}. \tag{10}$$

If $0 < \varepsilon_0 < 2$, $D_0 << 1$, and $D_1 << 1$, $\tau(\varepsilon)$ obtains a sharp maximum for narrow strip of states around ε_0. For p-type materials $\gamma_1 > 0$, and maximum of $\tau(\varepsilon)$ appears in lower half of conduction band. For n-type materials $\gamma_1 < 0$, and maximum of $\tau(\varepsilon)$ occurs in upper half of conduction band. As a consequence of such dependence of $\tau(\varepsilon)$, charge carriers occupying states close to this maximum will show an increased mobility. Figures 3 and 4 show dependences $\tau(\varepsilon, \gamma_1)$ for p- and n-type crystals at $T = 300$ K and $D_0 = 0.1$.

Thus, parameter γ_1 plays an important role in determining maximum of $\tau(\varepsilon, \gamma_1)$. This behavior of energy relaxation time is a consequence of mutual compensation of two above-mentioned hole-phonon interactions for narrow strip of states in conduction band close to ε_0. The maximum of relaxation time is limited by impurity and interchain scattering rate.

In order to determine parameters d_1 and d_2, we have calculated electrical conductivity components σ_{yy} and σ_{zz} in transversal to TTT chain directions. Because of small overlap of electronic wave functions in these directions, charge carriers transport becomes of hopping type and charge carriers are behaving like small polarons. The values of σ_{yy} and σ_{zz} were calculated numerically. By comparing those with experimentally reported data $\sigma_{yy} \sim \sigma_{zz} = 3.3$ S/cm [26], it was determined that $w_2 = w_3 = 0.015 w_1$. The values of w_2 and w_3 are the same because lattice constants a and c in y and z directions are very close to each other. Unfortunately, transport integrals (Eq. (9)) can be calculated only numerically.

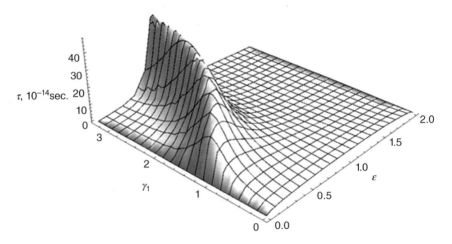

Fig. 3 Energy relaxation time of charge carriers as a function of γ_1 and $\varepsilon = E(k_x)/2w_1$ for p-type

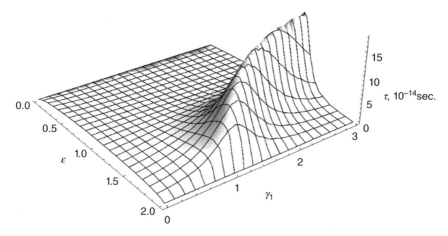

Fig. 4 Energy relaxation time of charge carriers as a function of γ_1 and $\varepsilon = E(k_x)/2w_1$ for n-type

Fig. 5 The ratio of electrical conductivity at temperature T to room temperature one: rhombuses, experiment; lines, calculations, dashed for 2D, dotted for 1D models [31]

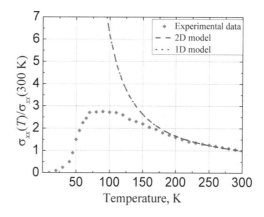

1.4   Results for Crystals of p-Type TTT₂I₃

For simplicity, first, we will consider 2D physical model of TTT_2I_3 crystals [31]. A small incommensurability between TTT chains and iodine chains produces disorder into TTT chains and defects [34] with small activation energy $E_a \sim 0.024$ eV. Thermal expansion coefficient β is not known in TTT_2I_3, but it was reported in anthracene $\beta = 14.5 \times 10^{-5}$ K^{-1}. Since internal structures are close, we have considered the same value for TTT_2I_3. Thermal expansion is considered only in parameter γ_1, because it contains b^{-5} and its value determines the resonance energy ε_0, which determines position and maximum value of relaxation time and plays important role in enhancement of TE properties. Figure 5 shows (rhombuses) the experimental data [35] for TTT_2I_3 crystal with almost stoichiometric hole concentration. It is seen that with decreasing temperature, electrical conductivity increases first, but for T lower than

~90 K, a smooth metal-dielectric transition takes place. Calculated data are represented by lines: dashed line is for 2D model and dotted for 1D (in 1D case, only Fermi energy was a little diminished by 0.0016 eV in order to have the same hole concentration). It is seen that for crystals with relatively high concentrations of impurity and defects, 1D and 2D models give the same results and it is possible to use the simple 1D model. Theoretical and measured results coincide very well in T interval from 180 K up to 300 K, the highest T when measuring. For lower T, a small gap is opened above Fermi level which becomes larger with decrease in T, determining slower increase in σ_{xx}, and leads to metal-dielectric transition near 90 K [36, 37]. This low T effect was not considered.

In Fig. 6 the results for thermopower S_{xx} are presented. It is seen that thermopower is less sensitive to appearance of the gap above Fermi level and theoretical results coincide with experimental ones [38] in larger interval, from 50 K up to 300 K. We can conclude that the theory describes rather well the experimental situation in large T interval.

Equations (6) and (7) have been calculated numerically for 3D physical model of Q1D organic crystals of TTT$_2$I$_3$ with different degrees of purity. The crystal parameters are $M = 6.5 \times 10^5 m_e$ (m_e is mass of free electron), $v_{s1} = 1.5 \times 10^3$ m/s, $w_1 = 0.16$ eV, $w_1' = 0.26$ eVÅ^{-1}, $r = 4$, $d_1 = d_2 = 0.015$, $\kappa_{xx}^L = 0.6$ W/(m $\times$ K), and $\alpha_0 = 45$ Å^{-3}. Parameter γ_1 was taken equal to 1.7, as it is shown in [36]. Parameters γ_2 and γ_3 were calculated after Eq. (5). At room T, sum $D_0 + D_1 e^{-Ea/kT}$ can be replaced by one parameter, D_0, for which three values were chosen: $D_0 = 0.1$ that corresponds to crystals grown by gas phase method [25] with stoichiometric electrical conductivity $\sigma_{xx} \sim 10^4$ S/cm, $D_0 = 0.02$ that corresponds to purer crystals grown also by gas phase method with higher $\sigma_{xx} \sim 3 \times 10^4$ S/cm, and $D_0 = 0.005$ that corresponds to still more perfect crystals with $\sigma_{xx} \sim 6.6 \times 10^4$ S/cm not obtained yet.

Figure 7 shows dependences of electrical conductivity σ_{xx} on hole concentration n. In stoichiometric crystals $n = 1.2 \times 10^{21}$ cm^{-3}, but crystals admit nonstoichiometric composition with surplus or deficiency of iodine as acceptors. It is seen that for not very pure crystals with $D_0 = 0.1$, the results of 3D model coincide with those of simpler 1D approximation for the entire interval of n variation. Even for purer crystals with $D_0 = 0.02$, deviation of 3D model from 1D one is still negligible. In these

Fig. 6 Thermopower S_{xx} as a function of T: rhombuses, experiment; dashed and dotted lines, calculations

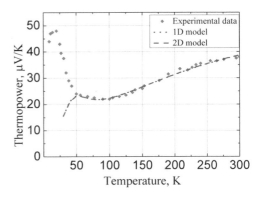

Fig. 7 Electrical
conductivity σ_{xx} along
chains as a function of hole
concentration n for $\gamma_1 = 1.7$

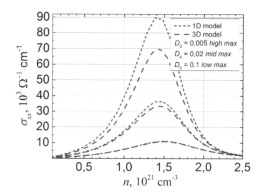

cases, mobility of charge carriers is limited by impurity scattering. The scattering on adjacent chains does not give important contribution to σ_{xx} and simpler 1D model may be used yet. For the purest crystals with $D_0 = 0.005$, deviation between 3D and 1D models achieves 20%. In this case, scattering of charge carriers on adjacent chains prevails scattering on impurities, and 3D model must be used.

Figure 8 shows dependences of Seebeck coefficient along chains S_{xx} on hole concentration n at room temperature. It is seen that results obtained with 3D and 1D models are very close in the whole interval of n variation. For ordinary crystals with $n \sim 1.3 \times 10^{21}$ cm^{-3}, S_{xx} depends weakly on crystal perfection and shows values between 40 and 35 μV/K as it is observed experimentally. For smaller n, S_{xx} achieves higher values which is favorable for improvement in TE properties.

Figures 7 and 8 show that with increase in n, electrical conductivity firstly increases and Seebeck coefficient decreases. At some values of n, σ_{xx} achieves maximum and $S_{xx} = 0$. With further increase in n, charge carriers change the sign and become electrons, σ_{xx} decreases and S_{xx} increases by absolute value staying negative. For free charge carriers there is a symmetry between holes and electrons and value of n for which S_{xx} changes the sign corresponds to half-filled conduction band. Of course, we will be interested in variation of n near stoichiometric value. We have used such large variations of n in order to show that when electron-phonon interaction is taken into account, the symmetry between electrons and holes does not exists. The interval of n for which charge carriers are holes is enlarged.

Figure 9 shows dependences of electronic thermal conductivity along chains κ_{xx}^e as a function of n at room temperature. It is seen that in low perfect crystals, the results of 1D and 3D models coincide practically. In ordinary crystals with $D_0 = 0.1$ and $n \sim 1.3 \times 10^{21}$ cm^{-3}, electronic thermal conductivity $\kappa_{xx}^e \sim 3$ W/(m × K) or 5 times larger than lattice thermal conductivity κ_{xx}^L. In purest crystals, electronic contribution to thermal conductivity is up to 20 times larger than phononic contribution. Thus, electrons realize almost all thermal conductivity. The maxima of κ_{xx}^e are displaced toward higher values of n with respect to σ_{xx}. This leads to violation of Wiedemann-Franz law and to decrease in Lorentz number [39], which is also favorable for improvement in TE properties.

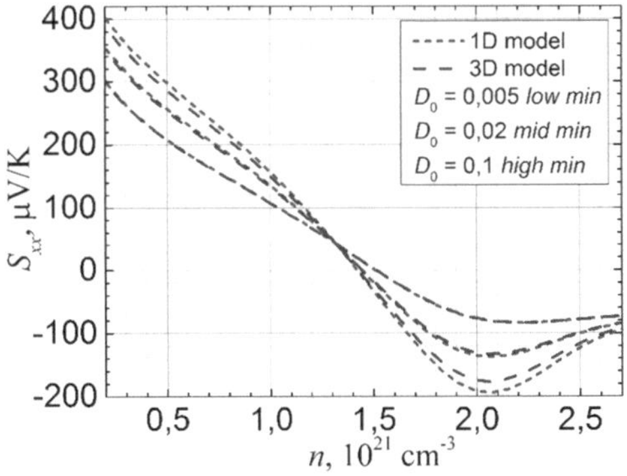

Fig. 8 Seebeck coefficient S_{xx} along chains as a function of hole concentration n for $\gamma_1 = 1.7$

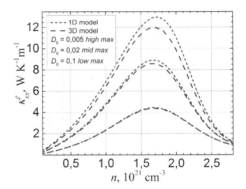

Fig. 9 Electronic thermal conductivity κ^e_{xx} along chains as a function of n for $\gamma_1 = 1.7$

Figure 10 shows dependences of PF along TTT chains P_{xx} as a function of n at room temperature. Deviation of 3D model from 1D one is more evident due to both contributions of deviations in σ_{xx} and in S_{xx}. In crystals with $D_0 = 0.1$, the maximum of P_{xx} achieves 80×10^{-4} W/(m × K²), or by two times higher than in Bi₂Te₃. However, in the purest crystals, maximum of P_{xx} is about 18 times higher than in Bi₂Te. Thus, we can conclude that enhancement in TE properties comes mainly from increase in power factor.

Figure 11 shows dependences of TE figure of merit along TTT chains ZT_{xx} as a function of n at room temperature. As-grown crystals usually have excess of iodine and $n \sim 1.3 \times 10^{21}$ cm⁻³. Figure 11 shows that in such crystals, ZT_{xx} is quite low, that is, ~0.1. In order to increase ZT_{xx}, it is necessary to decrease n. Thus, if n is decreased by two times, down to 0.65×10^{21} cm⁻³, ZT_{xx} increases up to 1 in existing crystals with $\sigma_{xx} \sim 10^4$ S/cm and $D_0 = 0.1$. In more perfect crystals not obtained yet with higher σ_{xx} and $D_0 = 0.02$, it is expected to obtain $ZT_{xx} \sim 2.8$ and even $ZT_{xx} \sim 5$ in still more perfect crystals with $D_0 = 0.005$.

Fig. 10 Power factor P_{xx} along chains as a function of concentration of charge carriers n for $\gamma_1 = 1.7$

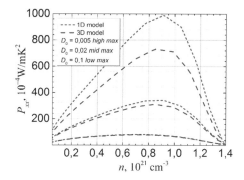

Fig. 11 ZT along chains as a function of hole concentration n for $\gamma_1 = 1.7$

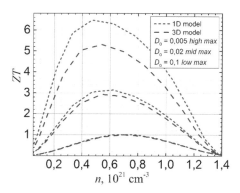

In the last case, crystal parameters are $\sigma_{xx} = 10 \times 10^4$ S/cm, $S_{xx} = 250$ μV/K, $P_{xx} = 6.2 \times 10^{-2}$ W/(m×K^2), and $\kappa_{xx}^e \sim 3$ W/(m×K), quite achievable parameters. It is seen that although electronic thermal conductivity is rather high, however, power factor P_{xx} achieves 6.2×10^{-2} W/(m×K^2) or 15 times higher than for Bi$_2$Te$_3$, very promising results.

1.5 Crystals of n-Type TTT(TCNQ)$_2$

Q1D organic crystals of TTT(TCNQ)$_2$ have the form of dark-violet needles of length of 3-6 mm and dimensions in transversal directions ~30-60 μm. The lattice constants are $a = 19.15$ Å, $b = 12.97$ Å, and $c = 3.75$ Å [28], showing pronounced Q1D structure. The internal crystalline structure, in perpendicular projection to TTT chain direction, c is shown in Fig. 12.

The structure is like that of TTT$_2$I$_3$ crystals. TTT molecules are donors and TCNQ ones are acceptors. Electrical conductivity of TCNQ chains is much higher than of TTT chains. Seebeck coefficient is negative, confirming that charge carriers are electrons. 3D physical model of the crystal is described in [40]. The crystal

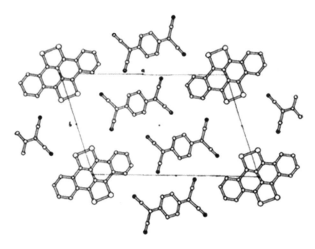

Fig. 12 Structure of TTT(TCNQ)$_2$ crystal projected along the **c** axis [28]

parameters are: mass of TCNQ molecule, $M = 3.72 \times 10^5 m_e$ (m_e is free electron mass), sound velocity along x direction, $v_{s1} = 2.8 \times 10^3$ m/s, transfer energy along x direcion $w_1 = 0.125$ eV, derivatives of transfer energies with respect to intermolecular distance, $w_1' = 0.22$ eVÅ^{-1}, $w_2'/w_1' = 0.015$, $w_3'/w_1' = 0.01$, number of TCNQ chains per elementary cell, $r = 2$, lattice thermal conductivity $\kappa_{xx}^L = 0.4$ W/(m×K), and parameter $\gamma_1 = 1.8$. Values of $D_0 = 0.1$, 0.04 and 0.02 were considered, describing crystals with different degrees of perfection. Unlike crystals of TTT$_2$I$_3$, where in order to achieve optimal concentration, it was necessary to diminish hole concentration, in TTT(TCNQ)$_2$ electron concentration needs to be increased. This can be achieved by additional doping. Consequently, the number of dislocations in crystalline lattice will increase, leading to larger values of D_0.

In stoichiometric TTT(TCNQ)$_2$ crystals, the electron concentration amounts to $n = 1.1 \times 10^{21}$ cm^{-3}. For such crystals, σ_{xx} is very small, ~0.46×10^3 S/cm, but if n is increased up to 2.5×10^{21} cm^{-3}, then values of $\sigma_{xx} \approx 10 \times 10^3$ S/cm, $\sigma_{xx} \approx 15 \times 10^3$ S/cm, and $\sigma_{xx} \approx 20 \times 10^3$ S/cm are expected (Fig. 13) in crystals with a given degree of perfection. Increase in n can be achieved by additional doping.

The dependences of Seebeck coefficient S_{xx} on n are presented in Fig. 14. It is seen that for $n = 2.2 \times 10^{21}$ cm^{-3}, it is expected to obtain $S_{xx} = -110$ μV/K, -140 μV/K, and -160 μV/K, rather high values.

Figures 13 and 14 show that with increase in electron concentration n, σ_{xx} firstly increases and S_{xx} decreases by absolute value. At $n \sim 3 \times 10^{21}$ cm^{-3}, σ_{xx} achieves maximum and $S_{xx} = 0$. At higher n, charge carriers become holes in the conduction band and $S_{xx} > 0$ (not indicated in Fig. 14). This region is not shown in Fig. 14, but the interval of electron conduction is enlarged.

Figure 15 shows that in stoichiometric crystals, electronic thermal conductivity κ_{xx}^e is small. However, if n increases up to 2.2×10^{21} cm^{-3}, κ_{xx}^e achieves values of 3.5,

Fig. 13 Electrical
conductivity σ_{xx} along
chains as a function of
concentration of charge
carriers n

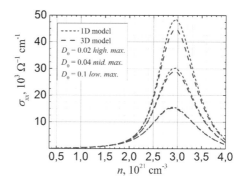

Fig. 14 Seebeck
coefficient S_{xx} along chains
as a function of
concentration of charge
carriers n

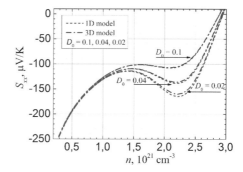

4.5, and 6.0 W/(m×K) for D_0 = 0.1, 0.4, and 0.02, respectively. This means that electrons provide almost all crystal thermal conductivity.

One can observe that unlike p-type crystals, maximum of κ^e_{xx} is displaced to smaller electron concentrations in comparison with maximum of σ_{xx}. This leads also to violation of Wiedemann-Franz law and to increase in Lorentz number. It is not favorable for improvement in TE properties. Probably, this situation explains why TE properties of n-type organic materials are usually worse than that of p-type materials.

In stoichiometric crystals, power factor P_{xx} is small (Fig. 16). However, if n increases up to 2.5 × 10^{21} cm^{-3}, P_{xx} achieves 5.1 × 10^{-3}, 17 × 10^{-3} and 32 × 10^{-3} W/(m×K^2), up to eight times higher than in Bi$_2$Te$_3$. This means that improvement in TE properties is determined mainly by increase in P_{xx}.

As can be seen from Fig. 17, in stoichiometric crystals, $(ZT)_{xx}$ is small, of about ~0.02, and does not strongly depend on the sample's perfection. If n is increased up to 2.5 × 10^{21} cm^{-3} (by 2.2 times), then $(ZT)_{xx}$ is expected to achieve 1.1 for crystals with D_0 = 0.02. In this case, TE parameters are $\sigma_{xx} \approx 20 \times 10^3$ S/cm, κ^e_{xx} = 8.3 W/(m×K), $S_{xx} \approx$ -140 μV/K, and P_{xx} = 32 × 10^{-3} W/(m×K). It is seen that κ^e_{xx} is rather high and the total thermal conductivity is realized by electrons. At the same time, P_{xx} is very high and increase in $(ZT)_{xx}$ is determined, namely, by P_{xx}.

Fig. 17 ZT along chains
as a function of
concentration of charge
carriers n

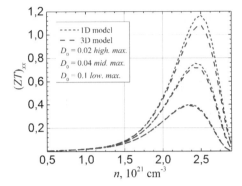

Fig. 16 Power factor P_{xx}
along chains as a function
of concentration of charge
carriers n

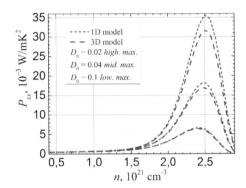

Fig. 15 Electronic thermal
conductivity κ^e_{xx} along
chains as a function of n

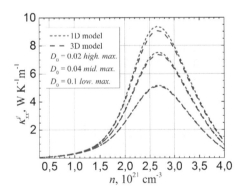

1.6 Efficiency of TE Module

The maximum efficiency of thermoelectric converter for electricity generation is
determined by parameter η_{max}, defined as:

$$\eta_{max} = \frac{T_h - T_c}{T_h} \frac{\sqrt{1 + ZT_{av}} - 1}{\sqrt{1 + ZT_{av}} + T_c / T_h},$$ (11)

where T_h is temperature of hot side, T_c is temperature of cold side, and T_{av} is average temperature $T_{av} = (T_h + T_c)/2$. ZT_{av} is average dimensionless figure of merit of TE device, which takes into consideration thermoelectric figures of merit of n- and p-type materials. After geometrical optimization with respect to legs sections, it is obtained [41]:

$$ZT_{av} = \frac{\left(S_p - S_n\right)^2 T_{av}}{\left[\left(\sigma_n^{-1}\kappa_n\right)^{1/2} + \left(\sigma_p^{-1}\kappa_p\right)^{1/2}\right]^2},$$ (12)

where σ_n, σ_p, S_n, S_p, κ_n, and κ_p are, respectively, electrical conductivities, Seebeck coefficients, and thermal conductivities of n- and p-type materials. The first factor in (11) is Carnot efficiency.

Thermoelectric converter consists of a series of thermocouples, each made of two materials of n-and p-type. For example, HZ-14 TE module by Hi-Z Company has sizes 6.27 cm × 6.27 cm and thickness of about 5 mm and includes 49 p-n pairs of legs made of Bi_2Te_3 based semiconductor [42]. The converter provides power output 25 W (5% efficiency) at temperature difference of 300 °C. Because now TE properties of commercial materials based on Bi_2Te_3 or PbTe are low, TE devices have still limited applications. Nevertheless, one can mention mass production of miniature thermoelectric modules designed to maintain constant temperatures in laser diodes [43], of climate control seats installed in hundreds of thousands of vehicles each year [44], of portable beverage coolers [45] and other applications, including cosmic ones.

Let us consider TE module constructed of n-type leg from TTT(TCNQ)$_2$ and p-leg from TTT$_2$I$_3$, working in power generation regime [46]. In Table 1, the numerical data for σ_n, σ_p, S_n, S_p, κ_n, and κ_p are extracted from the previous calculations as σ_{xx}, S_{xx}, and κ_{xx}^e, for n- and p-type, respectively. The figure of merit ZT_{av} of TE module and maximum efficiency η_{max} were calculated after (12) and (11).

Table 1 Calculations of ZT_{av} and of maximum efficiency η_{max} of TE module

σ_n, S/m	ε_F n-leg	S_n μV/K	$\kappa_n = \kappa^e + \kappa^L$ W/(m×K)	σ_p S/cm	ε_F p-leg	S_p μV/K	$\kappa_p = \kappa^e + \kappa^L$ W/(m×K)	ZT_{av}	η_{max} %
3.0×10^5	0.90	-170	4.4	7.5×10^5	0.09	300	3.1	2.53	13
7.0×10^5	1.00	-181	6.5	8.0×10^5	0.10	280	3.3	3.2	14
7.5×10^5	1.05	-183	7.8	12×10^5	0.15	226	4.3	2.5	13
11×10^5	1.10	-182	8.5	21×10^5	0.20	200	5.8	2.9	14
11×10^5	1.10	-182	8.5	8.0×10^5	0.10	280	3.3	3.6	15
13×10^5	1.15	-177	9.2	8.0×10^5	0.10	280	3.3	3.7	15.7

The efficiency is not big because we have chosen $T_h = 480$ K, the highest T admitted by these organic materials, and $T_c = 300$ K. The temperature difference ΔT is only 180 K. For such small ΔT, ideal Carnot efficiency is only 37.5%. Nevertheless, predicted generator efficiency ~13-15% for conversion of low-potential waste heat is very good result. These values are much higher than those realized in TE module on Bi_2Te_3 based materials for $\Delta T = 300$ K (see [42]). From Table 1, it is also seen that the highest efficiency is achieved when both n- and p-type legs have optimal parameters. Considered here organic materials could be used in low-temperature cascade of TE generators working at larger ΔT in order to enhance the overall efficiency.

1.7 Coefficient of Performance of TE Module

The cooling efficiency of TE converter is determined by coefficient of performance (COP), which is defined as the ratio of heat Q removed from cold side of TE cooler in Watts to power input $I \times V$:

$$COP = \frac{Q}{IV}, \tag{13}$$

where I is electric current and V is applied electrical voltage. The following equation determines Q:

$$Q = ST_c I - K\Delta T - \frac{1}{2}I^2 R, \tag{14}$$

where S is TE device Seebeck coefficient, T_c is cold side temperature, K is thermal conductance, ΔT is temperature difference across TE device, and R is electrical resistance of TE device.

For optimal current, we have:

$$COP_{max} = \frac{T_c}{\Delta T} \frac{\sqrt{1 + ZT_{av}} - T_{h/T_c}}{\sqrt{1 + ZT_{av}} + 1}, \tag{15}$$

where $T_{av} = (T_c + T_h)/2$, and T_c and T_h are temperatures of cold side and of hot side, respectively. The term $T_c/\Delta T$ in (15) is thermodynamic maximum of the coefficient of performance (COP of Carnot cycle).

Let us consider TE module constructed of n-type leg from TTT(TCNQ)$_2$ and p-leg from TTT$_2$I$_3$, working in cooler regime. The calculations of COP_{max} after Eq. (15) as a function of temperature of hot side T_h when temperature of cold side T_c is taken at 300 K are presented in Fig. 18. The values of ZT_{av} for different T_h have been calculated after Eq. (12). The numerical data for σ_n, σ_p, S_n, S_p, κ_n, and κ_p are taken from Table 1.

Fig. 18 COP_{max} as a function of T_h when temperature at cold surface $T_c = 300$ K

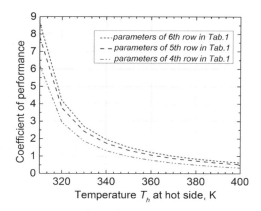

It is seen that for $T_h = 310$ K ($\Delta T = 10$ K), COP_{max} achieves rather high values: 6.0, 7.5, and 8.5 for parameters from fourth, fifth, and sixth rows (Table 1), respectively. Even for $\Delta T = 20$ K, COP_{max} takes also high values: 2.9, 3.8, and 4.2, respectively. From Eq. (13), $Q = COP \times IV$, it results that cooling, for example, 2 W with TEC having COP of 8.0 requires only 0.25 W of input power. Even for COP of 4, it requires only 0.5 W. Therefore, such modules are very efficient for applications in air-conditioning systems or in refrigerating systems for decreasing in temperature on $\Delta T \sim 20$ C with respect to room temperature.

Some experimental results are presented in [47, 48]. TTT_2I_3 crystals were synthesized and TE parameters were measured. For the first time, single thermocouple TEG was fabricated of n-type $(DCNQI)_2Cu$ and p-type TTT_2I_3 legs. TEG demonstrated unprecedented specific power output of a few mW $\times$ cm^{-2} at room temperature [47].

Single thermocouple thin-film TEGs were also prepared via vacuum deposition of both p-type TTT_2I_3 and n-type $TTT(TCNQ)_2$ materials on the same substrate by two consecutive sublimation cycles [48]. Power per degree Kelvin of 5.5 pW $\times$ K^{-1} was measured for fabricated single thermocouple TEG close to room temperature. This value is mainly limited by low electrical conductivity of polycrystalline thin films. However, simple fabrication process proposed allows easy duplication of such TEG modules; therefore, power output of TEG could be multiplied several times. Besides, it remains to optimize concentration of charge carriers in order to improve TE properties of these structures.

2 Conclusions

TE properties of highly conducting Q1D organic crystals of p-type TTT_2I_3 and n-type $TTT(TCNQ)_2$ are investigated. The physical model considers simultaneously two of the most important electron-phonon interactions: one is of deformation potential type and the other is like of polaron. Scattering of charge carriers on point-like impurity and on thermally activated defects is also considered. In TTT_2I_3 crystals, axis x is directed along conductive TTT chains, axis b. In $TTT(TCNQ)_2$ crystals, axis x is directed along conductive TCNQ chains, axis c. TE coefficients are calculated for the case when weak electrical field and low temperature gradient are applied along axis x and charge and thermal transport is calculated along the same axis x. Because the physical model contains many interactions, it is convenient to apply the method of two-particle Green functions in order to deduce linearized kinetic equation. Mass operator of two-particle Green function (inverse of charge carrier's relaxation time) is calculated in second approximation of perturbation theory. Electrical conductivity σ_{xx}, Seebeck coefficient S_{xx}, electronic thermal conductivity κ_{xx}^e, and TE figure of merit $(ZT)_{xx}$ were expressed through transport integrals and calculated numerically for p-type TTT_2I_3 and n-type $TTT(TCNQ)_2$ crystals. It was shown that due to partial compensation of both electron-phonon interactions, relaxation time of charge carriers as a function of carrier energy obtains a sharp maximum for narrow strip of states in the conduction band. The value of this maximum is limited by interchain scattering of charge carriers and by scattering on impurities and defects and may be rather high, if the crystal has pronounced Q1D properties and is sufficiently purified. Charge carriers close to this maximum will have increased mobility. If concentration of charge carriers is optimized so that Fermi level is displaced near energetic states that correspond to the maximum of relaxation time, then significant improvement in TE properties is expected. Thus, in TTT_2I_3 crystals, values of $(ZT)_{xx} \sim 4\text{-}5$ are predicted. For $TTT(TCNQ)_2$ crystals, values of $(ZT)_{xx} \sim 1.2\text{-}1.5$ are expected, according to our calculations. The efficiency of TE module as electric generator for conversion of low-potential waste heat is estimated $\sim 13\text{-}15\%$. The efficiency is not high because temperature difference $\Delta T = T_h - T_c = 480 \text{ K} - 300 \text{ K} = 180 \text{ K}$ is not big and is determined by the highest temperature allowed for these materials. Nevertheless, these values are much higher than those realized in TEG module based on Bi_2Te_3 materials for $\Delta T = 300$ K. Of course, organic materials considered in this chapter could be used in low-temperature cascade of TEG working at larger ΔT in order to enhance the overall efficiency. The coefficient of performance of TE module in regime of cooler for $T_h = 310$ K ($\Delta T = 10$ K) achieves rather high values: 6.0, 7.5, and 8.5. Even for $\Delta T = 20$ K, COP_{max} takes high values: 2.9, 3.8, and 4.2. Some recent experimental results are analyzed too.

References

1. A.F. Ioffe, Semiconductor thermo-elements and thermoelectric cooling, in *Infosearch Ltd*, (London, 1958)
2. A. Casian, in *Thermoelectric Handbook, Macro to Nano, Chapter 36*, ed. by D. M. Rowe, (CRC Press, Boca Raton, 2006)
3. A. Casian, Prospects of the thermoelectricity based on organic materials. J. Thermoelectricity **3**(45) (2007)

4. Y. Wang, J. Zhou, R. Yang, Thermoelectric properties of molecular nanowires. J. Phys. Chem. C **115**, 24418 (2011)
5. Z. Fan, J. Quyang, Thermoelectric properties of PEDOT:PSS. Adv. Electron. Mater. **2019**, 1800769
6. G.-H. Kim, L. Shao, K. Zhang, K.P. Pipe, Engineered doping of organic semiconductors for enhanced thermoelectric efficiency. Nat. Mater. **12**, 719 (2013)
7. K. Zhang, J. Qiud, S. Wang, Thermoelectric properties of PEDOT nanowire/ PEDOT hybrids. Nanoscale **8**, 8033 (2016)
8. B. Liu, J. Hu, J. Zhou, R. Yang, Thermoelectric transport in nanocomposites. Materials **10**, 418 (2017). https://doi.org/10.3390/ma10040418
9. S.P. Ashby, J. Garcia-Canadas, G. Min, Y. Chao, Measurement of thermoelectric properties of phenyl acetylene-capped silicon nanoparticles and their potential in fabrication of thermoelectric materials. J. Electron. Mater. **42**, 1495 (2013)
10. M. Romero, D. Mombru, R. Faccio, A. Mombru, Thermoelectric properties and thermal stability of conducting polymer nanocomposites: A review, in *Advanced Thermoelectric Materials*, ed. by C. R. Park, (Scrivener Publishing LLC, Hoboken, 2019), pp. 467–492
11. C.J. Boile et al., Tuning charge transport dynamics via clustering of doping in organic semiconductor thin films. Nat. Commun. **10**, 2827 (2019). https://doi.org/10.1038/s41467-019-10567-5
12. M. Upadhayaya, Z. Aksamija, C.J. Boile, *Venkataraman. Effects of Disorder on Thermoelectric Properties of Semiconducting Polymers*, vol arXiv (2019), p. 1901.03370v1. [cond-mat. mtrl-sci]
13. I. Petsagkourakis et al., Correlating the Seebeck coefficient of thermoelectric polymer thin films to their charge transport mechanism. Org. Electron.. https://doi.org/10.1016/j.orgel.2017.11.018
14. J.-S. Kim, W. Jang, D.H. Wang, The investigation of the Seebeck effect of the poly(3,4-Ethylenedioxythiophene)-Tosylate with the various concentrations of an oxidant. Polymers **11**, 21 (2019). https://doi.org/10.3390/polym11010021
15. R.M.W. Wolfe, A.K. Menon, T.R. Fletcher, et al., Simultaneous enhancement in electrical conductivity and Thermopower of n-type NiETT/PVDF composite films by annealing. Adv. Funct. Mater. **28**, 1803275 (2018)
16. C.-J. Yao, H.-L. Zhang, Q. Zhang, Recent Progress in thermoelectric materials based on conjugated polymers. Polymers **11**, 107 (2019). https://doi.org/10.3390/polym11010107
17. M. Culebras, K. Choi, C. Cho, Recent Progress in flexible organic Thermoelectrics. Micromachines **9**, 638 (2018)
18. H. Li et al., Dopant-dependent increase in Seebeck coefficient and electrical conductivity in blended polymers with offset carrier energies. Adv. Electron. Mater. **5**, 1800618 (2019)
19. N. Roland et al., Understanding morphology-mobility dependence in PEDOT:Tos. Phys. Rev. Materials **2**, 045605 (2018). https://doi.org/10.1103/PhysRevMaterials.2.045605
20. K.W. Shah, S. Wang, D. Soo, J. Xu, One-dimensional nanostructure engineering of conducting polymers for thermoelectric applications. Appl. Sci. **9**, 1422 (2019). https://doi.org/10.3390/app9071422
21. J.J. Urban, A.K. Menon, A. Jain, Z. Tian, K. Hippalgaonkar, Correlated electrons, organic transport, machine learning, and more. J. Appl. Phys. **125**, 180902 (2019). https://doi.org/10.1063/1.5092525
22. Z. Li et al., A free-standing high-output power density thermoelectric device based on structure-ordered PEDOT:PSS. Adv. Electron. Mater., **4**(2), 1700496 (2018). https://doi.org/10.1002/aelm.201700496
23. S. Mortazavinatanzi, A. Rezaniakolaei, A. Rosendahl, Printing and folding: A solution for high-throughput processing of organic thin-film thermoelectric devices. Sensors (Basel) **18**(4), 989 (2018). https://doi.org/10.3390/s18040989
24. W. Shi, D. Wang, Z. Shuai, High-performance organic thermoelectric materials: Theoretical insights and computational design. Adv. Electron. Mater. **5**, 1800882 (2019). https://doi.org/10.1002/aelm.201800882

25. B. Hilti, C.W. Mayer, Electrical properties of the organic metallic compound bis (tetrathiotetracene)-triiodide, $(TTT)_2I_3$. Helv. Chim. Acta **61**(40), 501–511 (1978)
26. L.C. Isett, Magnetic susceptibility, electrical resistivity, and thermoelectric power measurements of bis (tetrathiotetracene)-triiodide. Phys. Rev. B **18**, 439 (1978)
27. I. Shchegolev, E. Yagubskii, *Extended Linear Chain Compounds*, vol 2 (JS Miller, Plenum Press, New York, 1982), pp. 385–434
28. L. Buravov, O. Eremenko, R. Lyubovskii, E. Yagubskii, Structure and electromagnetic properties of a new high-conductivity complex $TTT(TCNQ)_2$. JETP **20**(7), 208–209 (1974)
29. N. Ueno, Electronic Structure of Molecular Solids: Bridge to the Electrical Conduction, Chapter 3, in *Physics of Organic Semiconductors*, ed. by W. Brutting, C. Adachi, 2nd edn., (Wiley-VCH Verlag GmbH & Co. KGaA, Weinheim, 2012), pp. 65–89
30. A. Casian, I. Sanduleac, Effect of interchain interaction on electrical conductivity in quasi-one-dimensional organic crystals of tetrathiotetracene-iodide. J. Nanoelectron. Optoelectron. **7**(7), 706–711 (2012). https://doi.org/10.1166/jno.2012.1408
31. A. Casian, I. Sanduleac, Thermoelectric properties of tetrathiotetracene iodide crystals: Modeling and experiment. J. Electron. Mater. **43**(10), 3740–3745 (2014). https://doi.org/10.1007/s11664-014-3105-6
32. A.I. Casian, I.I. Sanduleac, Organic thermoelectric materials: New opportunities. J. Thermoelectricity **3**, 11 (2013)
33. A. Casian, J. Pflaum, I. Sanduleac, Prospects of low dimensional organic materials for thermoelectric applications. J. Thermoelectricity **1**, 16–26 (2015)
34. A. Casian, V. Dusciac, I. Coropceanu, Phys. Rev. B **66**, 165404 (2002)
35. V.F. Kaminskii, M.L. Khidekel', R.B. Lyubovskii et al. Phys. Status Solidi A 44, 77 (1977)
36. S. Andronic, A. Casian, Adv. Mat. Physic. Chem. **7**, 212–222 (2017)
37. S. Andronic, A. Casian, Peierls structural transition in organic crystals of TTT_2I_3 type in 2D approximation. Mold. J. Phys. Scien. **18**, 21 (2019)
38. P.M. Chaikin, G. Gruner, I.F. Shchegolev, E.B. Yagubskii, Sol. State Commun. **32**, 1211 (1979)
39. A. Casian, Violation of Wiedemann-Franz law in quasi-one-dimensional organic crystals. Phys. Rev. B **81**, 155415 (2010)
40. I. Sanduleac, A. Casian, Nanostructured $TTT(TCNQ)_2$ organic crystals as promising thermoelectric *n*-type materials: 3D modeling. J. Electron. Mater. **45**, 1316–1320 (2015)
41. G.J. Snyder, M. Soto, R. Alley, D. Koester, B. Conner, *Hot Spot Cooling Using Embedded Thermoelectric Coolers, 22nd IEEE SEMI-THERM Symposium* (Dallas, TX, 2006), pp. 135–143
42. http://www.hi-z.com/hz-14.html
43. MARLOW   INC.., http://www.marlow.com/industries/telecommunications/transmission-lasers-dwdm.html
44. GENTHERM, "Climate Seats", http://www.gentherm.com/page/climateseats
45. "KOOLATRON", http://www.koolatron.com/
46. A.I. Casian, I.I. Sanduleac, Thermoelectric efficiency of a *p-n*-module formed from organic materials. J. Thermoelectricity **1**, 42 (2017)
47. F. Huewe, A. Steeger, K. Kostova, L. Burroughs, I. Bauer, P. Strohriegl, V. Dimitrov, S. Woodward, J. Pflaum, Low-cost and sustainable organic thermoelectrics based on low-dimensional molecular metals. Adv. Mater. **29**, 1605682 (2017)
48. K. Pudzs, A. Vembris, M. Rutkis, S. Woodward, Adv. Electron. Mater. **3**, 1600429 (2017)

Numerical Understanding of Thermal Properties of Dusty Plasmas

Aamir Shahzad and He Mao-Gang

1 Introduction

This chapter is focused on transport properties of dusty (complex) plasma liquids using new approach of molecular dynamics (MD) simulations and discloses the exact nature and cause of the deep understanding present in different types of complex fluid materials. Novel molecular simulation methods for investigating dynamical properties in microelectronic (thermoelectric and photoelectronic) devices are needed for improving device design and for understanding device physics. Such approaches are also needed for studying new transport phenomena in nanoscale devices made of novel nanostructures, such as carbon electronics or semiconductors nanowires, nanometer size sensors to detect proteins or single DNA and nanolevel fluidic flow, nanopowder production, nanocrystalline solar cells, and polymer coatings with embedded nanoparticles. In this chapter, we discuss the thermophysical properties of fluids [simple liquids, dense liquids, and complex liquids including dusty plasma liquids (DPLs)], and role and applications of transport properties in daily life and industrial processes are also overviewed. This chapter provides brief information about the dusty plasma liquids (DPLs), its ideal and nonideal behaviors, and also about the classification of DPLs in terms of density and temperature of constituents. Great details of dusty plasma and transport properties of dusty

A. Shahzad (✉)
Molecular Modeling and Simulation Laboratory, Department of Physics, Government College University Faisalabad (GCUF), Faisalabad, Pakistan

Key Laboratory of Thermo-Fluid Science and Engineering, Ministry of Education (MOE), Xi'an Jiaotong University, Xi'an, P. R. China
e-mail: aamir.awan@gcuf.edu.pk

H. Mao-Gang
Key Laboratory of Thermo-Fluid Science and Engineering, Ministry of Education (MOE), Xi'an Jiaotong University, Xi'an, P. R. China

© Springer Nature Switzerland AG 2021
S. Skipidarov, M. Nikitin (eds.), *Thin Film and Flexible Thermoelectric Generators, Devices and Sensors*, https://doi.org/10.1007/978-3-030-45862-1_12

plasma, based on different plasmas parameters, are also included. Moreover, applications of dusty plasma in daily life, laboratories, and in industries are mentioned in detail. Different molecular simulation techniques including MD simulation are presented herein with advantages and disadvantages of these methods. Motivations, aims, and objectives of this study from the previous literature are also taken into account. Plasma conductivity, heat process, and those nonlinear effects (non-Newtonian behaviors) are the basic data of transport property of DPLs, and are applied for studying the flow, and heat and mass transport individuality. We start here with some estimations of the thermal conductivity of strongly coupled complex (dusty) plasmas (SCCDPs) using the novel homogenous nonequilibrium MD (HNEMD) method, which is important parameter used in the heat designing system.

1.1 Thermophysical Properties

Thermophysical properties (chemical properties remain unaffected and physical properties of material fluctuate with variable temperature, composition, and pressure) of fluids, that is, simple liquids and complex liquids (gaseous, liquids, and crystal state can be complex liquids) explain the phase transition [1]. These fluids can be investigated by experimental, theoretical, and simulation techniques. Thermophysical properties (thermodynamics and transport coefficients) include thermal conductivity, thermal expansion, thermal radiative properties, thermal diffusivity, enthalpy internal energy, Joule–Thomson coefficients, and heat capacity, as well as thermal diffusion coefficients, mass coefficients, viscosity, speed of sound, interfacial, and surface tension. Thermophysical properties of gases and liquids such as hydrogen H_2, oxygen O_2, nitrogen N_2, and water H_2O at high pressure and low temperature are dissimilar from ideal gas, especially at phase boundary. To calculate these properties in the widest range of pressure and temperature, specific models are required. In different power plants, different fluids like gases and liquids are used as a power generation source. Nuclear power plants, gas turbine plants, and internal combustion plants use heavy water, steam, air, and different gases for power generation. In refrigerators and fast nuclear reactors, ammonia and sodium in liquid phase are used as a cooling agent. Some gases in certain limit act as an ideal gas, for example, helium and hydrogen are ideal gases in a closed cycle. Detailed diagrams and tables are needed for the calculation of thermodynamic properties because these properties cannot be calculated by the hypothesis for ideal and incompressible fluids. If details about thermodynamic properties are not available, then suitable equations are used for precise calculation of thermodynamic properties of fluids [2]. Transport properties played a crucial role in industrial and laboratory applications for the optimization and heat designing system. The industrial applications and huge existence of dusty plasmas, combined with numerous general and standard thermophysical properties, make dusty plasmas extremely attractive and exciting field for most of the researchers and attract attention toward the development of science and technology. When dealing with thermal conductivity at the industrial scale, many details of information are required for different system size [3, 4]. Transport coefficients (diffusion, shear viscosity, bulk viscosity, electrical, and thermal conductivity) of plasma can be depicted through experiments, theoretical study, and by computer simulation.

1.2 Historical View of Plasma

The three well-known types of matter are solid, liquid, and gas but another state exists in the universe and it is called plasma. Plasma is simply defined as "the fourth state of matter." It is ionized gas containing electrons, ions, and neutral particles. An ordinary gas cannot be plasma until it obeys certain conditions ($L > \lambda_D$, $N_D >> 1$, and $\omega_\tau > 1$), where L is edge length of the system, λ_D is Debye length, and τ is time period or inverse of plasma frequency; here, it is given as $\omega_p = (n_0 e^2/\varepsilon_0 m)^{1/2}$. Basically, this big difference between ordinary gas and plasma exists due to temperature. In reality, plasma is generated when ordinary gas is heated up more and more at a high temperature until atoms and molecules of gas become ionized or ions and electrons are separated from neutral atoms or molecules. Irving Langmuir, defined the plasma as "it is a quasineutral gas of charged and neutral particles which exhibit collective behavior" and he won Noble prize in 1927 for being the first to use the term plasma [5]. There are two things in the definition: quasineutral and collective behavior. The first term is quasineutral; the quasineutrality confirms that plasma is electrically neutral and has ion density approximately equal to electron density ($n_i \approx n_e$). So, quasineutrality disappears over Debye length. The second term is collective behavior, wherein plasma particles collide with each other by Coulomb potential or electric field and, if the distance increases between charged particles, then particles collide with each other and all those particles appear in the way [6].

1.2.1 Classifications of Plasmas

Relativistic Plasmas and Quantum Plasmas

In relativistic plasma, the colliding speed of particles is comparable to the speed of light like in the core of the supernova; inside the core of a supernova, it is so hot that electrons gain additional energy than massive particles and travel with the speed of light. The quark–gluon plasma (QGP) and electron–positron pair plasma in supernova are universal examples of such plasmas. In quantum plasmas, De-Broglie wavelength of particles is approximately equal to or larger than the average distance between those. The white dwarf is the example of quantum plasma.

Nonrelativistic Plasmas and Classical Plasma

The nonrelativistic plasma is opposite to the relativistic plasma. In this plasma collision speed of particles is very less than the speed of light. In fact, ideal and nonideal plasmas are also nonrelativistic plasmas. In this chapter, only classical and nonrelativistic plasma is studied.

Ideal Plasma

The ideal plasma has a very low temperature and high density, kinetic energy is much larger than potential energy in such kind of plasmas. In some cases, interaction potential energy can be ignored due to large average distance between the interacting particles; this interaction potential energy is often ignored because average distance between the particles is less. Due to large distance between the particles, ideal plasma doesn't possess any arrangement of particles or self-organization like crystal plasma.

Nonideal (Complex) Plasmas

The nonideal (complex) plasma can be characterized on the basis of Coulomb coupling parameter into two types.

(a) *Weakly Coupled Plasma*

The ratio of potential energy to kinetic energy is called Coulomb coupling parameter. When the ratio is small, then average kinetic energy of particles dominates over interaction potential energy, and that type of plasma is called weakly coupled plasma ($\Gamma < 1$). Specifically, we can say that in nonideal weakly coupled plasma, Coulomb coupling parameter has a range of $0.1 < \Gamma < 1$. It is also called hot plasma having low density. Due to weak interactions between the particles, particles do not carry specific arrangement or position, and as a result, particles remain in a dynamical motion like molecular gas. At high temperature and low density, particles gain high speed and then acquire high kinetic energy.

(b) *Strongly Coupled Plasma*

When the ratio of potential to kinetic energy is larger than in weakly couples plasma, it is called a strongly coupled plasma ($\Gamma > 1$) and also called cold, dense, or low-temperature plasma. It comes into view in many physical systems such as dusty plasma, electrons levitate on the surface of helium liquid, and condensed matter systems, that is, liquid metals and molten salts and astrophysical systems such as neutron stars, giant planetary interiors, ions in white dwarf interior, layered semiconductor nanostructure, and in cryogenic trap [7]. The strongly coupled plasma is not hard to generate in the laboratory, and it can easily be produced at low temperature and high density. If the ratio of inter-particle potential energy to kinetic energy

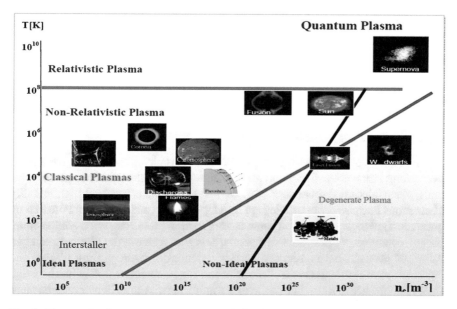

Fig. 1 Electron density versus temperature graph for the different plasmas is shown. Strongly coupled plasma exists above the red line at ($\Gamma \geq 1$)

is small, then the particles will be in gaseous state; if the ratio is larger, then particles will form liquid, crystal, or solid structure. This demonstrates Coulomb coupling parameter (determines the structure of matter). The strongly coupled plasma is shown in Fig. 1 versus temperature.

Characteristics and properties of complex (dusty) plasma can be calculated from experimental data and theoretically; however, recently, computational techniques are widely used for the most precise and fast calculation of different properties of various materials [6, 7].

1.2.2 Complex (Dusty) Plasmas

As we know, 99% of visible materials are formed by the plasma in the universe [5]. The dusty plasma is a modified form of simple plasma as discussed earlier, so it can be defined as quasineutral gas having charged particles, dust particles, and neutral particles, which exhibit a collective behavior. In the twentieth century, pioneers of plasma physics, Irving Langmuir, Hannes Alfve'n, and Lyman Spitzer, have discussed the role of dust in laboratory and cosmic plasma and found that dynamics of dust particles can be controlled by gravitational and electromagnetic force [8]. Dust particles can be incorporated intentionally or unintentionally into the plasma. The complex plasma is a dense plasma described by a strong interaction between the existing molecules and atoms; it is also called the strongly coupled complex plasma.

The dusty plasma is a complex multicomponent plasma including ions, electrons, neutral particle, and dust particles. Few features of dust particles are as follows:

(a) *Size of Dust Particle*

The size of a dust particle is measured in micron or tens of nanometer. So, the microscope is not necessary to find out the dust particles in plasma; it can be seen with naked eyes.

(b) *Charge on Dust Particle*

Due to the collective nature of plasma, dust particles interact with each other over a long range collectively and acquire high negative charge due to plasma currents, ultraviolet rays, secondary emission, photoelectric effect, and high mobility of electrons because dust particles are heavier than ions and electrons. Potential on grain (dust) particle depends on some factors such as grain size, grain composition, plasma condition, velocity, and temperature. Figure 2 shows the plot between two values of plasma, that is, reduced surface potential of hydrogen ion plasma and argon ion plasma and the ratio of electron and ion temperature T_e/T_i. So, the graph shows that for heavy ions, dust particles are more negative than lighter ones because electrons are more mobile for heavy ions [9].

(c) *Nature of Dust Particle*

The dust particle may be solid, crystal, dielectric, etc.

(d) *Force on Charged Particle*

To get information about the dynamics of dust particles, it is necessary to get the knowledge of forces through which dust particle is experienced in plasma environment. The force value depends on the size of particles and that is why the force remains balanced.

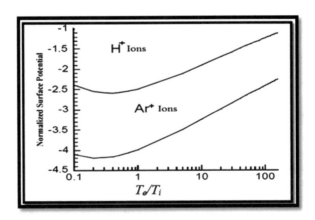

Fig. 2 Ratio of electrons and ions temperature T_e/T_i versus normalized surface potential argon and hydrogen ions

1.2.3 Role of Dust Particle

The existence of dust particles is mostly investigated in physics, chemistry, medicine, environmental science, astrophysics, and nuclear energy, etc. The basic reason to outstanding interest in this growing field is that dust particles allow to study at the microscopic or atomistic level. The existence of dust particles in the plasma controls properties such as plasma temperature, density, and potential energy. In the same way, efficiency of solar cell can be also enhanced by specific number of dust particles incorporated in thin film. In some cases, contamination destroys the microchips during improvement in chip properties. So, the dusty plasma is extended or modified form of plasma due to the addition of the fourth particle in plasma, that is, grain or dust particle, and it has a size in microns. In the recent two decades, there are many reports or papers on transport properties of dusty plasma, and that is why, plasma physics becomes an important area of research with the existence of dust particles.

1.2.4 Composition of Dusty Plasma

In dusty plasma, there may be a possibility of particles self-ordering like crystal due to strong electric field, and hence that type of plasma is called plasma crystal, which is a fascinating property of dusty plasma. Actually, 2D crystals are formed because of gravity, most of heavy particles move down to the plasma, and are levitated on horizontal electrodes in space charge sheath. In the same manner, three-dimensional crystals can also be formed under thermophoretic levitation or micro gravity. Plasma crystals are illustration of strongly coupled plasma. These crystals are usually used to study the phase transition and wave phenomena. In dusty plasma, for the higher Coulomb coupling parameter value ($\Gamma > 170$), liquid–solid phase transition is found [10]. Dust particles carry charge and interact with neighboring dust particles with Coulomb potential, so when thermal energy exceeds it, then plasma becomes strongly coupled. Kinetics, dynamics, and weak fractional damping of each particle can be observed by video microscopy, so from this technique, one can investigate plasma crystal, solid, plasma liquids, waves, phase transition, and many other phenomena at a microscopic level [11].

1.3 Applications of Complex (Dusty) Plasmas

1.3.1 Dusty Plasma in Nature

Dusty plasmas have a well-known use in industrial (laser, surface modification, lighting, etc.) polymers, textile, health, and biomedical applications. But environmental science is interested also in use of dusty plasma. In nature, there is existence of dusty plasma in space surroundings such as nebulae, solar wind, ionosphere, planetary rings, interstellar clouds, supernova remnants, and white dwarf [12].

1.3.2 Dusty Plasma in Laboratory

In experiment, this type of plasma is found in fluorescent neon lamps, rail guns, arcs, and inertial confinement fusion laser-generated plasmas. Plasma is used to make progress in the world by invention of solar cell, semiconductor chip, micro-electronic devices, storage devices, coating, energy production, biomedical tools, flat panel displays, and fusion devises, thermonuclear fireball, dusty plasma devices, plasma-enhanced chemical vapor deposition (PECVD), etc. are laboratory examples of dusty plasma. The presence of these microparticles (dust particle) in plasma enhances the property of whole dusty plasma by combining with the basic properties of plasma.

1.3.3 Dusty Plasma in Daily Life

After valuable application of plasmas in nature and in textile, the role of plasma is not sustained or ended at the above-mentioned fields or in industries. Plasma also plays a crucial role in manufacturing of many products which we use directly at room temperature or moderate temperature in daily life. These products may be automobile bumpers, plastic bags, artificial joints, airplane turbine blades, and semiconductor chips or circuits. At manufacture of transistors and capacitors for integrated circuits (ICs), etching, deposition, masking, and stripping take place a number of times. Plasma TV, plasma foam, and plasma are also used to enhance the properties of medical tools [5]. When instrument or tool (neurosurgical, vascular, endoscope, and bone saw blade, etc.) is used for the purpose of surgery or dental treatment, these tools may be harmful for the immune system because of the direct contact with human body. To protect immune defense system from any kind of inflammation or infection, tools should be sterilized for the reuse purpose. During the reuse of instrument, microorganisms contaminate the surface of the instrument; for decontamination or sterilization of germs, a nonequilibrium discharge plasma is used. Plasma processing at low temperature is harmless for the human body, which may be that of operator or patient, and for environment. Medical supporting materials are made of natural and synthetic materials like joints, artificial heart, vascular grafts, sutures, and intraocular lenses, etc. and for extracorporeal therapeutic and other devices, for example, blood oxygenation, hemoperfusion, intravenous lines, hemodialysis, blood-bags, and needle catheters. Material should be biocompatible if used for the drug delivery at the targeted position purpose like microcapsule. Plasma modifies characteristic of surface by bombardments of electrons and free radicals which collide with the surface atom and break the bond and change the chemical properties of that surface [13]. Moving from microworld to nanoworld requires details for the phase transition, mass, and heat transport. It is a valuable challenge for the progress in nanoworld and it widens also the scale (spatial and temporal) of macroscopic phenomena for heat transport.

1.4 Weakness of Experimental Work over Simulation

Experimental results have a limited range of calculations and for precise results, there are lots of difficulties in handling large parameters simultaneously. Mostly experiments conduct values at macroscopic level and do not conduct results of transport coefficients at microscopic or atomistic level and for dynamical point of view it has become more difficult and uncertain [14]. So, simulation technique is the best option for the researchers for fast and precise calculations at atomistic level with a wide or desired range of parameters. Another advantage of simulation is that it simultaneously saves cost and time for the researcher.

1.5 Molecular Simulation

Simulation is a tool through which macroscopic system can be studied by microscopic model, and this model is specified in terms of intermolecular interaction and molecular structure. Computer simulation is used to find how accurate is the model by comparing with analytical or theoretical results. Complex model cannot be studied analytically, but it can be tested by computer simulation. Simulation is also used to study at the microscopic level which cannot be studied by experiments [15].

1.5.1 Computational Techniques

There are several computational techniques which have advantages in respective fields. Monte Carlo and molecular dynamics are influential tools to study transport properties of dusty plasma. Transport properties can also be calculated by Langevin dynamics (LD), Monte Carlo (MC), Path integral MC (PIMC), and molecular dynamics (MD) [16–18]. A disadvantage of MC technique is that it cannot evaluate transport properties of dynamic systems and cannot solve the equation of motion.

Molecular Dynamics Simulation

Molecular dynamics simulation (MDS) is a favorable technique over the other computational techniques for the dynamic study of materials (gases, liquids, solids, and plasmas) and complex systems. MD plays a central role for simulation in all progressing fields, for example, material sciences, engineering, environmental sciences, plasma physics, astrophysics, life sciences or biological sciences, and chemical industries. Computer simulations become more important to study complex and dynamical systems. The properties, structure, and behavior of complex and large system may be explored by using a faster and powerful computer simulation (MDS) modeling tool. The first MD simulation was carried out on liquid argon in the 1950s by Alder, Wainwright, and Rahman. Molecular dynamics has two families according to the model; one is approach to "classical" mechanics, which tells the dynamics of system and in the 1980s, the second family approach to quantum MD simulation started. Quantum MD simulation is improved approach over classical approach and it is helpful for solving many problems in complex systems, but as many resources are required for the purpose of computational simulation, at present, classical approach is used in practice. Simulations have changed the relationship between experiment and theory [19]. If simulation can be depicted from experiments, that type of simulation is called experimental simulation and that which is depicted theoretically is called theoretical simulation.

Purpose of Computer Simulation

The main purpose of computer simulation is to reduce the cost of time of computation and uncertainty in results when going from small scale (laboratory) to industrial (commercial) scale. For the study of classical N-body systems such as dusty plasmas, Newton's equation of motion (by integration, trajectories of particles can be find) and its model are the base of molecular dynamics. For the large system sizes, simple models and choice of technique are required. To study the system at molecular level by molecular dynamics technique, up to thousand particles can be studied. As we know that, it is difficult to handle a large number of particles analytically and theoretically and the study of fluid in dynamical motion becomes more complex and impossible from analytical point of view. Therefore, molecular simulations are required.

1.6   Types of Molecular Dynamics Simulation

There are two types of molecular dynamics simulation to study thermophysical properties of system.

1.6.1   Equilibrium Molecular Dynamics Simulation

Mostly, Green-Kubo relations (GKRs) are used in equilibrium molecular dynamics simulation (EMDS) for the calculation of transport coefficients in determining spontaneous fluctuation decay. Einstein relation is equivalent to GKR, but it does not require derivation for heat current [20]. Required transport properties computed by GKR involve equilibrium autocorrelation function for related current. EMDS can compute thermophysical properties of materials (solids, liquids, gases, and plasmas) in microcanonical, canonical thermal, or isobaric ensembles, but EMD has the disadvantage that it consumes more time for the calculation of lengthy portion of time correlation function.

1.6.2 Nonequilibrium Molecular Dynamics Simulation

Sometimes, thermophysical properties of fluids do not remain in equilibrium position, so properties of that type of fluids can also be found by nonequilibrium molecular dynamics simulation (NEMDS) technique [21]. Perturbed force is applied to the system and after applying, it remains constant during the simulation. In 1973, Gosling et al. used NEMDS technique to calculate the shear viscosity; in 1979, Ciccotti et al. used this technique to calculate diffusion and thermal conductivity by this technique. In NEMDS technique, transport coefficient can be determined by linear response theory and system perturbed due to external fields; this technique is faster than EMDS. The entire transport coefficient cannot compute simultaneously in NEMDS rather than in EMDS.

1.7 Advantages of NEMDS over EMDS

EMDS takes lots of time to calculate transport coefficients of dynamic systems. Time correlation function caused statistical error in EMDS. EMDS does not give a good response to minimize these statistical errors or fluctuations. In contrast, NEMDS technique is quick and efficient and simulates more than one transport coefficients at a time. It gives a good response in order to minimize statistical errors. It can examine the response of large perturbation which is artificially imposed from external source [22].

2 Numerical Model and Design Parameters

We consider a cubic box of edge length L in dusty plasma having N number of particles or millions of atoms, but the range of particles in molecular dynamics is only in thousands [23]. The size or dimension of the box is selected by periodic boundary conditions (PBCs). The periodic boundary condition avoids the surface size effects. In box, particles interact with each other with some potential. This potential may be Yukawa or Coulomb and or Lennard-Jones (LJ) potential depending on the type of system.

2.1 Algorithm for HNEMD Simulation

Evans has proposed a new method for computing thermal conductivity of 3D Yukawa liquids called HNEMD. In Yukawa systems, molecules or charge particles interact with neighboring particles with a long-range Yukawa potential. This potential is different from Coulomb potential because of dust particles. It can be determined not only by electrostatic potential, but also by the variety of charges and number of collective effects. Yukawa potential depends on the distance between interacting particles denoted by r and Debye length λ_D. Plasma becomes purely Columbic when distance between interacting particles is very less than Debye length $(r < <\lambda_D)$; when distance r approaches Debye length $(r \sim \lambda_D)$, the screening becomes important. When distance is exceeding Debye length $(r> > \lambda_D)$ then potential follows the inverse law and that type of potential is called screened Coulomb or Yukawa potential [24].

$$\phi_Y \; \mathbf{r} = \frac{1}{4\pi\varepsilon_o} \frac{Z_d^2 e^2}{\mathbf{r}} e^{-r/\lambda_D}. \tag{1}$$

In Yukawa system, three dimensionless parameters can describe the system completely. First, the screening parameter is defined as the ratio of Wigner-Seitz (WS) radius and Debye screening length:

$$\kappa = \frac{a_{ws}}{\lambda_D}, \tag{2}$$

where a_{ws} is WS radius and λ_D is Debye screening length. WS radius is defined as radius of sphere (e.g., dust particle) which is equal to average volume per atom:

$$\frac{4}{3}\pi r^3 = \frac{V}{N}. \tag{3}$$

Consider that $r = a$ then for 3D, WS radius is equal to:

$$a_{ws} = \left(\frac{3V}{4\pi N}\right)^{1/3} \approx \left(\frac{3}{4\pi n}\right)^{1/3}, \tag{4}$$

where n is the number density $(n = N/V)$ and the second parameter is Coulomb coupling parameter:

$$\Gamma = \frac{Ze^2}{4\pi\varepsilon} \frac{1}{ak_B T}, \tag{5}$$

where T is temperature of the system in terms of energy unit and k_B is Boltzmann constant. The third reduced parameter is the external force field used to perturb or disturb the system from an equilibrium state:

$$F^* = \frac{F_z a_{ws}}{J_{Q=Ze}}. \tag{6}$$

In the above Eq. $F^* = (0\ \hat{x}, 0\ \hat{y}, F\ \hat{z})$ means F is applied along z-axis only to perturb the system and $\mathbf{J}_{Q\,=\,Ze}$ is heat flux vector of charge particle. In 3D systems, for charge-free particle, thermal conductivity formula according to GKR is expressed as:

$$\lambda = \frac{1}{3Vk_B T^2} \int_\infty^0 \mathbf{J}_{Q=Ze}(t) \cdot \mathbf{J}_{Q=Ze}(0)\,dt. \tag{7}$$

Here, $\mathbf{J}_{Q\,=\,Ze}$ is vector of heat flux, V is volume, and T is temperature of system. At a microscopic level, $\mathbf{J}_{Q\,=\,Ze}$ has value:

$$\mathbf{J}_{Q=Ze} = \frac{1}{V} \sum_N^{i=1} E_i \frac{\mathbf{p}_i}{m} - \frac{1}{2} \sum_{i\neq j} \mathbf{r}_{ij} \left(\frac{\mathbf{p}_i}{m} . \mathbf{F}_{ij} \right). \tag{8}$$

As discussed earlier, $\mathbf{r}_{ij}$ is the distance between two interacting particles and E_i is the energy of interacting particle which is the combination of kinetic $(p_i^2/2m)$ and potential $(1/2\sum\phi_{ij})$ part:

$$E_i = \frac{p_i^2}{2m} + \frac{1}{2} \sum_{i\neq j} \phi_{ij}. \tag{9}$$

The general non-Hamiltonian theory of linear response, modified by using Newton's equation of motion, is proposed by Evans for Yukawa system:

$$\dot{\mathbf{r}}_i = \frac{\mathbf{p}_i}{m}, \tag{10}$$

$$\dot{\mathbf{p}}_i = \sum_N^{j=1} \mathbf{F}_i(t) + E_i F_e(t) - \frac{1}{2} \sum_N^{j=1} \mathbf{F}_{ij}\left(\mathbf{r}_{ij}.\mathbf{F}_e(t)\right) + \frac{1}{2N} \sum_N^{j,k} \mathbf{F}_{jk}\left(\mathbf{r}_{jk}.\mathbf{F}_e(t)\right) - \alpha \mathbf{p}_i, \tag{11}$$

$$\mathbf{F}_i = -\frac{\partial \phi_{ij}}{\partial r_i}, \tag{12}$$

$$\alpha = \frac{\sum_N^{i=1} \mathbf{p}_i/m_i.(\mathbf{F}_i(t) + E_i F_e(t) - \frac{1}{2}\sum_N^{j=1}\mathbf{F}_{ij}\left(\mathbf{r}_{ij}.\mathbf{F}_e(t)\right) + \frac{1}{2N}\sum_N^{j,k}\mathbf{F}_{jk}\left(\mathbf{r}_{jk}.\mathbf{F}_e(t)\right)}{\sum_N^{i=1} p_i^2/m_i}, \tag{13}$$

where $\mathbf{F}_i(t)$ is the total internal force influenced by i^{th} charge particle, $\mathbf{F}_e(t)$ is the external force applied at any time t, and α is Gaussian thermostat that keeps the system in equilibrium state (at constant temperature):

$$\mathbf{F}_i' = \sum_N^{j=1} \mathbf{F}_i(t) + E_i F_e(t) - \frac{1}{2} \sum_N^{j=1} \mathbf{F}_{ij} \left(\mathbf{r}_{ij}.\mathbf{F}_e(t) \right) + \frac{1}{2N} \sum_N^{j,k} \mathbf{F}_{jk} \left(\mathbf{r}_{jk}.\mathbf{F}_e(t) \right). \quad (14)$$

$\mathbf{F}_i$ is added term of force on each particle and by substituting this value in Eq. (15), Gaussian thermostat is reduced to:

$$\alpha = \frac{\sum_N^{i=1} p_i / m_i.\left(\mathbf{F}_i + \mathbf{F}_i' \right)}{\sum_N^{i=1} p_i^2 / m_i}, \quad (15)$$

$$\lambda = \frac{1}{3Vk_B T^2} \int_\infty^0 \mathbf{J}_{Q_z}(t) \mathbf{J}_{Q_z}(0) dt. \quad (16)$$

When the external force field is applied along z-axis, thermal conductivity is reduced to the following term:

$$\lambda = \lim_{F_z \to 0} \lim_{t \to \infty} \frac{-\mathbf{J}_{Q_z}(t)}{TF_z}. \quad (17)$$

This is the final formula for thermal conductivity having only z-factor. In simulation of Yukawa 3D systems, the number of particles $N = 256–864$ is to be selected to study the size effect of system. It comes to be known that system size does not affect thermal conductivity or any other properties significantly under some statistical uncertainties. The force exerted by particles on each other or in minimum image convection is numerically calculated by negative divergence of Yukawa potential $\mathbf{F} = (-\nabla\phi)$ and this Newton's equation of motion is integrated by the predictor corrector algorithm. The handling of infinite system remains a difficult task in every aspect; it may be theoretical, experimental, or simulation. In simulation, periodic boundary conditions are imposed on the cell, which is L/a of cell size to handle the infinite system. Ewald sums method is used to ensure long-range interaction; this method is divided into two parts: one touches the real space and the second touches the reciprocal space [25, 26]. For a high screening parameter value $\kappa > 0$, only real space can give a precise and accurate result, but when screening parameter is less than zero $\kappa < 0$, then Fourier part cannot be ignored; it becomes important because Yukawa system at this state behaves like a gas. Here, HNEMD simulation is performed in canonical ensemble and when the system attains the equilibrium state, then the thermostat continues to maintain a constant temperature of the system. The desired time step as discussed earlier is time step $\Delta\tau = 0.001/\omega_p$, time step is the inverse of plasma frequency [$\omega_p = (Q^2/2\pi\varepsilon_0 ma^3)^{1/2}$] in Yukawa system have mass m.

All computational data for normalized thermal conductivity (λ) is selected in between $2 \times 10^5/\omega_p$ and $5 \times 10^4/\omega_p$ range of time unit. Here, thermal conductivity (λ) of 3D Yukawa systems are computed for a wide range of Coulomb coupling parameter ($1 \leq \Gamma \leq 300$) and screening parameter ($1 \leq \kappa \leq 4$) and normalized by Einstein's frequency $\lambda^* = \lambda/nm\omega_E a^2$. Einstein's frequency increased with decreases in coupling parameter $\kappa \to 0$ so $\omega_{E\to}\omega_p/\sqrt{3}$.

3 Numerical Outcomes and Discussion

In this section, we discuss the lattice correlation obtained by applying external force field for 3D Yukawa system. A graphical representation of thermal conductivity and comparison of simulated results with the earlier published results are presented. We have checked out how the smaller size of the system influences thermal conductivity or not. In addition, we have observe that the potential energy, kinetic energy, and total energy vary with a variation in temperature.

3.1 Lattice Correlation

Lattice correlation gives information about the structure of considered systems. The ordered structure of particles shows that it may be in solid or crystal form depending on the density of material at point r and can be written as:

$$\rho(\mathbf{r}) = \sum_{N}^{j=1} \delta(\mathbf{r} - \mathbf{r}_j). \tag{18}$$

Fourier transform gives the lattice correlation equation:

$$\Psi = \frac{1}{N} \sum_{N}^{i=1} \exp(-i\mathbf{k}.\mathbf{r}_j). \tag{19}$$

The computation of the above equation is used to know about ordered or disordered arrangement of the system. If $|\Psi| \approx 1$, then the system is in an ordered state; if its value is less than 1, then the system lies in a gaseous state (nonideal gaseous state), and then it follows $|\Psi| \approx O(N^{-1/2})$. Figure 3 shows lattice correlation $|\Psi(t)|$ versus simulation time step $\Delta\tau$ for the whole range of screening parameters ($1 \leq \kappa \leq 4$) with different numbers of particles ($N = 256$–864). In Eq. (19), $\mathbf{k}$ is lattice correlation vector for ordered state and its value is different for the different lattice structure. Its value for face cantered-cubic (FCC) is $\mathbf{k} = 2\pi/(1,-1, 1)l$, body-centered cubic (BCC) is $\mathbf{k} = 2\pi/(1, 0, 1)l$, and for simple cubic (SC) is $\mathbf{k} = 2\pi/(1, 0, 0)l$, here l is the edge length [18].

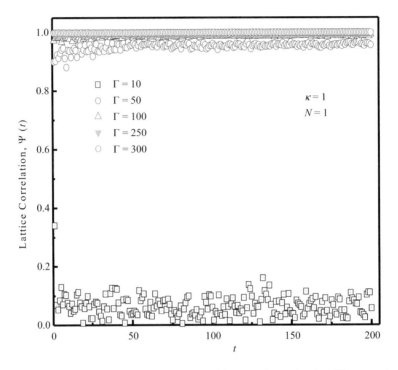

Fig. 3 Dependence of lattice correlation function $|\Psi(t)|$ on time t for five different values of Coulomb coupling parameters $\Gamma = 10, 50, 100, 250$, and 300 at $\kappa = 1$, when external force field $F^* = 0.003$ is applied to Yukawa system and number of particles is $N = 256$

3.1.1 Lattice Correlation Analysis

The above figure shows graphical representation of lattice correlation (long-range order) of Yukawa system for a fixed number of particles $N = 256$, but at different screening parameters. For higher Γ, it is observed that the lattice correlation value is close to one, (i.e., $|\Psi(t)| \leq 1$). At a higher Γ, particles are partially or completely fixed in its position and orderly arranged; however, when lattice correlation value is less than one (such as $|\Psi(t)| << 1$), then the system becomes disordered. This behavior shows that Yukawa system remains in strongly coupled states for a whole range of plasma coupling (Γ, κ). Figure 3 shows the lattice correlation of 3D Yukawa systems versus time steps t for five different coupling parameters ($\Gamma = 10, 50, 100, 200,$ and 300) at $\kappa = 1$. As shown in Fig. 3, at higher Γ (low temperature), the system is completely in ordered state, but in contrast, at lower Γ (high temperature), the system becomes disordered. As we know, Debye screening ($\kappa = a_{ws}/\lambda_D$) is inverse of Debye length [$\lambda_D = (\varepsilon_0 K T_e/ne^2)^{1/2}$]; therefore, density would be larger at higher κ. At this stage, the system is close to disordered liquid state and behaves as nonideal strongly coupled plasma for both lower values of $\Gamma = 10$ and 50 [18]. As discussed earlier, strongly coupled plasma has high density and low temperature, so this type of plasma is also called cold plasma. Moreover, at high temperatures ($\Gamma < 1$) and low screening value ($\kappa < 1$), particles attain high kinetic energy and the system has low density. The kinetic energy dominated over potential energy throughout the system. The potential energy in this state is very low that can be ignored; therefore, the system will behave as a gas and act as weakly coupled plasma (ideal complex plasma).

3.1.2 Normalized Thermal Conductivity

HNEMD method is applied to investigate thermal conductivity normalized by plasma frequency (ω_p) of 3D SCCDPs, at a reduced external force field $F^* = a_{ws}F_z/J_Q$. Normalization of thermal conductivity is widely used in earlier studies of one component strongly coupled plasma (OCCP) [30] and complex dusty plasma liquids (CDPLs) [27, 28]. λ can be reduced by plasma frequency ω_p, $\lambda_0 = \lambda/nk_B\omega_p a_{ws}$, or by Einstein frequency (ω_E), $\lambda^* = \lambda/\sqrt{3}nk_B\omega_E a_{ws}$. In order to employ HNEMD technique, F^* is used to establish near equilibrium value of normalized thermal conductivity. It is observed that at $F^* = 0.003$, the present results have a good agreement with HNEMD simulation of Shahzad and He [2], the homogenous perturbed molecular dynamics (HPMD) simulation of Shahzad and He [17], the equilibrium molecular dynamics (EMD) of Salin and Caillol [27], NEMD of Donko and Hartmann [28], and variational procedure (VP) of Faussurier and Murillo [29].

The simulations are executed for $N = 256$ and the present results, for the case of $\kappa = 1$, are shown in Fig. 4a. This figure shows that for lower values of Γ, present results of λ_0 are higher than obtained by theoretical approach of Faussurier and Murillo and NEMD results of Donko and Hartmann, but lower (within ~13–25%) than HNEMD results of Shahzad and He, and EMD investigation of Salin and Caillol, for $5 \leq \Gamma \leq 20$. A comparison with the reference data is shown in Fig. 4b for $\kappa = 1$. HNEMD simulations are carried out for different numbers of particles ($N = 256$, 500, and 864) over a large domain of plasma parameters (Γ, κ) and obtained results are used as reference data (λ_{REF}) points in our study (in all figures), at $F^* = 0.003$. It should be noted that most of data are taken for $N = 256$. The deviations of earlier results are calculated as λ_0/λ_{REF} and taken from reference data (λ_{REF}). It is observed from Fig. 4b that presented results for λ_0 show the realistic agreement with the earlier results calculated by different numerical techniques such as EMD, NEMD, HPMD, and HNEMD: λ_0 results are within ~1–30% for EMD, ~ 5–30% for HNEMD, ~ 3–35% for NEMD, and ~1 –48% for HPMD. For the case of $\kappa = 2$, present results of λ_0 give satisfactory agreement with the earlier results based on different numerical techniques. It is noticed from Fig. 4c, for lower values of Γ, obtained results of λ_0 are comparatively higher than Donko and Hartmann NEMD results, while relatively less than EMD, HNEMD, and HPMD results within the range of $5 \leq \Gamma \leq 50$. Thermal conductivity is shown well in agreement with the earlier simulation data in Fig. 4d, within the range ~ 1–12% for EMD, ~ 1–25% for HNEMD, and ~ 1–10% for VP theoretical approach. These results are also significant because

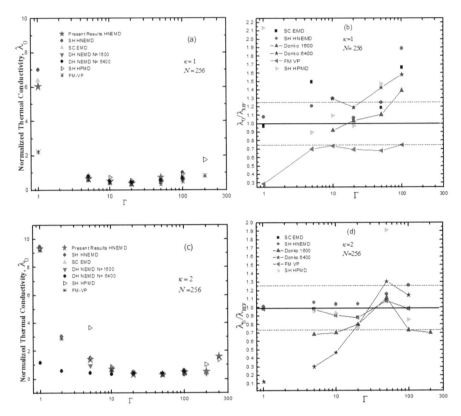

Fig. 4 Variation of present results of normalized (by plasma frequency) λ_0 obtained through HNEMD with Shahzad and He for HNEMD: SH HNEMD [2], and Shahzad and He homogeneous perturbed MD: SH HPMD [17], Salin and Caillol for EMD: SC EMD [27], Donko and Hartmann for NEMD: DH NEMD [28], and Faussurier and Murillo for Variational Procedure: FM VP [29] results, for $N = 256$ and wide range of Coulomb coupling parameters $(1 \leq \Gamma \leq 300)$ and force filed $F^* = 0.003$ at (**a**) $\kappa = 1$ (**c**) $\kappa = 2$. Panels (**b**) $\kappa = 1$ (**d**) $\kappa = 2$ show the comparison obtained by λ_0/λ_{REF} with the earlier results. The present data shown in (a) and (c) are taken as a reference data, λ_{REF}. The horizontal dotted lines show ±25% deviation from the reference data (λ_{REF})

computations take place at the same values of Γ where $\lambda_{min} = 0.38$ at $\kappa = 1$ and $\lambda_{min} = 0.32$ at $\kappa = 2$ were earlier shown to have the minimum values of λ_0.

For the case $\kappa = 3$ and 4 in Fig. 5a and c, it is examined that the present data illustrate the excellent agreement with the earlier numerical results at high values of κ. Our results are slightly higher than EMD results but slightly less than the earlier numerical results of EMD, NEMD, HPMD, and HNEMD within the range $5 \leq \Gamma \leq 100$ for $\kappa = 3$ and $10 \leq \Gamma \leq 100$ for $\kappa = 4$. Comparison of data with the reference results (λ_{REF}), in Fig. 5b and d, is within the range ~ 1–25% (~3–27%) for EMD, ~3–26% (~2–19%) for HNEMD, ~ 6–23% for NEMD, and ~ 1–32% (~3–20%) for HPMD, for the case $\kappa = 3$ (and 4). There are two possibilities for these differences in λ_0, and both may be arising from Gaussian thermostat and

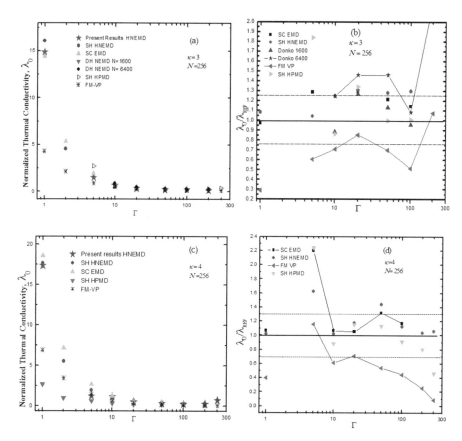

Fig. 5 Reduced thermal conductivity λ_0 for a wide range of Coulomb coupling parameter ($1 \leq \Gamma \leq 300$) for $N = 256$ and force filed $F^* = 0.003$ at (**a**) $\kappa = 3$ (**c**) $\kappa = 4$, and the comparison of obtained results with earlier numerical results is shown in panels (**b**) $\kappa = 3$ and (**d**) $\kappa = 4$. For more detail, see the caption of Fig. 4

homogeneity in HNEMD algorithm that are absent in EMD and inhomogeneous NEMD (InH-NEMD) simulations, respectively. In HNEMD, Gaussian thermostat exists; as a result, we get uniform values of Γ and λ_0 while EMD simulation also gives the uniform values at constant energy. This difference may be caused by limited system size in EMD simulations. The homogenous simulation is in nonequilibrium state, with employment of small external force in specific direction so that all atoms possess similar environment. In contrast, inhomogeneous simulation is in nonequilibrium state when large external force is applied to the system with relatively large momentum to drive heat flux.

3.1.3   Energies Analysis of SCCDPs

The simulations are carried out in canonical ensemble (*NVT*), for fixed number of particles, volume, and temperature. Figure 6a shows that kinetic energy of each Γ remains the same for the entire range of time steps, but its value varies for different Γ, $N = 256$, and $\kappa = 1$. The graph represents that kinetic energy is higher for a lower value of Γ while its value decreases with increase in Γ. The kinetic energy is not high enough at $\Gamma = 1$, it is about ~1.4832 and for $\Gamma = 300$, kinetic energy is less than or near to zero (such as ~0.0049). It is observed from Fig. 6b that potential energy varies with time steps $\Delta\tau$ as well as with Γ. In contrast to kinetic energy, potential energy increases with increase in Γ, and vice versa. For the same case of κ, the maximum value of potential energy is ~ 57.984 at higher Γ. The combination of kinetic and potential energy is shown in Fig. 6c. The graph seems to be analogous to the potential energy graph because kinetic energy is relatively very small (~0.005). The total energy for the higher Γ is nearly ~57.989.

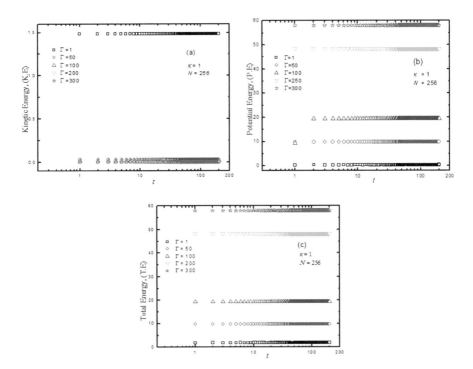

Fig. 6 Variation of SCCDP system energies with respect to simulation time steps, for five different Coulomb coupling parameters ($\Gamma = 10$, 50, 100, 200 and 300) at $\kappa = 1$ and $N = 256$, (**a**) kinetic energy (K.E.), (**b**) potential energy (P.E.), and (**c**) total energy (T.E)

4 Summary

The purpose of this chapter is basic as well as applied in the field of applied plasma field including material science. The intended MD methods for complex materials, when industrial at a commercially viable level, have a strong potential to offer in an attractive way to achieve new technologies in industries and highly advanced academic problems. The major impact of progress of transport properties is to minimize operating or modifying and growing cost in new industries or system designing in view of energy saving, reducing time, and shortening of the uncertainty in going from process innovation at a small laboratory scale to a commercial industrial scale. An improved HNEMD method is used to evaluate thermal conductivity of 3D SCCDPs system for a wide domain of Coulomb coupling parameter and screening parameter by applying external force field ($F^* = 0.003$). This HNEMDS method demonstrates that the present results have a good agreement with the earlier results obtained through different numerical techniques for Yukawa system. Furthermore, HNEMDS method can be used to explore non-Newtonian behaviors of 3D dusty plasmas with the use of linear regimes of external force field F^*. HNEMDS gives a reasonable accuracy for a lower system size and good signal-to-noise ratio for complete range of plasma parameters. Investigation shows that the minimum value of λ_0 shifts toward Γ with increase in κ, as expected in earlier simulation results. Our new simulation results are generally overpredicted (1–30%), but within limited uncertainty statistical errors, for different numbers of particle N. The system size does not affect the behavior of lattice correlation, while lattice correlation decreases with increase in κ and Γ. The kinetic energy is not affected by the system size and independent also of time steps and κ; however, it depends on system Γ. In future work, quantum effects in λ_0 can be added to see how these effects influence λ_0 of Yukawa systems and in other fields of science such as chemical, biological, ionic systems, and material sciences.

Acknowledgments The authors thank Z. Donkó (Hungarian Academy of Sciences) for providing his thermal conductivity data of Yukawa Liquids for the comparisons with our simulation results, and for useful discussions. We are grateful to the National Advanced Computing Center of National Center of Physics (NCP), Pakistan, for allocating computer time to test and run our MD code.

References

1. V.E. Fortov, A.V. Ivlev, S.A. Khrapak, A.G. Khrapak, G.E. Morfill, Complex (dusty) plasmas: Current status, open issues, perspectives. Phys. Rep. **421**, 1–104 (2005)
2. A. Shahzad, M.-G. He, Thermal conductivity of three dimensional Yukawa liquids (dusty plasma). Contrib. Plasma Physics **52**, 667–675 (2012)
3. A. Shahzad, M.-G. He, Interaction contributions in thermal conductivity of three-dimensional complex liquids. AIP Conf. Proc. **1547**, 173 (2013)
4. A. Shahzad, *Impact of Thermal Conductivity on Energy Technologies* (InTech, Rijeka: Croatia, 2018). https://doi.org/10.5772/intechopen.72471
5. F.F. Chen, *Introduction to Plasma Physics and Controlled Fusion*, 2nd edn. (Springer verlag, New York, 2010)

6. A. Shahzad, M.-G. He, Diffusion motion of two-dimensional weakly coupled complex (dusty) plasmas. Phys. Scr. **87**, 035501 (2013)

7. G.J. Kalman, J.M. Rommel, K. Blagoev, *Strongly Coupled Coulomb Systems* (Plenum, New York, 1998)

8. R.L. Merlino, j.A. Goree, Dusty plasma in laboratory, industry, and space. Phys. Today **57**, 32–38 (2004)

9. R.L. Merlino, *A Dusty Plasma Is an Ionized Gas Containing Dust Particles Plasma Physics Applied* (2006), pp. 73–110

10. Arp, O.,Block, D., and Piel, Alexander, Dust coulomb balls: Three-dimensional plasma crystals. PRL **93**, 165004 (2004)

11. A. Melzer, M. Himpel, C. Carsten Killer, M. Mulsow, Stereoscopic imaging of dusty plasmas. Aust. J. Plant Physiol. **82**, 615820102 (2016)

12. M. Slimullah, M.R. Amin, M. Salahuddin, A.R. Chowdhury, Ultra-low-frequency electrostatic modes in a magnetized dusty plasma. Phys. Scr. **58**, 76 (1998)

13. A. Shahzad, M.-G. He, Homogeneous nonequilibrium molecular dynamics evaluation of thermal conductivity in 2D Yukawa liquids. Int. J. Thermophys. **36**, 2565 (2015)

14. B. Liu, J. Goree, Superdiffusion and non-Gaussian statistics in a driven-dissipative 2D dusty plasma. Phys. Rev. Lett. **100**, 055003 (2008)

15. A. Shahzad, M.G. He, Thermoelectrics for power generation-a look at trends in the technology, in *Thermal Conductivity and Non-Newtonian Behavior of Complex Plasma Liquids*, ed. by D. M. Nikitin , (InTech, Rijeka, Croatia, 2016). https://doi.org/10.5772/65563Chp 13

16. A. Shahzad, M.G. He, Numerical experiment of thermal conductivity in two-dimensional Yukawa liquids. Physic. Plasmas **22**(12), 123707 (2015). https://doi.org/10.1063/1.4938275

17. A. Shahzad, M.-G. He, Thermal conductivity calculation of complex (dusty) plasmas. Physic. Plasmas **19**(8), 083707 (2012). https://doi.org/10.1063/1.4748526

18. A. Shahzad, M.-G. He, Structural order and disorder in strongly coupled Yukawa liquids. Physic. Plasmas **23**, 093708 (2016). https://doi.org/10.1063/1.4963390

19. A. Shahzad, S.I. Haider, M. Kashif, M.S. Shifa, T. Munir, M.-G. He, Thermal conductivity of complex plasmas using novel Evan-Gillan approach. Commun. Theor. Phys. **69**, 704–710 (2018)

20. A. Kinaci, J.B. Haskins, C. Tahir, On calculation of thermal conductivity from Einstein relation in equilibrium MD. Phys.Chem **137**, 01410 (2012)

21. D.J. Evans, G.P. Morriss, *Statistical Mechanics of Non-equilibrium Liquids* (London Academic press, 1990)

22. A. Shahzad, S. Maryam, A. Arfa, M.-G. He, Thermal conductivity measurements of 2D complex liquids using nonequilibrium molecular dynamics simulations. Appl Sci Technol (IBCAST), 11th International Bhurban Conference, Jan. 14–18, Proceeding of the IEEE Transaction **1**, 212–217 (2014)

23. Toukmaji, A. Y., Board, Jr. John. A, Ewald summation techniques in perspective: A survey. Comput. Phys. Commun. **95**, 73–92 (1996)

24. V.E. Fortov, A.G. Khrapak, S.A. Khrapak, V.I. Molotkov, O.F. Petrov, Dusty plasmas. Physics – Uspekhi **47**, 447–492 (2004)

25. A. Shahzad, M.-G. He, Thermodynamics characteristics of dusty plasma by using molecular dynamics simulations. Plasma Sci. Technol **14**, 771–777 (2012)

26. A. Shahzad, M.-G. He, Calculations of thermal conductivity of complex (dusty) plasmas using homogenous nonequilibrium molecular simulations. Radiat Eff Defects Solids **170**(9), 758–770 (2015). https://doi.org/10.1080/10420150.2015.1108316

27. G. Salin, J.-M. Caillol, Equilibrium molecular dynamics simulations of the transport coefficients of the Yukawa one component plasma. Phys Plasmas **10**, 1220 (2003)

28. Z. Donkó, P. Hartmann, Thermal conductivity of strongly coupled Yukawa liquids. Phys. Rev. E **69**, 016405 (2004). https://doi.org/10.1103/PhysRevE.69.016405

29. G. Faussurier, M.S. Murillo, Gibbs-Bogolyubov inequality and transport properties for strongly coupled Yukawa fluids. Phys. Rev. E **67**, 046404 (2003)

30. C. Pierleoni, G. Ciccotti, B. Bernu, Thermal conductivity of the classical one-component plasma by nonequilibrium molecular dynamics. Europhys. Lett. **4**, 1115 (1987)

Index

© Springer Nature Switzerland AG 2021
S. Skipidarov, M. Nikitin (eds.), *Thin Film and Flexible Thermoelectric
Generators, Devices and Sensors*, https://doi.org/10.1007/978-3-030-45862-1

Printed in the United States
by Baker & Taylor Publisher Services